Principles of Mathematics in Operations Research

Recent titles in the **INTERNATIONAL SERIES IN**
OPERATIONS RESEARCH & MANAGEMENT SCIENCE
Frederick S. Hillier, Series Editor, *Stanford University*

* A list of the early publications in the series is at the end of the book *

Levent Kandiller

Principles of Mathematics in Operations Research

 Springer

Levent Kandiller
Middle East Technical University
Ankara, Turkey

ISBN-10: 0-387-37735-2 (e-book)

ISBN-13: 978-1-4419-4250-0   ISBN-13: 978-0387-37735-3 (e-book)

Printed on acid-free paper.

springer.com

To my daughter, Deniz

Preface

The aim of this book is to provide an overview of mathematical concepts and their relationships not only for graduate students in the fields of Operations Research, Management Science and Industrial Engineering but also for practitioners and academicians who seek to refresh their mathematical skills.

The contents, which could broadly be divided into two as linear algebra and real analysis, may also be more specifically categorized as linear algebra, convex analysis, linear programming, real and functional analysis. The book has been designed to include fourteen chapters so that it might assist a 14–week graduate course, one chapter to be covered each week.

The introductory chapter aims to introduce or review the relationship between Operations Research and mathematics, to offer a view of mathematics as a language and to expose the reader to the art of proof–making. The chapters in Part 1, linear algebra, aim to provide input on preliminary linear algebra, orthogonality, eigen values and vectors, positive definiteness, condition numbers, convex sets and functions, linear programming and duality theory. The chapters in Part 2, real analysis, aim to raise awareness of number systems, basic topology, continuity, differentiation, power series and special functions, and Laplace and z–transforms.

The book has been written with an approach that aims to create a snowball effect. To this end, each chapter has been designed so that it adds to what the reader has gained insight into in previous chapters, and thus leads the reader to the broader picture while helping establish connections between concepts.

The chapters have been designed in a reference book style to offer a concise review of related mathematical concepts embedded in small examples. The remarks in each section aim to set and establish the relationship between concepts, to highlight the importance of previously discussed ones or those currently under discussion, and to occasionally help relate the concepts under scrutiny to Operations Research and engineering applications. The problems at the end of each chapter have been designed not merely as simple exercises requiring little time and effort for solving but rather as in–depth problem solving tasks requiring thorough mastery of almost all of the concepts provided within that chapter. Various Operations Research applications from deterministic (continuous, discrete, static, dynamic) modeling, combinatorics, regression, optimization, graph theory, solution of equation systems as well as geometric and conceptual visualization of abstract mathematical concepts have been included.

As opposed to supplying the readers with a reference list or bibliography at the end of the book, active web resources have been provided at the end of each chapter. The rationale behind this is that despite the volatility of Internet sources, which has recently proven to be less so with the necessary solid maintenance being ensured, the availability of web references will enable the ambitious reader to access materials for further study without delay at the end of each chapter. It will also enable the author to keep this list of web materials updated to exclude those that can no longer be accessed and to include new ones after screening relevant web sites periodically.

I would like to acknowledge all those who have contributed to the completion and publication of this book. Firstly, I would like to extend my gratitude to Prof. Fred Hillier for agreeing to add this book to his series. I am also indebted to Gary Folven, Senior Editor at Springer, for his speedy processing and encouragement.

I owe a great deal to my professors at Bilkent University, Mefharet Kocatepe, Erol Sezer and my Ph.D. advisor Mustafa Akgül, for their contributions to my development. Without their impact, this book could never have materialized. I would also like to extend my heartfelt thanks to Prof. Çağlar Güven and Prof. Halim Doğrusöz from Middle East Technical University for the insight that they provided as regards OR methodology, to Prof. Murat Köksalan for his encouragement and guidance, and to Prof. Nur Evin Özdemirel for her mentoring and friendship.

The contributions of my graduate students over the years it took to complete this book are undeniable. I thank them for their continuous feedback, invaluable comments and endless support. My special thanks go to Dr. Tevhide Altekin, former student current colleague, for sharing with me her view of the course content and conduct as well as for her suggestions as to the presentation of the material within the book.

Last but not least, I am grateful to my family, my parents in particular, for their continuous encouragement and support. My final words of appreciation go to my local editor, my wife Sibel, for her faith in what started out as a far-fetched project, and most importantly, for her faith in me.

Ankara, Turkey,

June 2006   *Levent Kandiller*

Contents

1

Introduction

Operations Research, in a narrow sense, is the application of scientific models, especially mathematical and statistical ones, to decision making problems. The present course material is devoted to parts of mathematics that are used in Operations Research.

1.1 Mathematics and OR

In order to clarify the understanding of the relation between two disciplines, let us examine Figure 1.1. The scientific inquiry has two aims:

- *cognitive:* knowing for the sake of knowing
- *instrumental:* knowing for the sake of doing

If *A is B* is a proposition, and if *B* belongs to *A*, the proposition is *analytic*. It can be validated logically. All analytic propositions are *a priori*. They are tautologies like "all husbands are married". If *B* is outside of *A*, the proposition is synthetic and cannot be validated logically. It can be *a posteriori* like "all African-Americans have dark skin" and can be validated *empirically*, but there are difficulties in establishing necessity and *generalizability* like "Fenerbahçe beats Galatasaray".

Mathematics is purely analytical and serves cognitive inquiry. Operations Research is (should be) instrumental, hence closely related to engineering, management sciences and social sciences. However, like scientific theories, Operations Research

- refers to idealized models of the world,
- employs theoretical concepts,
- provides explanations and predictions using empirical knowledge.

The purpose of this material is to review the related mathematical knowledge that will be used in graduate courses and research as well as to equip the student with the above three tools of Operations Research.

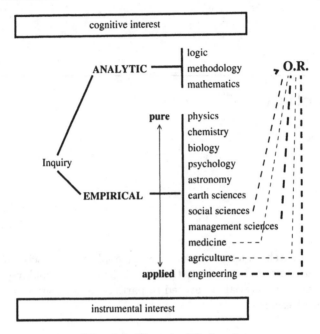

Fig. 1.1. The scientific inquiry.

1.2 Mathematics as a language

The main objective of mathematics is to state certainty. Hence, the main role of a mathematician is to communicate truths but usually in its own language. One example is

$$\forall i \in S, \exists j \in T \ni i \bot j \ \Rightarrow \ \forall j \in T, \exists i \in S \ni i \bot j \ \Longleftrightarrow \ S \bot T.$$

That is, if for all i in S there exists an element j of T such that i is orthogonal to j then for all elements j of T there is an element j of S such that j is orthogonal to i; if and only if, S is orthogonal to T.

To help the reader appreciate the expressive power of modern mathematical language, and as a tribute to those who achieved so much without it, a few samples of (original but translated) formulation of theorems and their equivalents have been collected below.

$$(a + b)^2 = a^2 + b^2 + 2ab$$

If a straight line be cut at random, the square on the whole is equal to the squares on the segments and twice the rectangle contained by the segments (Euclid, Elements, II.4, 300B.C.).

$$1 + 2 + \cdots + 2^n \ is \ prime \ \Rightarrow \ 2^n(1 + 2 + \cdots + 2^n) \ is \ perfect$$

If as many numbers as we please beginning from a unit be set out continuously in double proportion, until the sum of all becomes prime, and if the sum multiplied into the last make some number, the product will be perfect (Euclid, Elements, IX.36, 300B.C.).

$$A = \frac{2\pi r \cdot r}{2} = \pi r^2$$

The area of any circle is equal to a right-angled triangle in which one of the sides about the right angle is equal to the radius, and the other to the circumference, of the circle (Archimedes, Measurement of a Circle, 225B.C.).

$$S = 4\pi r^2$$

The surface of any sphere is equal four times the greatest circle in it (Archimedes, On the Sphere and the Cylinder, 220B.C.).

$$x = \sqrt[3]{\frac{n}{2} + \sqrt{\frac{n^2}{4} + \frac{m^3}{27}}} - \sqrt[3]{-\frac{n}{2} + \sqrt{\frac{n^2}{4} + \frac{m^3}{27}}}$$

Rule to solve $x^3 + mx = n$: Cube one-third the coefficient of x; add to it the square of one-half the constant of the equation; and take the square root of the whole. You will duplicate this, and to one of the two you add one-half the number you have already squared and from the other you subtract one-half the same... Then, subtracting the cube root of the first from the cube root of the second, the remainder which is left is the value of x (Gerolamo Cardano, Ars Magna, 1545).

However, the language of mathematics does not consist of formulas alone. The definitions and terms are verbalized often acquiring a meaning different from the customary one. In this section, the basic grammar of mathematical language is presented.

Definition 1.2.1 *Definition is a statement that is agreed on by all parties concerned. They exist because of mathematical concepts that occur repeatedly.*

Example 1.2.2 *A prime number is a natural integer which can only be (integer) divided by itself and one without any remainder.*

Proposition 1.2.3 *A Proposition or Fact is a true statement of interest that is being attempted to be proven.*

Here are some examples:

Always true Two different lines in a plane are either parallel or they intersect at exactly one point.
Always false $-1 = 0$.
Sometimes true $2x = 1$, $5y \leq 1$, $z \geq 0$ and $x, y, z \in \mathbb{R}$.

Needs proof! There is an angle t such that $\cos t = t$.

Proof. Proofs should not contain ambiguity. However, one needs creativity, intuition, experience and luck. The basic guidelines of proof making is tutored in the next section. Proofs end either with Q.E.D. ("Quod Erat Demonstrandum"), means "which was to be demonstrated" or a square such as the one here. □

Theorem 1.2.4 *Theorems are important propositions.*

Lemma 1.2.5 *Lemma is used for preliminary propositions that are to be used in the proof of a theorem.*

Corollary 1.2.6 *Corollary is a proposition that follows almost immediately as a result of knowing that the most recent theorem is true.*

Axiom 1.2.7 *Axioms are certain propositions that are accepted without formal proof.*

Example 1.2.8 *The shortest distance between two points is a straight line.*

Conjecture 1.2.9 *Conjectures are propositions that are to date neither proven nor disproved.*

Remark 1.2.10 *A remark is an important observation.*

There are also *quantifiers:*

∃ there is/are, exists/exist
∀ for all, for each, for every
∈ in, element of, member of, choose
∋ such that, that is
: member definition

An example to the use of these delimiters is

$$\forall y \in S = \{x \in \mathbb{Z}^+ : x \text{ is odd }\}, \quad y^2 \in S,$$

that is the square of every positive odd number is also odd.

Let us concentrate on $A \Rightarrow B$, i.e. if A is true, then B is true. This statement is the main structure of every element of a proposition family which is to be proven. Here, statement A is known as a *hypothesis* whereas B is termed as a *conclusion*. The operation table for this logical statement is given in Table 1.1. This statement is incorrect if A is true and B is false. Hence, the main aim of making proofs is to detect this case or to show that this case cannot happen.

Table 1.1. Operation table for $A \Rightarrow B$

A	B	$A \Rightarrow B$
True	True	True
True	False	False
False	True	True
False	False	True

Formally speaking, $A \Rightarrow B$ means

1. whenever A is true, B must also be true.
2. B follows from A.
3. B is a necessary consequence of A.
4. A is sufficient for B.
5. A only if B.

There are related statements to our primal assertion $A \Rightarrow B$:

$B \Rightarrow A$: converse
$\bar{A} \Rightarrow \bar{B}$: inverse
$\bar{B} \Rightarrow \bar{A}$: contrapositive

where $\bar{A}$ is negation (complement) of A.

1.3 The art of making proofs

This section is based on guidelines of how to read and make proofs. Our pattern here is once again $A \Rightarrow B$. We are going to start with the forward–backward method. After discussing the special cases defined in A or B in terms of quantifiers, we will see proof by Contradiction, in particular contraposition. Finally, we will investigate uniqueness proofs and theorem of alternatives.

1.3.1 Forward–Backward method

If the statement $A \Rightarrow B$ is proven by showing that B is true after assuming A is true $(A \rightarrow B)$, the method is called full *forward* technique. Conversely, if we first assume that B is true and try to prove that A is true $(A \leftarrow B)$, this is the full *backward* method.

Proposition 1.3.1 *If the right triangle XYZ with sides x, y and hypotenuse of length z has an area of $\frac{z^2}{4}$ (A), then the triangle XYZ is isosceles (B). See Figure 1.2.*

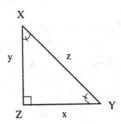

Fig. 1.2. Proposition 1.3.1

Proof. Backward:
B: $x = y$ $(x - y = 0)$ $\Leftrightarrow$ $\widehat{YXZ} \equiv \widehat{XYZ}$ (triangle is equilateral)
  Forward:
A–(i) Area: $\frac{1}{2}xy = \frac{z^2}{4}$
A–(ii) Pythagorean Theorem: $x^2 + y^2 = z^2$
$\Leftrightarrow \frac{1}{2}xy = \frac{x^2+y^2}{4} \Leftrightarrow x^2 - 2xy + y^2 = 0 \Leftrightarrow (x - y)^2 = 0 \Leftrightarrow x - y = 0.$ $\square$

The above proof is a good example of how forward-backward combination can be used. There are special cases defined by the forms of A or B with the use of quantifiers. The first three out of four cases are based on conditions on statement B and the last one arises when A has a special form.

Construction ($\exists$)

If there is an object ($\exists x \in \mathbb{N}$) with a certain property($x > 2$) such that something happens ($x^2 - 5x + 6 = 0$), this is a construction. Our objective here is to first construct the object so that it possesses the certain property and then to show that something happens.

Selection ($\forall$)

If something ($\exists x \in \mathbb{R} \ni 2^x = y$) happens for every object ($\forall y \in \mathbb{R}_+$) with a certain property ($y > 0$), this is a selection. Our objective here is to first make a list (set) of all objects in which something happens ($T = \{y \in \mathbb{R}_+ : \exists x \in \mathbb{R} \ni 2^x = y\}$) and show that this set is equivalent to the set whose elements has the property ($S = \mathbb{R}_+$). In order to show an equivalence of two sets ($S = T$), one usually has to show ($S \subseteq T$) and ($T \subseteq S$) by choosing a generic element in one set and proving that it is in the other set, and vice versa.

Specialization

If A is of the form "for all objects with a certain property such that something happens", then the method of specialization can be used. Without loss of generality, we can fix an object with the property. If we can show that something happens for this particular object, we can generalize the result for all the objects with the same property.

Proposition 1.3.2 *Let $T \subseteq S \subseteq \mathbb{R}$, and u be an upper bound for S; i.e. $\forall x \in S$, $x \le u$. Then, u is an upper bound for T.*

Proof. Let u be an upper bound for S, so $\forall x \in S$, $x \le u$. Take any element y of T. $T \subseteq S \Rightarrow y \in S \Rightarrow y \le u$. Thus, $\forall y \in T$, $y \le u$. Then, u is an upper bound for T. $\square$

Uniqueness

When statement B has the word *unique* in it, the proposition is more restrictive. We should first show the existence then prove the uniqueness. The standard way of showing uniqueness is to assume two different objects with the property and to conclude that they are the same.

Proposition 1.3.3

$$\forall r \in \mathbb{R}_+, \ \exists \ unique \ x \in \mathbb{R} \ni x^3 = r.$$

Proof. Existence: Let $y = r^{\frac{1}{3}}$, $y \in \mathbb{R}$.
Uniqueness: Let $x, y \in \mathbb{R} \ni x \ne y$, $x^3 = r = y^3 \Rightarrow x^3 - y^3 = 0 \Rightarrow (x - y)(x^2 + xy + y^2) = 0 \Rightarrow (x^2 + xy + y^2) = 0$, since $x \ne y$. The roots of the last equation (if we take y as parameter and solve for x) are

$$\frac{-y \pm \sqrt{y^2 - 4y^2}}{2} = \frac{-y \pm \sqrt{-3y^2}}{2}.$$

Hence, $y = 0 \Rightarrow y^3 = 0 = r \notin \mathbb{R}_+$. Contradiction. Thus, $x = y$. $\square$

1.3.2 Induction Method

Proofs of the form "for every integer $n \ge 1$, something happens" is made by induction. Formally speaking, *induction* is used when B is true for each integer beginning with an initial one (n_0). If the base case ($n = n_0$) is true, it is assumed that something happens for a generic intermediate case ($n = n_k$). Consequently, the following case ($n = n_{k+1}$) is shown, usually using the properties of the induction hypothesis ($n = n_k$). In some instances, one may relate any previous case (n_l, $0 \le l \le k$). Let us give the following example.

Theorem 1.3.4

$$1 + 2 + \cdots + n = \sum_{k=1}^{n} k = \frac{n(n+1)}{2}.$$

Proof. Base: $n = 1 = \frac{1 \cdot 2}{2}$.

Hypothesis: $n = j$, $\sum_{k=1}^{j} k = \frac{j(j+1)}{2}$.

Conclusion: $n = j + 1$, $\sum_{k=1}^{j+1} k = \frac{(j+1)(j+2)}{2}$.

$$\sum_{k=1}^{j+1} k = (j+1) + \sum_{k=1}^{j} k = (j+1) + \frac{j(j+1)}{2} = (j+1)\left[1 + \frac{j}{2}\right] = \frac{(j+1)(j+2)}{2}.$$

Thus, $1 + 2 + \cdots + n = \sum_{k=1}^{n} k = \frac{n(n+1)}{2}$. $\square$

1.3.3 Contradiction Method

When we examine the operation table for $A \Rightarrow B$ in Table 1.2, we immediately conclude that the only circumstance under which $A \Rightarrow B$ is not correct is when A is true and B is false.

Contradiction

Proof by Contradiction assumes the condition (A is true B is false) and tries to reach a legitimate condition in which this cannot happen. Thus, the only way $A \Rightarrow B$ being incorrect is ruled out. Therefore, $A \Rightarrow B$ is correct. This proof method is quite powerful.

Proposition 1.3.5

$$n \in \mathbb{N}, \ n^2 \ is \ even \ \Rightarrow n \ is \ even.$$

Proof. Let us assume that $n \in \mathbb{N}$, n^2 is even but n is odd. Let $n = 2k - 1$, $k \in \mathbb{N}$. Then, $n^2 = 4k^2 - 4k + 1$ which is definitely odd. Contradiction. $\square$

Contraposition

In contraposition, we assume A and $\bar{B}$ and go forward while we assume $\bar{A}$ and come backward in order to reach a Contradiction. In that sense, contraposition is a special case of Contradiction where all the effort is directed towards a specific type of Contradiction (A vs. $\bar{A}$). The main motivation under contrapositivity is the following:

$$A \Rightarrow B \equiv \bar{A} \vee B \equiv (\bar{A} \vee B) \vee \bar{A} \equiv (A \wedge \bar{B}) \Rightarrow \bar{A}.$$

One can prove the above fact simply by examining Table 1.2.

Table 1.2. Operation table for some logical operators.

A	$\bar{A}$	B	$\bar{B}$	$A \Rightarrow B$	$\bar{A} \vee B$	$A \wedge \bar{B}$	$A \wedge \bar{B} \Rightarrow \bar{A}$
T	F	T	F	T	T	F	T
T	F	F	T	F	F	T	F
F	T	T	F	T	T	F	T
F	T	F	T	T	T	F	T

Proposition 1.3.6

$$p, q \in \mathbb{R}_+ \ni \sqrt{pq} \neq \frac{p+q}{2} \Rightarrow p \neq q.$$

Proof. A: $\sqrt{pq} \neq \frac{p+q}{2}$ and hence $\bar{A}$: $\sqrt{pq} = \frac{p+q}{2}$. Similarly, B: $p \neq q$ and $\bar{B}$: $p = q$. Let us assume $\bar{B}$ and go forward $\frac{p+q}{2} = p = \sqrt{p^2} = \sqrt{pq}$. However, this is nothing but $\bar{A}$: $\sqrt{pq} = \frac{p+q}{2}$. Contradiction.   □

1.3.4 Theorem of alternatives

If the pattern of the proposition is $A \Rightarrow$ either C or (else) D is true (but not both), we have a theorem of alternatives. In order to prove such a proposition, we first assume A and $\bar{C}$ and try to reach D. Then, we should interchange C and D, do the same operation.

Proposition 1.3.7 *If $x^2 - 5x + 6 \geq 0$, then either $x \leq 2$ or $x \geq 3$.*

Proof. Let $x > 2$. Then,

$$x^2 - 5x + 6 \geq 0 \Rightarrow (x-2)(x-3) \geq 0 \Rightarrow (x-3) \geq 0 \Rightarrow x \geq 3.$$

Let $x < 3$. Then,

$$x^2 - 5x + 6 \geq 0 \Rightarrow (x-2)(x-3) \geq 0 \Rightarrow (x-2) \leq 0 \Rightarrow x \leq 2.   □$$

Problems

1.1. Prove the following two propositions:
(a) If f and g are two functions that are continuous [1] at x, then the function $f + g$ is also continuous at x, where $(f + g)(y) = f(y) + g(y)$.
(b) If f is a function of one variable that (at point x) satisfies

$$\exists\, c > 0,\ \delta > 0 \text{ such that } \forall y \ni |x - y| < \delta,\ |f(x) - f(y)| \leq c|x - y|$$

then f is continuous at x.

1.2. Assume you have a chocolate bar consisting, as usual, of a number of squares arranged in a rectangular pattern. Your task is to split the bar into small squares (always breaking along the lines between the squares) with a minimum number of breaks. How many will it take? Prove[2].

---

[1] A function f of one variable is *continuous at point x* if
$\forall \epsilon > 0,\ \exists \delta > 0$ such that $\forall y \ni |x - y| < \delta \Rightarrow |f(x) - f(y)| < \epsilon$.
[2] www.cut-the-knot.org/proofs/chocolad.shtml

1.3. Prove the following:

(a) $\binom{n}{r} = \binom{n}{n-r}$.

(b) $\binom{n}{r} = \binom{n-1}{r} + \binom{n-1}{r-1}$.

(c) $\binom{n}{0} + \binom{n}{1} + \cdots + \binom{n}{n} = 2^n$.

(d) $\binom{n}{m}\binom{m}{r} = \binom{n}{r}\binom{n-r}{m-r}$.

(e) $\binom{n}{0} + \binom{n+1}{1} + \cdots + \binom{n+r}{r} = \binom{n+r+1}{r}$.

Web material

http://acept.la.asu.edu/courses/phs110/si/chapter1/main.html
http://cas.umkc.edu/math/MathUGcourses/Math105.htm
http://cresst96.cse.ucla.edu/Reports/TECH429.pdf
http://descmath.com/desc/language.html
http://economictimes.indiatimes.com/articleshow/1024184.cms
http://en.wikipedia.org/wiki/Mathematical_proof
http://en.wikipedia.org/wiki/Mathematics_as_a_language
http://fcis.oise.utoronto.ca/~ghanna/educationabstracts.html
http://fcis.oise.utoronto.ca/~ghanna/philosophyabstracts.html
http://germain.umemat.maine.edu/faculty/wohlgemuth/DMAltIntro.pdf
http://interactive-mathvision.com/PaisPortfolio/CKMPerspective/
    Constructivism(1998).html
http://mathforum.org/dr.math/faq/faq.proof.html
http://mathforum.org/library/view/5758.html
http://mathforum.org/mathed/mtbib/proof.methods.html
http://mtcs.truman.edu/~thammond/history/Language.html
http://mzone.mweb.co.za/residents/profmd/proof.pdf
http://online.redwoods.cc.ca.us/instruct/mbutler/BUTLER/
    mathlanguage.pdf
http://pass.maths.org.uk/issue7/features/proof1/index.html
http://pass.maths.org.uk/issue8/features/proof2/index.html
http://plus.maths.org/issue9/features/proof3/index.html
http://plus.maths.org/issue10/features/proof4/
http://research.microsoft.com/users/lamport/pubs/
    lamport-how-to-write.pdf
http://serendip.brynmawr.edu/blog/node/59
http://teacher.nsrl.rochester.edu/phy_labs/AppendixE/
    AppendixE.html
http://weblog.fortnow.com/2005/07/understanding-proofs.html
http://www-didactique.imag.fr/preuve/ICME9TG12
http://www-didactique.imag.fr/preuve/indexUK.html
http://www-leibniz.imag.fr/DIDACTIQUE/preuve/ICME9TG12
http://www-logic.stanford.edu/proofsurvey.html
http://www-personal.umich.edu/~tappen/Proofstyle.pdf
http://www.4to40.com/activities/mathemagic/index.asp?
    article=activities_mathemagic_mathematicalssigns
http://www.ams.org/bull/pre-1996-data/199430-2/thurston.pdf

```
http://www.answers.com/topic/mathematics-as-a-language
http://www.bisso.com/ujg_archives/000158.html
http://www.bluemoon.net/~watson/proof.htm
http://www.c3.lanl.gov/mega-math/workbk/map/mptwo.html
http://www.cal.org/ericcll/minibibs/IntMath.htm
http://www.chemistrycoach.com/language.htm
http://www.cis.upenn.edu/~ircs/mol/mol.html
http://www.crystalinks.com/math.html
http://www.culturaleconomics.atfreeweb.com/Anno/Boulding
    %20Limitations%20of%20Mathematics%201955.htm
http://www.cut-the-knot.com/language/index.shtml
http://www.cut-the-knot.org/ctk/pww.shtml
http://www.cut-the-knot.org/language/index.shtml
http://www.cut-the-knot.org/proofs/index.shtml
http://www.education.txstate.edu/epic/mellwebdocs/
    SRSUlitreview.htm
http://www.ensculptic.com/mpg/fields/webpages/GilaHomepage/
    philosophyabstracts.html
http://www.fdavidpeat.com/bibliography/essays/maths.htm
http://www.fiz-karlsruhe.de/fiz/publications/zdm/zdm985r2.pdf
http://www.iigss.net/
http://www.indiana.edu/~mfl/cg.html
http://www.isbe.state.il.us/ils/math/standards.htm
http://www.lettredelapreuve.it/ICME9TG12/index.html
http://www.lettredelapreuve.it/TextesDivers/ICMETGProof96.html
http://www.maa.org/editorial/knot/Mathematics.html
http://www.maa.org/reviews/langmath.html
http://www.math.csusb.edu/notes/proofs/pfnot/node10.html
http://www.math.csusb.edu/notes/proofs/pfnot/pfnot.html
http://www.math.lamar.edu/MELL/index.html
http://www.math.montana.edu/math151/
http://www.math.rochester.edu/people/faculty/rarm/english.html
http://www.math.toronto.edu/barbeau/hannajoint.pdf
http://www.mathcamp.org/proofs.php
http://www.mathematicallycorrect.com/allen4.htm
http://www.mathmlconference.org/2002/presentations/naciri/
http://www.maths.ox.ac.uk/current-students/undergraduates/
    study-guide/p2.2.6.html
http://www.mtholyoke.edu/courses/rschwart/mac/writing/language.shtml
http://www.nctm.org/about/position_statements/
    position_statement_06.htm
http://www.nwrel.org/msec/science_inq/
http://www.quotedb.com/quotes/3002
http://www.righteducation.org/id28.htm
http://www.sciencemag.org/cgi/content/full/307/5714/1402a
http://www.sciencemag.org/sciext/125th/
http://www.southwestern.edu/~sawyerc/math-proofs.htm
http://www.theproofproject.org/bibliography
http://www.uoregon.edu/~moursund/Math/language.htm
```

http://www.utexas.edu/courses/bio301d/Topics/Scientific.method/
    Text.html
http://www.w3.org/Math/
http://www.warwick.ac.uk/staff/David.Tall/themes/proof.html
http://www.wmich.edu/math-stat/people/faculty/chartrand/proofs
http://www2.edc.org/makingmath/handbook/Teacher/Proof/Proof.asp
http://www2.edc.org/makingmath/mathtools/contradiction/
    contradiction.asp
http://www2.edc.org/makingmath/mathtools/proof/proof.asp
https://www.theproofproject.org/bibliography/

2

Preliminary Linear Algebra

This chapter includes a rapid review of basic concepts of Linear Algebra. After defining fields and vector spaces, we are going to cover bases, dimension and linear transformations. The theory of simultaneous equations and triangular factorization are going to be discussed as well. The chapter ends with the fundamental theorem of linear algebra.

2.1 Vector Spaces

2.1.1 Fields and linear spaces

Definition 2.1.1 *A set* $\mathbb{F}$ *together with two operations*

$$\begin{cases} + : \mathbb{F} \times \mathbb{F} \mapsto \mathbb{F} \ Addition \\ \cdot : \mathbb{F} \times \mathbb{F} \mapsto \mathbb{F} \ Multiplication \end{cases}$$

is called a field if

1. *a)* $\alpha + \beta = \beta + \alpha$, $\forall \alpha, \beta \in \mathbb{F}$ *(Commutative)*
   b) $(\alpha + \beta) + \gamma = \alpha + (\beta + \gamma)$, $\forall \alpha, \beta, \gamma \in \mathbb{F}$ *(Associative)*
   c) $\exists$ *a distinguished element denoted by* $0 \ni \forall \alpha \in \mathbb{F}$, $\alpha + 0 = \alpha$ *(Additive identity)*
   d) $\forall \alpha \in \mathbb{F} \exists - \alpha \in \mathbb{F} \ni \alpha + (-\alpha) = 0$ *(Existence of an inverse)*
2. *a)* $\alpha \cdot \beta = \beta \cdot \alpha$, $\forall \alpha, \beta \in \mathbb{F}$ *(Commutative)*
   b) $(\alpha \cdot \beta) \cdot \gamma = \alpha \cdot (\beta \cdot \gamma)$, $\forall \alpha, \beta, \gamma \in \mathbb{F}$ *(Associative)*
   c) $\exists$ *an element denoted by* $1 \ni \forall \alpha \in \mathbb{F}$, $\alpha \cdot 1 = \alpha$ *(Multiplicative identity)*
   d) $\forall \alpha \neq 0 \in \mathbb{F} \exists \alpha^{-1} \in \mathbb{F} \ni \alpha \cdot \alpha^{-1} = 1$ *(Existence of an inverse)*
3. $\alpha \cdot (\beta + \gamma) = (\alpha \cdot \beta) + (\alpha \cdot \gamma)$, $\forall \alpha, \beta, \gamma \in \mathbb{F}$ *(Distributive)*

Definition 2.1.2 *Let* $\mathbb{F}$ *be a field. A set* V *with two operations*

$$\begin{cases} + : V \times V \mapsto V \quad \textit{Addition} \\ \cdot : \mathbb{F} \times V \mapsto V \quad \textit{Scalar multiplication} \end{cases}$$

is called a vector space (linear space) over the field $\mathbb{F}$ *if the following axioms are satisfied:*

1. a) $u + v = u + v$, $\forall u, v \in V$
    b) $(u + v) + w = u + (v + w)$, $\forall u, v, w \in V$
    c) $\exists$ *a distinguished element denoted by* $\theta \ni \forall v \in V$, $v + \theta = v$
    d) $\forall v \in V \exists$ *unique* $-v \in V \ni v + (-v) = \theta$
2. a) $\alpha \cdot (\beta \cdot u) = (\alpha \cdot \beta) \cdot u$, $\forall \alpha, \beta \in \mathbb{F}$, $\forall u \in V$
    b) $\alpha \cdot (u + v) = (\alpha \cdot u) + (\alpha \cdot v)$, $\forall \alpha \in \mathbb{F}$, $\forall u, v \in V$
    c) $(\alpha + \beta) \cdot u = (\alpha \cdot u) + (\beta \cdot u)$, $\forall \alpha, \beta \in \mathbb{F}$, $\forall u \in V$
    d) $1 \cdot u = u$, $\forall u \in V$, *where 1 is the multiplicative identity of* $\mathbb{F}$

Example 2.1.3 $\mathbb{R}^n = \{(\alpha_1, \alpha_2, \dots, \alpha_n)^T : \alpha_1, \alpha_2, \dots, \alpha_n \in \mathbb{R}\}$ *is a vector space over* $\mathbb{R}$ *with* $(\alpha_1, \alpha_2, \dots, \alpha_n) + (\beta_1, \beta_2, \dots, \beta_n) = (\alpha_1 + \beta_1, \alpha_2 + \beta_2, \dots, \alpha_n + \beta_n)$; $c \cdot (\alpha_1, \alpha_2, \dots, \alpha_n) = (c\alpha_1, c\alpha_2, \dots, c\alpha_n)$; *and* $\theta = (0, 0, \dots, 0)^T$.

Example 2.1.4 *The set of all m by n complex matrices is a vector space over* $\mathbb{C}$ *with usual addition and multiplication.*

Proposition 2.1.5 *In a vector space* V,

i. θ *is unique.*
ii. $0 \cdot v = \theta$, $\forall v \in V$.
iii. $(-1) \cdot v = -v$, $\forall v \in V$.
iv. $-\theta = \theta$.
v. $\alpha \cdot v = \theta \Leftrightarrow \alpha = 0$ *or* $v = \theta$.

Proof. Exercise. $\square$

2.1.2 Subspaces

Definition 2.1.6 *Let* V *be a vector space over* $\mathbb{F}$, *and let* $W \subset V$. W *is called a subspace of* V *if* W *itself is a vector space over* $\mathbb{F}$.

Proposition 2.1.7 W *is a subspace of* V *if and only if it is closed under vector addition and scalar multiplication, that is*

$$w_1, w_2 \in W, \ \alpha_1, \alpha_2 \in \mathbb{F} \Leftrightarrow \alpha_1 \cdot w_1 + \alpha_2 \cdot w_2 \in W.$$

Proof. (Only if: $\Rightarrow$) Obvious by definition.
(If: $\Leftarrow$) we have to show that $\theta \in W$ and $\forall w \in W$, $-w \in W$.

i. Let $\alpha_1 = 1$, $\alpha_2 = -1$, and $w_1 = w_2$. Then,

$$1 \cdot w_1 + (-1) \cdot w_1 = w_1 + (-w_1) = \theta \in W.$$

ii. Take any w. Let $\alpha_1 = -1$, $\alpha_2 = 0$, and $w_1 = w$. Then,

$$(-1) \cdot w + (0) \cdot w_2 = -w \in W. \quad \square$$

Example 2.1.8 $S \subset \mathbb{R}^{2 \times 3}$, consisting of the matrices of the form $\begin{bmatrix} 0 & \beta & \gamma \\ \alpha & \alpha - \beta & \alpha + 2\gamma \end{bmatrix}$ is a subspace of $\mathbb{R}^{2 \times 3}$.

Proposition 2.1.9 If W_1, W_2 are subspaces, then so is $W_1 \cap W_2$.

Proof. Take $w_1, w_2 \in W_1 \cap W_2$, $\alpha_1, \alpha_2 \in \mathbb{F}$.

i. $w_1, w_2 \in W_1 \Rightarrow \alpha_1 \cdot w_1 + \alpha_2 \cdot w_2 \in W_1$
ii. $w_1, w_2 \in W_2 \Rightarrow \alpha_1 \cdot w_1 + \alpha_2 \cdot w_2 \in W_2$

Thus, $\alpha_1 w_1 + \alpha_2 w_2 \in W_1 \cap W_2$. $\quad \square$

Remark 2.1.10 If W_1, W_2 are subspaces, then $W_1 \cup W_2$ is not necessarily a subspace.

Definition 2.1.11 Let V be a vector space over $\mathbb{F}$, $X \subset V$. X is said to be linearly dependent if there exists a distinct set of $x_1, x_2, \ldots, x_k \in X$ and scalars $\alpha_1, \alpha_2, \ldots, \alpha_k \in \mathbb{F}$ not all zero $\ni \sum_{i=1}^{k} \alpha_i x_i = \theta$. Otherwise, for any subset of size k,

$$x_1, x_2, \ldots, x_k \in X, \ \sum_{i=1}^{k} \alpha_i x_i = \theta \Rightarrow \alpha_1 = \alpha_2 = \cdots = \alpha_k = 0.$$

In this case, X is said to be linearly independent.

We term an expression of the form $\sum_{i=1}^{k} \alpha_i x_i$ as linear combination. In particular, if $\sum_{i=1}^{k} \alpha_i = 1$, we call it affine combination. Moreover, if $\sum_{i=1}^{k} \alpha_i = 1$ and $\alpha_i \geq 0$, $\forall i = 1, 2, \ldots, k$, it becomes convex combination. On the other hand, if $\alpha_i \geq 0$, $\forall i = 1, 2, \ldots, k$; then $\sum_{i=1}^{k} \alpha_i x_i$ is said to be canonical combination.

Example 2.1.12 In $\mathbb{R}^n$, let $E = \{e_i\}_{i=1}^{n}$ where $e_i^T = (0, \cdots 0, 1, 0, \cdots, 0)$ is the ith canonical unit vector that contains 1 in its ith position and 0s elsewhere. Then, E is an independent set since

$$\theta = \alpha_1 e_1 + \cdots + \alpha_n e_n = \begin{bmatrix} \alpha_1 \\ \vdots \\ \alpha_n \end{bmatrix} \Rightarrow \alpha_i = 0, \forall i.$$

Let $X = \{x_i\}_{i=1}^{n}$ where $x_i^T = (0, \cdots 0, 1, 1, \cdots, 1)$ is the vector that contains 0s sequentially up to position i, and it contains 1s starting from position i onwards. X is also linearly independent since

$$\theta = \alpha_1 x_1 + \cdots + \alpha_n x_n = \begin{bmatrix} \alpha_1 \\ \alpha_1 + \alpha_2 \\ \vdots \\ \alpha_1 + \cdots + \alpha_n \end{bmatrix} \Rightarrow \alpha_i = 0, \ \forall i.$$

Let $Y = \{y_i\}_{i=1}^n$ where $y_i^T = (0, \cdots 0, -1, 1, 0, \cdots, 0)$ is the vector that contains -1 in i^{th} position, 1 in $(i+1)^{st}$ position, and 0s elsewhere. Y is not linearly independent since $y_1 + \cdots + y_n = \theta$.

Definition 2.1.13 Let $X \subset V$. The set

$$Span(X) = \left\{ v = \sum_{i=1}^k \alpha_i x_i \in V : x_1, x_2, \ldots, x_k \in X; \ \alpha_1, \alpha_2, \ldots, \alpha_k \in \mathbb{F}; \ k \in \mathbb{N} \right\}$$

is called the span of X. If the above linear combination is of the affine combination form, we will have the affine hull of X; if it is a convex combination, we will have the convex hull of X; and finally, if it is a canonical combination, what we will have is the cone of X. See Figure 2.1.

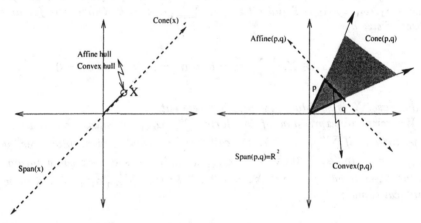

Fig. 2.1. The subspaces defined by $\{x\}$ and $\{p, q\}$.

Proposition 2.1.14 $Span(X)$ is a subspace of V.

Proof. Exercise. □

2.1.3 Bases

Definition 2.1.15 A set X is called a basis for V if it is linearly independent and spans V.

Remark 2.1.16 *Since $Span(X) \subset V$, in order to show that it covers V, we only need to prove that $\forall v \in V$, $v \in Span(X)$.*

Example 2.1.17 *In $\mathbb{R}^n$, $E = \{e_i\}_{i=1}^n$ is a basis since E is linearly independent and $\forall \alpha = (\alpha_1, \alpha_2, \ldots, \alpha_n)^T \in \mathbb{R}^n$, $\alpha = \alpha_1 e_1 + \cdots + \alpha_n e_n \in Span(E)$.*
$X = \{x_i\}_{i=1}^n$ is also a basis for $\mathbb{R}^n$ since $\forall \alpha = (\alpha_1, \alpha_2, \ldots, \alpha_n)^T \in \mathbb{R}^n$, $\alpha = \alpha_1 x_1 + (\alpha_2 - \alpha_1)x_2 + \cdots + (\alpha_n - \alpha_{n-1})x_n \in Span(X)$.

Proposition 2.1.18 *Suppose $X = \{x_i\}_{i=1}^n$ is a basis for V over $\mathbb{F}$. Then,*

a) $\forall v \in V$ can be expressed as $v = \sum_{i=1}^n \alpha_i x_i$ where α_i 's are unique.
b) Any linearly independent set with exactly n elements forms a basis.
c) All bases for V contain n vectors, where n is the dimension of V.

Remark 2.1.19 *Any vector space V of dimension n and an n–dimensional field $\mathbb{F}^n$ have an isomorphism.*

Proof. Suppose $X = \{x_i\}_{i=1}^n$ is a basis for V over $\mathbb{F}$. Then,

a) Suppose v has two different representations: $v = \sum_{i=1}^n \alpha_i x_i = \sum_{i=1}^n \beta_i x_i$. Then, $\theta = v - v = \sum_{i=1}^N (\alpha_i - \beta_i)x_i \Rightarrow \alpha_i = \beta_i$, $\forall i = 1, 2, \ldots, n$. Contradiction, since X is independent.

b) Let $Y = \{y_i\}_{i=1}^n$ be linearly independent. Then, $y_1 = \sum \delta_i x_i$ ($\spadesuit$), where at least one $\delta_i \neq 0$. Without loss of generality, we may assume that $\delta_1 \neq 0$. Consider $X_1 = \{y_1, x_2, \ldots, x_n\}$. X_1 is linearly independent since $\theta = \beta_1 y_1 + \sum_{i=2}^n \beta_i x_i = \beta_1(\sum \delta_i x_i)^{(\spadesuit)} + \sum_{i=2}^n \beta_i x_i = \beta_1 \delta_1 x_1 + \sum_{i=2}^n (\beta_1 \delta_i + \beta_i)x_i \Rightarrow \beta_1 \delta_1 = 0$; $\beta_1 \delta_i + \beta_i = 0$, $\forall i = 2, \ldots, n \Rightarrow \beta_1 = 0$ ($\delta_1 \neq 0$); and $\beta_i = 0$, $\forall i = 2, \ldots, n$. Any $v \in V$ can be expressed as $v = \sum_{i=1}^n \gamma_i x_i = \gamma_1 x_1 + \sum_{i=2}^n \gamma_i x_i$
$v = \gamma_1(\delta_1^{-1} y_1 - \sum_{i=2}^n \delta_1^{-1} \delta_i x_i)^{(\spadesuit)} = (\gamma_1 \delta_1^{-1})y_1 + \sum_{i=2}^n (\gamma_i - \gamma_1 \delta_1^{-1} \delta_i)x_i$.
Thus, $Span(X_1) = V$.
Similarly,
$X_2 = \{y_1, y_2, x_3, \ldots, x_n\}$ is a basis.
$\vdots$
$X_n = \{y_1, y_2, \ldots, y_n\} = Y$ is a basis.
c) Obvious from part b). $\square$

Remark 2.1.20 *Since bases for V are not unique, the same vector may have different representations with respect to different bases. The aim here is to find the best (simplest) representation.*

2.2 Linear transformations, matrices and change of basis

2.2.1 Matrix multiplication

Let us examine another operation on matrices, matrix multiplication, with the help of a small example. Let $A \in \mathbb{R}^{3 \times 4}$, $B \in \mathbb{R}^{4 \times 2}$, $C \in \mathbb{R}^{3 \times 2}$

$$\begin{bmatrix} c_{11} \ c_{12} \\ c_{21} \ c_{22} \\ c_{31} \ c_{32} \end{bmatrix} = C = AB = \begin{bmatrix} a_{11} \ a_{12} \ a_{13} \ a_{14} \\ a_{21} \ a_{22} \ a_{23} \ a_{24} \\ a_{31} \ a_{32} \ a_{33} \ a_{34} \end{bmatrix} \begin{bmatrix} b_{11} \ b_{12} \\ b_{21} \ b_{22} \\ b_{31} \ b_{32} \\ b_{41} \ b_{42} \end{bmatrix}$$

$$= \begin{bmatrix} a_{11}b_{11} + a_{12}b_{21} + a_{13}b_{31} + a_{14}b_{41} \ \ a_{11}b_{12} + a_{12}b_{22} + a_{13}b_{32} + a_{14}b_{42} \\ a_{21}b_{11} + a_{22}b_{21} + a_{23}b_{31} + a_{24}b_{41} \ \ a_{21}b_{12} + a_{22}b_{22} + a_{23}b_{32} + a_{24}b_{42} \\ a_{31}b_{11} + a_{32}b_{21} + a_{33}b_{31} + a_{34}b_{41} \ \ a_{31}b_{12} + a_{32}b_{22} + a_{33}b_{32} + a_{34}b_{42} \end{bmatrix}$$

Let us list the properties of this operation:

Proposition 2.2.1 *Let A, B, C, D be matrices and x be a vector.*

1. $(AB)x = A(Bx)$.
2. $(AB)C = A(BC)$.
3. $A(B+C) = AB + AC$ and $(B+C)D = BD + CD$.
4. $AB = BA$ does not hold (usually $AB \neq BA$) in general.
5. Let I_n be a square n by n matrix that has 1s along the main diagonal and 0s everywhere else, called identity matrix. Then, $AI = IA = A$.

2.2.2 Linear transformation

Definition 2.2.2 *Let $A \in \mathbb{R}^{m \times n}$, $x \in \mathbb{R}^n$. The map $x \mapsto Ax$ describing a transformation $\mathbb{R}^n \mapsto \mathbb{R}^m$ with property (matrix multiplication)*

$$\forall x, y \in \mathbb{R}^n; \ \forall a, b \in \mathbb{R}, \ A(bx + cy) = b(Ax) + c(Ay)$$

is called linear.

Remark 2.2.3 *Every matrix A leads to a linear transformation $\mathcal{A}$. Conversely, every linear transformation $\mathcal{A}$ can be represented by a matrix A. Suppose the vector space V has a basis $\{v_1, v_2, \ldots, v_n\}$ and the vector space W has a basis $\{w_1, w_2, \ldots, w_m\}$. Then, every linear transformation $\mathcal{A}$ from V to W is represented by an m by n matrix A. Its entries a_{ij} are determined by applying $\mathcal{A}$ to each v_j, and expressing the result as a combination of the w's:*

$$\mathcal{A}v_j = \sum_{i=1}^{m} a_{ij} w_i, \ j = 1, 2, \ldots, n.$$

Example 2.2.4 *Suppose $\mathcal{A}$ is the operation of integration of special polynomials if we take $1, t, t^2, t^3, \cdots$ as a basis where v_j and w_j are given by t^{j-1}. Then,*

$$\mathcal{A}v_j = \int t^{j-1} \, dt = \frac{t^j}{j} = \frac{1}{j} w_{j+1}.$$

For example, if dim $V = 4$ and dim $W = 5$ then $A = \begin{bmatrix} 0 & 0 & 0 & 0 \\ 1 & 0 & 0 & 0 \\ 0 & \frac{1}{2} & 0 & 0 \\ 0 & 0 & \frac{1}{3} & 0 \\ 0 & 0 & 0 & \frac{1}{4} \end{bmatrix}$. *Let us try*

to integrate $v(t) = 2t + 8t^3 = 0v_1 + 2v_2 + 0v_3 + 8v_4$:

$$\begin{bmatrix} 0 & 0 & 0 & 0 \\ 1 & 0 & 0 & 0 \\ 0 & \frac{1}{2} & 0 & 0 \\ 0 & 0 & \frac{1}{3} & 0 \\ 0 & 0 & 0 & \frac{1}{4} \end{bmatrix} \begin{bmatrix} 0 \\ 2 \\ 0 \\ 8 \end{bmatrix} = \begin{bmatrix} 0 \\ 0 \\ 1 \\ 0 \\ 2 \end{bmatrix} \Leftrightarrow \int (2t + 8t^3)\, dt = t^2 + 2t^4 = w_3 + 2w_5.$$

Proposition 2.2.5 *If the vector x yields coefficients of v when it is expressed in terms of basis $\{v_1, v_2, \ldots, v_n\}$, then the vector $y = Ax$ gives the coefficients of Av when it is expressed in terms of the basis $\{w_1, w_2, \ldots, w_m\}$. Therefore, the effect of A on any v is reconstructed by matrix multiplication.*

$$Av = \sum_{i=1}^{m} y_i w_i = \sum_{i,j} a_{ij} x_j w_i.$$

Proof.

$$v = \sum_{j=1}^{n} x_j v_j \Rightarrow Av = A\left(\sum_{1}^{n} x_j v_j\right) = \sum_{1}^{n} x_j Av_j = \sum_{j} x_j \sum_{i} a_{ij} w_i. \quad \square$$

Proposition 2.2.6 *If the matrices A and B represent the linear transformations A and B with respect to bases $\{v_i\}$ in V, $\{w_i\}$ in W, and $\{z_i\}$ in Z, then the product of these two matrices represents the composite transformation BA.*

Proof. $A : v \mapsto Av \quad B : Av \mapsto BAv \Rightarrow BA : v \mapsto BAv. \quad \square$

Example 2.2.7 *Let us construct 3×5 matrix that represents the second derivative $\frac{d^2}{dt^2}$, taking P_4 (polynomial of degree four) to P_2.*

$$t^4 \mapsto 4t^3, \; t^3 \mapsto 3t^2, \; t^2 \mapsto 2t, \; t \mapsto 1$$

$$\Rightarrow B = \begin{bmatrix} 0 & 1 & 0 & 0 & 0 \\ 0 & 0 & 2 & 0 & 0 \\ 0 & 0 & 0 & 3 & 0 \\ 0 & 0 & 0 & 0 & 4 \end{bmatrix}, \; A = \begin{bmatrix} 0 & 1 & 0 & 0 \\ 0 & 0 & 2 & 0 \\ 0 & 0 & 0 & 3 \end{bmatrix} \Rightarrow AB = \begin{bmatrix} 0 & 0 & 2 & 0 & 0 \\ 0 & 0 & 0 & 6 & 0 \\ 0 & 0 & 0 & 0 & 12 \end{bmatrix}.$$

Let $v(t) = 2t + 8t^3$, *then*

$$\frac{d^2 v(t)}{dt^2} = \begin{bmatrix} 0 & 0 & 2 & 0 & 0 \\ 0 & 0 & 0 & 6 & 0 \\ 0 & 0 & 0 & 0 & 12 \end{bmatrix} \begin{bmatrix} 0 \\ 2 \\ 0 \\ 8 \\ 0 \end{bmatrix} = \begin{bmatrix} 0 \\ 48 \\ 0 \end{bmatrix} = 48t.$$

Proposition 2.2.8 *Suppose* $\{v_1, v_2, \ldots, v_n\}$ *and* $\{w_1, w_2, \ldots, w_n\}$ *are both bases for the vector space* V, *and let* $v \in V$, $v = \sum_1^n x_j v_j = \sum_1^n y_j w_j$. *If* $v_j = \sum_1^n s_{ij} w_i$, *then* $y_i = \sum_1^n s_{ij} x_j$.

Proof.

$$\sum_j x_j v_j = \sum_j \sum_i x_j s_{ij} w_i \text{ is equal to } \sum_i y_i w_i \sum_i \sum_j s_{ij} x_j w_i. \quad \square$$

Proposition 2.2.9 *Let* $A : V \mapsto V$. *Let* A_v *be the matrix form of the transformation with respect to basis* $\{v_1, v_2, \ldots, v_n\}$ *and* A_w *be the matrix form of the transformation with respect to basis* $\{w_1, w_2, \ldots, w_n\}$. *Assume that* $v_j = \sum_i s_{ij} w_j$. *Then,*

$$A_v = S^{-1} A_w S.$$

Proof. Let $v \in V$, $v = \sum x_j v_j$. Sx gives the coefficients with respect to w's, then $A_w S x$ yields the coefficients of Av with respect to original w's, and finally $S^{-1} A_w S x$ gives the coefficients of Av with respect to original v's. $\square$

Remark 2.2.10 *Suppose that we are solving the system* $Ax = b$. *The most appropriate form of* A *is* I_n *so that* $x = b$. *The next simplest form is when* A *is diagonal, consequently* $x_i = \frac{b_i}{a_{ii}}$. *In addition, upper-triangular, lower-triangular and block-diagonal forms for* A *yield easy ways to solve for* x. *One of the main aims in applied linear algebra is to find a suitable basis so that the resultant coefficient matrix* $A_v = S^{-1} A_w S$ *has such a simple form.*

2.3 Systems of Linear Equations

2.3.1 Gaussian elimination

Let us take a system of linear m equations with n unknowns $Ax = b$. In particular,

$$\begin{aligned} 2u + v + w &= 1 \\ 4u + v &= -2 \\ -2u + 2v + w &= 7 \end{aligned} \Leftrightarrow \begin{bmatrix} 2 & 1 & 1 \\ 4 & 1 & 0 \\ -2 & 2 & 1 \end{bmatrix} \begin{bmatrix} u \\ v \\ w \end{bmatrix} = \begin{bmatrix} 1 \\ -2 \\ 7 \end{bmatrix}.$$

Let us apply some elementary row operations:

S1. Subtract 2 times the first equation from the second,
S2. Subtract -1 times the first equation from the third,
S3. Subtract -3 times the second equation from the third.

The result is an equivalent but simpler system, $Ux = c$ where U is upper-triangular:

$$\begin{bmatrix} 2 & 1 & 1 \\ 0 & -1 & -2 \\ 0 & 0 & -4 \end{bmatrix} \begin{bmatrix} u \\ v \\ w \end{bmatrix} = \begin{bmatrix} 1 \\ -4 \\ -4 \end{bmatrix}.$$

Definition 2.3.1 *A matrix U (L) is upper(lower)-triangular if all the entries below (above) the main diagonal are zero. A matrix D is called diagonal if all the entries except the main diagonal are zero.*

Remark 2.3.2 *If the coefficient matrix of a linear system of equations is either upper or lower triangular, then the solution can be characterized by backward or forward substitution. If it is diagonal, the solution is obtained immediately.*

Let us name the matrix that accomplishes S1 (E_{21}), subtracting twice the first row from the second to produce zero in entry $(2,1)$ of the new coefficient matrix, which is a modified I_3 such that its $(2,1)$st entry is -2. Similarly, the elimination steps S2 and S3 can be described by means of E_{31} and E_{32}, respectively.

$$E_{21} = \begin{bmatrix} 1 & 0 & 0 \\ -2 & 1 & 0 \\ 0 & 0 & 1 \end{bmatrix}, \ E_{31} = \begin{bmatrix} 1 & 0 & 0 \\ 0 & 1 & 0 \\ 1 & 0 & 1 \end{bmatrix}, \ E_{32} = \begin{bmatrix} 1 & 0 & 0 \\ 0 & 1 & 0 \\ 0 & 3 & 1 \end{bmatrix}.$$

These are called *elementary matrices*. Consequently,

$$E_{32} E_{31} E_{21} A = U \text{ and } E_{32} E_{31} E_{21} b = c,$$

where $E_{32} E_{31} E_{21} = \begin{bmatrix} 1 & 0 & 0 \\ -2 & 1 & 0 \\ -5 & 3 & 1 \end{bmatrix}$ is lower triangular. If we undo the steps of *Gaussian elimination* through which we try to obtain an upper-triangular system $Ux = c$ to reach the solution for the system $Ax = b$, we have

$$A = E_{32}^{-1} E_{31}^{-1} E_{21}^{-1} U = LU,$$

where

$$L = E_{21}^{-1} E_{31}^{-1} E_{32}^{-1} = \begin{bmatrix} 1 & 0 & 0 \\ 2 & 1 & 0 \\ 0 & 0 & 1 \end{bmatrix} \begin{bmatrix} 1 & 0 & 0 \\ 0 & 1 & 0 \\ -1 & 0 & 1 \end{bmatrix} \begin{bmatrix} 1 & 0 & 0 \\ 0 & 1 & 0 \\ 0 & 3 & 1 \end{bmatrix} = \begin{bmatrix} 1 & 0 & 0 \\ 2 & 1 & 0 \\ -1 & -3 & 1 \end{bmatrix}$$

is again lower-triangular. Observe that the entries below the diagonal are exactly the multipliers 2, -1, and -3 used in the elimination steps. We term L as the matrix form of the Gaussian elimination. Moreover, we have $Lc = b$. Hence, we have proven the following proposition that summarizes the Gaussian elimination or triangular factorization.

Proposition 2.3.3 *As long as pivots are nonzero, the square matrix A can be written as the product LU of a lower triangular matrix L and an upper triangular matrix U. The entries of L on the main diagonal are 1s; below the main diagonal, there are the multipliers l_{ij} indicating how many times of row j is subtracted from row i during elimination. U is the coefficient matrix, which appears after elimination and before back–substitution; its diagonal entries are the pivots.*

In order to solve $x = A^{-1}b = U^{-1}c = U^{-1}L^{-1}b$ we never compute inverses that would take n^3–many steps. Instead, we first determine c by forward-substitution from $Lc = b$, then find x by backward-substitution from $Ux = c$. This takes a total of n^2 operations. Here is our example,

$$\begin{bmatrix} 1 & 0 & 0 \\ 2 & 1 & 0 \\ -1 & -3 & 1 \end{bmatrix} \begin{bmatrix} c_1 \\ c_2 \\ c_3 \end{bmatrix} = \begin{bmatrix} 1 \\ -2 \\ 7 \end{bmatrix} \Rightarrow \begin{bmatrix} c_1 \\ c_2 \\ c_3 \end{bmatrix} = \begin{bmatrix} 1 \\ -4 \\ -4 \end{bmatrix} \Longrightarrow$$

$$\begin{bmatrix} 2 & 1 & 1 \\ 0 & -1 & -2 \\ 0 & 0 & -4 \end{bmatrix} \begin{bmatrix} x_1 \\ x_2 \\ x_3 \end{bmatrix} = \begin{bmatrix} 1 \\ -4 \\ -4 \end{bmatrix} \Rightarrow \begin{bmatrix} x_1 \\ x_2 \\ x_3 \end{bmatrix} = \begin{bmatrix} -1 \\ 2 \\ 1 \end{bmatrix}.$$

Remark 2.3.4 *Once factors U and L have been computed, the solution x' for any new right hand side b' can be found in the similar manner in only n^2 operations. For instance*

$$b' = \begin{bmatrix} 8 \\ 11 \\ 3 \end{bmatrix} \Rightarrow \begin{bmatrix} c'_1 \\ c'_2 \\ c'_3 \end{bmatrix} = \begin{bmatrix} 8 \\ -5 \\ -4 \end{bmatrix} \Rightarrow \begin{bmatrix} x'_1 \\ x'_2 \\ x'_3 \end{bmatrix} = \begin{bmatrix} 2 \\ 3 \\ 1 \end{bmatrix}.$$

Remark 2.3.5 *We can factor out a diagonal matrix D from U that contains pivots, as illustrated below.*

$$U = \begin{bmatrix} d_1 & & & \\ & d_2 & & \\ & & \ddots & \\ & & & d_n \end{bmatrix} \begin{bmatrix} 1 & \frac{u_{12}}{d_1} & \frac{u_{13}}{d_1} & \cdots & \frac{u_{1n}}{d_1} \\ & 1 & \frac{u_{23}}{d_2} & \cdots & \frac{u_{2n}}{d_2} \\ & & 1 & \cdots & \vdots \\ & & & \ddots & \vdots \\ & & & & 1 \end{bmatrix}$$

Consequently, we have $A = LDU$, where L is lower triangular with 1s on the main diagonal, U is upper diagonal with 1s on the main diagonal and D is the diagonal matrix of pivots. LDU factorization is uniquely determined.

Remark 2.3.6 *What if we come across a zero pivot? We have two possibilities:*

Case (i) If there is a nonzero entry below the pivot element in the same column:

We interchange rows. For instance, if we are faced with

$$\begin{bmatrix} 0 & 2 \\ 3 & 4 \end{bmatrix} \begin{bmatrix} u \\ v \end{bmatrix} = \begin{bmatrix} b_1 \\ b_2 \end{bmatrix},$$

we will interchange row 1 and 2. The permutation matrix, $P_{12} = \begin{bmatrix} 0 & 1 \\ 1 & 0 \end{bmatrix}$, represents the exchange. A permutation matrix P_{kl} is the modified identity matrix of the same order whose rows k and l are interchanged. Note that $P_{kl} = P_{lk}^{-1}$ (exercise!). In summary, we have

$$PA = LDU.$$

Case (ii) If the pivot column is entirely zero below the pivot entry:
The current matrix (so was A) is singular. Thus, the factorization is lost.

2.3.2 Gauss-Jordan method for inverses

Definition 2.3.7 The left (right) inverse B of A exists if $BA = I$ $(AB = I)$.

Proposition 2.3.8 $BA = I$ and $AC = I \Leftrightarrow B = C$.

Proof. $B(AC) = (BA)C \Leftrightarrow BI = IC \Leftrightarrow B = C$. □

Proposition 2.3.9 If A and B are invertible, so is AB.

$$(AB)^{-1} = B^{-1}A^{-1}.$$

Proof.

$$(AB)(B^{-1}A^{-1}) = A(BB^{-1})A^{-1} = AIA^{-1} = AA^{-1} = I.$$

$$(B^{-1}A^{-1})AB = B^{-1}(A^{-1}A)B = B^{-1}IB = B^{-1}B = I. \quad □$$

Remark 2.3.10 Let $A = LDU$. $A^{-1} = U^{-1}D^{-1}L^{-1}$ is never computed. If we consider $AA^{-1} = I$, one column at a time, we have $Ax_j = e_j, \forall j$. When we carry out elimination in such n equations simultaneously, we will follow the Gauss-Jordan method.

Example 2.3.11 In our example instance,

$$[A|e_1 e_2 e_3] = \begin{bmatrix} 2 & 1 & 1 & 1 & 0 & 0 \\ 4 & 1 & 0 & 0 & 1 & 0 \\ -2 & 2 & 1 & 0 & 0 & 1 \end{bmatrix} \rightarrow \begin{bmatrix} 2 & 1 & 1 & 1 & 0 & 0 \\ 0 & -1 & -2 & -2 & 1 & 0 \\ 0 & 3 & 2 & 1 & 0 & 1 \end{bmatrix}$$

$$\rightarrow \begin{bmatrix} 2 & 1 & 1 & 1 & 0 & 0 \\ 0 & -1 & -2 & -2 & 1 & 0 \\ 0 & 0 & -4 & -5 & 3 & 1 \end{bmatrix} = [U|L^{-1}] \rightarrow \begin{bmatrix} 1 & 0 & 0 & \frac{1}{8} & \frac{1}{8} & -\frac{1}{8} \\ 0 & 1 & 0 & -\frac{1}{2} & \frac{1}{2} & \frac{1}{2} \\ 0 & 0 & 1 & \frac{5}{4} & -\frac{3}{4} & -\frac{1}{4} \end{bmatrix} = [I|A^{-1}].$$

2.3.3 The most general case

In this subsection, we are going to concentrate on the equation system, $Ax = b$, where we have n unknowns and m equations.

Axiom 2.3.12 *The system $Ax = b$ is solvable if and only if the vector b can be expressed as the linear combination of the columns of A (lies in Span[columns of A] or geometrically lies in the subspace defined by columns of A).*

Definition 2.3.13 *The set of non-trivial solutions $x \neq \theta$ to the homogeneous system $Ax = \theta$ is itself a vector space called the null space of A, denoted by $\mathcal{N}(A)$.*

Remark 2.3.14 *All the possible cases in the solution of the simple scalar equation $\alpha x = \beta$ are below:*

- $\alpha \neq 0$: $\forall \beta \in \mathbb{R}$, $\exists x = \frac{\beta}{\alpha} \in \mathbb{R}$ *(nonsingular case),*
- $\alpha = \beta = 0$: $\forall x \in \mathbb{R}$ *are the solutions (undetermined case),*
- $\alpha = 0, \beta \neq 0$: *there is no solution (inconsistent case).*

Let us consider a possible LU decomposition of a given $A \in \mathbb{R}^{m \times n}$ with the help of the following example:

$$A = \begin{bmatrix} 1 & 3\,3\,2 \\ 2 & 6\,9\,5 \\ -1 & -3\,3\,0 \end{bmatrix} \rightarrow \begin{bmatrix} 1\,3\,3\,2 \\ 0\,0\,3\,1 \\ 0\,0\,6\,2 \end{bmatrix} \rightarrow \begin{bmatrix} 1\,3\,3\,2 \\ 0\,0\,3\,1 \\ 0\,0\,0\,0 \end{bmatrix} = U.$$

The final form of U is upper-trapezoidal.

Definition 2.3.15 *An upper-triangular (lower-triangular) rectangular matrix U is called upper-(lower-)trapezoidal if all the nonzero entries u_{ij} lie on and above (below) the main diagonal, $i \leq j$ ($i \geq j$). An upper-trapezoidal matrices has the following "echelon" form:*

$$\begin{bmatrix} \odot & * & * & * & * & * & * & * \\ 0 & \odot & * & * & * & * & * & * \\ 0 & 0 & 0 & \odot & * & * & * & * \\ 0 & 0 & 0 & 0 & 0 & 0 & \odot & * \\ 0 & 0 & 0 & 0 & 0 & 0 & 0 & 0 \\ 0 & 0 & 0 & 0 & 0 & 0 & 0 & 0 \end{bmatrix}$$

In order to obtain such an U, we may need row interchanges, which would introduce a permutation matrix P. Thus, we have the following theorem.

Theorem 2.3.16 *For any $A \in \mathbb{R}^{m \times n}$, there is a permutation matrix P, a lower–triangular matrix L, and an upper-trapezoidal matrix U such that $PA = LU$.*

Definition 2.3.17 *In any system $Ax = b \Leftrightarrow Ux = c$, we can partition the unknowns x_i as basic (dependent) variables those that correspond to a column with a nonzero pivot $\odot$, and free (nonbasic,independent) variables corresponding to columns without pivots.*

We can state all the possible cases for $Ax = b$ as we did in the previous remark without any proof.

Theorem 2.3.18 *Suppose the m by n matrix A is reduced by elementary row operations and row exchanges to a matrix U in echelon form. Let there be r nonzero pivots; the last $m - r$ rows of U are zero. Then, there will be r basic variables and $n - r$ free variables as independent parameters. The null space, $\mathcal{N}(A)$, composed of the solutions to $Ax = \theta$, has $n - r$ free variables.*

If $n = r$, then null space contains only $x = \theta$.

Solutions exist for every b if and only if $r = m$ (U has no zero rows), and $Ux = c$ can be solved by back-substitution.

If $r < m$, U will have $m - r$ zero rows. If one particular solution $\hat{x}$ to the first r equations of $Ux = c$ (hence to $Ax = b$) exists, then $\hat{x} + \alpha\dot{x}$, $\forall \dot{x} \in \mathcal{N}(A) \setminus \{\theta\}$, $\forall \alpha \in \mathbb{R}$ is also a solution.

Definition 2.3.19 *The number r is called the rank of A.*

2.4 The four fundamental subspaces

Remark 2.4.1 *If we rearrange the columns of A so that all basic columns containing pivots are listed first, we will have the following partition of U:*

$$A = [B|N] \rightarrow U = \left[\frac{U_B|U_N}{O}\right] \rightarrow V = \left[\frac{I_r|V_N}{O}\right]$$

where $B \in \mathbb{R}^{m \times r}$, $N \in \mathbb{R}^{m \times (n-r)}$, $U_B \in \mathbb{R}^{r \times r}$, $U_N \in \mathbb{R}^{r \times (n-r)}$, O is an $(m - r) \times n$ matrix of zeros, $V_N \in \mathbb{R}^{r \times (n-r)}$, and I_r is the identity matrix of order r. U_B is upper-triangular, thus non-singular.

If we continue from U and use elementary row operations to obtain I_r in the U_B part, like in the Gauss-Jordan method, we will arrive at the reduced row echelon form V.

2.4.1 The row space of A

Definition 2.4.2 *The row space of A is the space spanned by rows of A. It is denoted by $\mathcal{R}(A^T)$.*

$$\mathcal{R}(A^T) = Span(\{a_i\}_{i=1}^m) = \left\{y \in \mathbb{R}^m : y = \sum_{i=1}^m \alpha_i a_i\right\}$$

$$= \left\{d \in \mathbb{R}^m : \exists y \in \mathbb{R}^m \ni y^T A = d^T\right\}.$$

Proposition 2.4.3 *The row space of A has the same dimension r as the row space of U and the row space of V. They have the same basis, and thus, all the row spaces are the same.*

Proof. Each elementary row operation leaves the row space unchanged.   □

2.4.2 The column space of A

Definition 2.4.4 *The column space of A is the space spanned by the columns of A. It is denoted by $\mathcal{R}(A)$.*

$$\mathcal{R}(A) = Span\left\{a^j\right\}_{j=1}^n = \left\{y \in \mathbb{R}^n : y = \sum_{j=1}^n \beta_j a^j\right\}$$

$$= \{b \in \mathbb{R}^n : \exists x \in \mathbb{R}^n \ni Ax = b\}.$$

Proposition 2.4.5 *The dimension of column space of A equals the rank r, which is also equal to the dimension of the row space of A. The number of independent columns equals the number of independent rows. A basis for $\mathcal{R}(A)$ is formed by the columns of B.*

Definition 2.4.6 *The rank is the dimension of the row space or the column space.*

2.4.3 The null space (kernel) of A

Proposition 2.4.7

$$\mathcal{N}(A) = \{x \in \mathbb{R}^n : Ax = 0 (Ux = 0, Vx = 0)\} = \mathcal{N}(U) = \mathcal{N}(V).$$

Proposition 2.4.8 *The dimension of $\mathcal{N}(A)$ is $n - r$, and a base for $\mathcal{N}(A)$ is the columns of $T = \begin{bmatrix} -V_N \\ I_{n-r} \end{bmatrix}$.*

Proof.

$$Ax = 0 \Leftrightarrow Ux = 0 \Leftrightarrow Vx = 0 \Leftrightarrow x_B + V_N x_N = 0.$$

The columns of $T = \begin{bmatrix} -V_N \\ I_{n-r} \end{bmatrix}$ is linearly independent because of the last $(n-r)$ coefficients. Is their span $\mathcal{N}(A)$?

Let $y = \sum_j \alpha_j T^j$, $Ay = \sum_j \alpha_j(-V_N^j + V_N^j) = 0$. Thus, $Span(\{T^j\}_{j=1}^{n-r}) \subseteq \mathcal{N}(A)$. Is $Span(\{T^j\}_{j=1}^{n-r}) \supseteq \mathcal{N}(A)$? Let $x = \begin{bmatrix} x_B \\ x_N \end{bmatrix} \in \mathcal{N}(A)$. Then,

$$Ax = 0 \Leftrightarrow x_B + V_N x_N = 0 \Leftrightarrow x = \begin{bmatrix} x_B \\ x_N \end{bmatrix} = \begin{bmatrix} -V_N \\ I_{n-r} \end{bmatrix} x_N \in Span(\{T^j\}_{j=1}^{n-r})$$

Thus, $Span(\{T^j\}_{j=1}^{n-r}) \supseteq \mathcal{N}(A)$.   □

2.4.4 The left null space of A

Definition 2.4.9 *The subspace of $\mathbb{R}^m$ that consists of those vectors y such that $y^T A = \theta$ is known as the left null space of A.*

$$\mathcal{N}(A^T) = \{y \in \mathbb{R}^m : y^T A = \theta\}.$$

Proposition 2.4.10 *The left null space $\mathcal{N}(A^T)$ is of dimension $m - r$, where the basis vectors are the last $m - r$ rows of $L^{-1}P$ of $PA = LU$ or $L^{-1}PA = U$.*

Proof.

$$\bar{A} = [A|I_m] \rightarrow \bar{V} = \left[\frac{I_r|V_N}{O}\Big|L^{-1}P\right]$$

Then, $(L^{-1}P) = \begin{bmatrix} S_I \\ S_{II} \end{bmatrix}$, where S_{II} is the last $m - r$ rows of $L^{-1}P$. Then, $S_{II}A = \theta.$ □

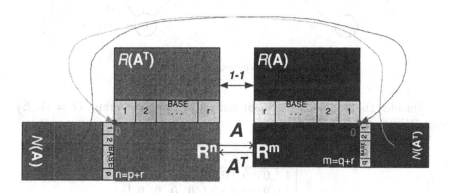

Fig. 2.2. The four fundamental subspaces defined by $A \in \mathbb{R}^{m \times n}$.

2.4.5 The Fundamental Theorem of Linear Algebra

Theorem 2.4.11 $\mathcal{R}(A^T) =$ *row space of A with dimension r;*
$\mathcal{N}(A) =$ *null space of A with dimension $n - r$;*
$\mathcal{R}(A) =$ *column space of A with dimension r;*
$\mathcal{N}(A^T) =$ *left null space of A with dimension $m - r$;*

Remark 2.4.12 *From this point onwards, we are going to assume that $n \geq m$ unless otherwise indicated.*

Problems

2.1. Graph spaces

Definition 2.4.13 *Let $GF(2)$ be the field with $+$ and $\times$ (addition and multiplication modulo 2 on $\mathbb{Z}^2$)*

$$
\begin{array}{c|cc}
+ & 0 & 1 \\
\hline
0 & 0 & 1 \\
1 & 1 & 0
\end{array}
\quad and \quad
\begin{array}{c|cc}
\times & 0 & 1 \\
\hline
0 & 0 & 0 \\
1 & 0 & 1
\end{array}
$$

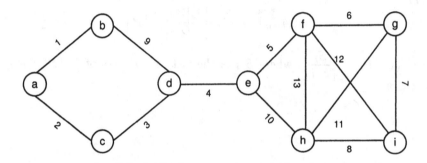

Fig. 2.3. The graph in Problem 2.1

Consider the node–edge incident matrix of the given graph $G = (V, E)$ over $GF(2)$, $A \in \mathbb{R}^{\|V\| \times \|E\|}$:

$$
A = \begin{array}{c}
\\ a \\ b \\ c \\ d \\ e \\ f \\ g \\ h \\ i
\end{array}
\begin{bmatrix}
1 & 2 & 3 & 4 & 5 & 6 & 7 & 8 & 9 & 10 & 11 & 12 & 13 \\
1 & 1 & 0 & 0 & 0 & 0 & 0 & 0 & 0 & 0 & 0 & 0 & 0 \\
1 & 0 & 0 & 0 & 0 & 0 & 0 & 0 & 1 & 0 & 0 & 0 & 0 \\
0 & 1 & 1 & 0 & 0 & 0 & 0 & 0 & 0 & 0 & 0 & 0 & 0 \\
0 & 0 & 1 & 1 & 0 & 0 & 0 & 0 & 1 & 0 & 0 & 0 & 0 \\
0 & 0 & 0 & 1 & 1 & 0 & 0 & 0 & 0 & 1 & 0 & 0 & 0 \\
0 & 0 & 0 & 0 & 1 & 1 & 0 & 0 & 0 & 0 & 0 & 1 & 1 \\
0 & 0 & 0 & 0 & 0 & 1 & 1 & 0 & 0 & 0 & 1 & 0 & 0 \\
0 & 0 & 0 & 0 & 0 & 0 & 0 & 1 & 0 & 1 & 1 & 0 & 1 \\
0 & 0 & 0 & 0 & 0 & 0 & 1 & 1 & 0 & 0 & 0 & 1 & 0
\end{bmatrix}
$$

The addition $+$ operator helps to point out the end points of the path formed by the added edges. For instance, if we add the first and ninth columns of A, we will have $[1, 0, 0, 1, 0, 0, 0, 0, 0]^T$, which indicates the end points (nodes a and d) of the path formed by edges one and nine.

(a) Find the reduced row echelon form of A working over $GF(2)$. Interpret the meaning of the bases.

(b) Let $T = \{1,2,3,4,5,6,7,8\}$ and $T^\perp = E \setminus T = \{9,10,11,12,13\}$.

Let $\bar{A} = \begin{bmatrix} I_8 & N \\ 0 & 0 \end{bmatrix}$. Let $Z = [I_8|N]$. For each row, $z_i, i \in T$, color the edges with non-zero entries. Interpret z_i

(c) Let $Y = \begin{bmatrix} N \\ I_5 \end{bmatrix}$. For each column y^j, $j \in T^\perp$, color the edges with non-zero entries. Interpret y_j.

(d) Find a basis for the four fundamental subspaces related with A.

2.2. Derivative of a polynomial

Let us concentrate on a $(n - k + 1) \times (n + 1)$ real valued matrix $A(n,k)$ that represents "taking k^{th} derivative of nth order polynomial"

$$P(t) = a_0 + a_1 t + \cdots + a_n t^n.$$

(a) Let $n = 5$ and $k = 2$. Characterize bases for the four fundamental subspaces related with $A(5,2)$.

(b) Find bases for and the dimensions of the four fundamental subspaces related with $A(n,k)$.

(c) Find $B(n,k)$, the right inverse of $A(n,k)$. Characterize the meaning of the underlying transformation and the four fundamental subspaces.

2.3. As in Example 2.1.12, let $Y = \{y_i\}_{i=1}^n$ be defined as

$$y_i^T = (0, \cdots 0, -1, 1, 0, \cdots, 0),$$

the vector that contains -1 in i^{th} position, 1 in $(i+1)^{st}$ position, and 0s elsewhere. Let $A = [y_1|y_2|\cdots|y_n]$. Characterize the four fundamental subspaces of A.

Web material

http://algebra.math.ust.hk/matrix_linear_trans/02_linear_transform/
    lecture5.shtml
http://algebra.math.ust.hk/vector_space/11_changebase/lecture4.shtml
http://archives.math.utk.edu/topics/linearAlgebra.html
http://calculusplus.cuny.edu/linalg.htm
http://ceee.rice.edu/Books/CS/chapter2/linear43.html
http://ceee.rice.edu/Books/CS/chapter2/linear44.html
http://dictionary.reference.com/search?q=vector%20space
http://distance-ed.math.tamu.edu/Math640/chapter1/node6.html
http://distance-ed.math.tamu.edu/Math640/chapter4/node2.html
http://distance-ed.math.tamu.edu/Math640/chapter4/node4.html
http://distance-ed.math.tamu.edu/Math640/chapter4/node6.html

```
http://en.wikibooks.org/wiki/Algebra/Linear_transformations
http://en.wikibooks.org/wiki/Algebra/Vector_spaces
http://en.wikipedia.org/wiki/Examples_of_vector_spaces
http://en.wikipedia.org/wiki/Fundamental_theorem_of_linear_algebra
http://en.wikipedia.org/wiki/Gauss-Jordan_elimination
http://en.wikipedia.org/wiki/Gaussian_elimination
http://en.wikipedia.org/wiki/Linear_transformation
http://en.wikipedia.org/wiki/Vector_space
http://encyclopedia.laborlawtalk.com/Linear_transformation
http://eom.springer.de/L/l059520.htm
http://eom.springer.de/t/t093180.htm
http://eom.springer.de/v/v096520.htm
http://euler.mcs.utulsa.edu/~class_diaz/cs2503/Spring99/lab7/
    node8.html
http://everything2.com/index.pl?node=vector%20space
http://graphics.cs.ucdavis.edu/~okreylos/ResDev/Geometry/
    VectorSpaceAlgebra.html
http://kr.cs.ait.ac.th/~radok/math/mat5/algebra12.htm
http://math.postech.ac.kr/~kwony/Math300/chapter2P.pdf
http://math.rice.edu/~hassett/teaching/221fall05/linalg5.pdf
http://mathforum.org/workshops/sum98/participants/sinclair/
    outline.html
http://mathonweb.com/help/backgd3e.htm
http://mathworld.wolfram.com/Gauss-JordanElimination.html
http://mathworld.wolfram.com/GaussianElimination.html
http://mathworld.wolfram.com/LinearTransformation.html
http://mathworld.wolfram.com/VectorSpace.html
http://mizar.uwb.edu.pl/JFM/Vol1/vectsp_1.html
http://planetmath.org/encyclopedia/GaussianElimination.html
http://planetmath.org/encyclopedia/
    ProofOfMatrixInverseCalculationByGaussianElimination.html
http://planetmath.org/encyclopedia/VectorField.html
http://planetmath.org/encyclopedia/VectorSpace.html
http://rkb.home.cern.ch/rkb/AN16pp/node101.html
http://thesaurus.maths.org/mmkb/entry.html?action=entryById&id=2243
http://triplebuffer.devmaster.net/file.php?id=5&page=1
http://tutorial.math.lamar.edu/AllBrowsers/2318/
    LinearTransformations.asp
http://uspas.fnal.gov/materials/3_LinearAlgebra.doc
http://vision.unige.ch/~marchand/teaching/linalg/
http://web.mit.edu/18.06/www/Video/video-fall-99.html
http://www-math.cudenver.edu/~wbriggs/5718s01/notes2/notes2.html
http://www-math.mit.edu/~djk/18_022/chapter16/section01.html
http://www.absoluteastronomy.com/v/vector_space
http://www.amath.washington.edu/courses/352-spring-2001/Lectures/
    lecture7_print.pdf
http://www.answers.com/topic/linear-transformation
http://www.answers.com/topic/vector-space
http://www.biostat.umn.edu/~sudiptob/pubh8429/
```

```
MatureLinearAlgebra.pdf
```
http://www.bookrags.com/sciences/mathematics/vector-spaces-wom.html
http://www.cap-lore.com/MathPhys/Vectors.html
http://www.cartage.org.lb/en/themes/Sciences/Mathematics/Algebra/
    foci/topics/transformations/transformations.htm
http://www.cee.umd.edu/menufiles/ence203/fall01/Chapter%205c%20
    (Simultaneous%20Linear%2http://www.sosmath.com/matrix/system1/
    system1.html
http://www.cs.berkeley.edu/~demmel/cs267/lectureSparseLU/
    lectureSparseLU1.html
http://www.cs.cityu.edu.hk/~luoyan/mirror/mit/ocw.mit.edu/18/
    18.013a/f01/required-readings/chapter04/section02.html
http://www.cs.nthu.edu.tw/~cchen/CS2334/ch4.pdf
http://www.cs.ut.ee/~toomas_l/linalg/lin1/node6.html
http://www.cs.ut.ee/~toomas_l/linalg/lin1/node7.html
http://www.cse.buffalo.edu/~hungngo/classes/2005/Expanders/notes/
    LA-intro.pdf
http://www.dc.uba.ar/people/materias/ocom/apunte1.doc
http://www.eas.asu.edu/~aar/classes/eee598S98/4vectorSpaces.txt
http://www.ee.ic.ac.uk/hp/staff/dmb/matrix/vector.html
http://www.ee.nchu.edu.tw/~minkuanc/courses/2006_01/LA/Lectures/
    Lecture%205%20-%202006.pdf
http://www.eng.fsu.edu/~cockburn/courses/eel5173_f01/four.pdf
http://www.everything2.com/index.pl?node_id=579183
http://www.fact-index.com/v/ve/vector_space_1.html
http://www.faqs.org/docs/sp/sp-129.html
http://www.fismat.umich.mx/~htejeda/aa/AS,L24.pdf
http://www.geometrictools.com/Books/GeometricTools/BookSample.pdf
http://www.krellinst.org/UCES/archive/classes/CNA/dir1.6/
    uces1.6.html
http://www.lehigh.edu/~brha/m43fall2004notes5_rev.pdf
http://www.library.cornell.edu/nr/bookcpdf/c2-2.pdf
http://www.ltcconline.net/greenl/courses/203/MatrixOnVectors/
    kernelRange.htm
http://www.ltcconline.net/greenl/courses/203/MatrixOnVectors/
    matrix_of_a_linear_transformatio.htm
http://www.ma.umist.ac.uk/tv/Teaching/Linear%20algebra%20B/
    Spring%202003/lecture2.pdf
http://www.math.byu.edu/~schow/work/GEnoP.htm
http://www.math.gatech.edu/~bourbaki/math2601/Web-notes/8.pdf
http://www.math.gatech.edu/~mccuan/courses/4305/notes.pdf
http://www.math.grin.edu/~stone/events/scheme-workshop/gaussian.html
http://www.math.harvard.edu/~elkies/M55a.99/field.html
http://www.math.hmc.edu/calculus/tutorials/lineartransformations/
http://www.math.hmc.edu/~su/pcmi/topics.pdf
http://www.math.jhu.edu/~yichen/teaching/2006spring/linear/
    review2.pdf
http://www.math.niu.edu/~beachy/aaol/fields.html
http://www.math.nps.navy.mil/~art/ma3046/handouts/Mat_Fund_Spa.pdf

http://www.math.poly.edu/courses/ma2012/Notes/GeneralLinearT.pdf
http://www.math.psu.edu/xu/451/HOMEWORK/computer6/node5.html
http://www.math.rutgers.edu/~useminar/basis.pdf
http://www.math.rutgers.edu/~useminar/lintran.pdf
http://www.math.sfu.ca/~lunney/macm316/hw05/node1.html
http://www.math.ubc.ca/~carrell/NB.pdf
http://www.math.uiuc.edu/documenta/vol-01/04.ps.gz
http://www.math.uiuc.edu/Software/magma/text387.html
http://www.math.uiuc.edu/~bergv/coordinates.pdf
http://www.mathcs.emory.edu/~rudolf/math108/summ1-2-3/node19.html
http://www.mathematik.uni-karlsruhe.de/mi2weil/lehre/stogeo2005s/
    media/cg.pdf
http://www.mathonweb.com/help/backgd3.htm
http://www.mathonweb.com/help/backgd3e.htm
http://www.mathreference.com/fld,intro.html
http://www.mathreference.com/la,1xmat.html
http://www.mathreference.com/la,xform.html
http://www.mathresource.iitb.ac.in/linear%20algebra/
    mainchapter6.2.html
http://www.maths.adelaide.edu.au/people/pscott/linear_algebra/lapf/
    24.html
http://www.maths.adelaide.edu.au/pure/pscott/linear_algebra/lapf/
    21.html
http://www.maths.nottingham.ac.uk/personal/sw/HG2NLA/gau.pdf
http://www.maths.qmul.ac.uk/~pjc/class_gps/ch1.pdf
http://www.mathwords.com/g/gaussian_elimination.htm
http://www.matrixanalysis.com/DownloadChapters.html
http://www.met.rdg.ac.uk/~ross/DARC/LinearVectorSpaces.html
http://www.numbertheory.org/courses/MP274/lintrans.pdf
http://www.phy.auckland.ac.nz/Staff/smt/453707/chap2.pdf
http://www.ping.be/~ping1339/lintf.htm
http://www.purplemath.com/modules/systlin5.htm
http://www.reference.com/browse/wiki/Vector_space
http://www.rsasecurity.com/rsalabs/node.asp?id=2370
http://www.sosmath.com/matrix/system1/system1.html
http://www.stanford.edu/class/ee387/handouts/lect07.pdf
http://www.swgc.mun.ca/~richards/M2051/M2051%20-March%2010%20-%
    20Vector%20Spaces%20and%20Subspaces.doc
http://www.ucd.ie/math-phy/Courses/MAPH3071/nummeth6.pdf
http://www.what-means.com/encyclopedia/Vector
http://www2.parc.com/spl/members/hhindi/reports/CvxOptTutPaper.pdf
http://xmlearning.maths.ed.ac.uk/eLearning/linear_algebra/
    binder.php?goTo=4-5-1-1
https://www.cs.tcd.ie/courses/baict/bass/4ict10/Michealmas2002/
    Handouts/12_Matrices.pdf

3

Orthogonality

In this chapter, we will analyze distance functions, inner products, projection and orthogonality, the process of finding an orthonormal basis, QR and singular value decompositions and conclude with a final discussion about how to solve the general form of $Ax = b$.

3.1 Inner Products

Following a rapid review of norms, an operation between any two vectors in the same space, inner product, is discussed together with the associated geometric implications.

3.1.1 Norms

Norms (distance functions, metrics) are vital in characterizing the type of network optimization problems like the Travelling Salesman Problem (TSP) with the rectilinear distance.

Definition 3.1.1 *A norm on a vector space V is a function that assigns to each vector, $v \in V$, a nonnegative real number $\|v\|$ satisfying*

i. $\|v\| > 0$, $\forall v \neq \theta$ and $\|\theta\| = 0$,
ii. $\|\alpha v\| = |\alpha|\,\|v\|$, $\forall \alpha \in \mathbb{R}$; $v \in V$.
iii. $\|u + v\| \leq \|u\| + \|v\|$, $\forall u, v \in V$ *(triangle inequality).*

Definition 3.1.2 $\forall x \in \mathbb{C}^n$, *the most commonly used norms,* $\|\cdot\|_1$, $\|\cdot\|_2$, $\|\cdot\|_\infty$, *are called the* l_1, l_2 *and* l_∞ *norms, respectively. They are defined as below:*

1. $\|x\|_1 = |x_1| + \cdots + |x_n|$,
2. $\|x\|_2 = (|x_1|^2 + \cdots + |x_n|^2)^{\frac{1}{2}}$,
3. $\|x\|_\infty = \max\{|x_1|, \ldots, |x_n|\}$.

Furthermore, we know the following relations:

$$\frac{\|x\|_2}{\sqrt{n}} \leq \|x\|_\infty \leq \|x\|_2\,,$$

$$\|x\|_2 \leq \|x\|_1 \leq \|x\|_2\,\sqrt{n},$$

$$\frac{\|x\|_1}{\sqrt{n}} \leq \|x\|_\infty \leq \|x\|_1\,.$$

Remark 3.1.3 *The good–old Euclidian distance is the l_2 norm that indicates the bird-flight distance. In Figure 3.1, for instance, a plane's trajectory between two points (given latitude and longitude pairs) projected on earth (assuming that it is flat!) is calculated by using the Pythagoras Formula. The rectilinear distance (l_1 norm) is also known as the Manhattan distance. It indicates the mere sum of the distances along the canonical unit vectors. It assumes the dependence of the movements along with the coordinate axes. In Figure 3.1, the length of the pathway restricted by blocks, of the car from the entrance of a district to the current location is calculated by adding the horizontal movement to the vertical. The Tchebychev's distance (l_∞) simply picks the maximum distance among all movements along the coordinate axes, and thus, assumes total independence. The forklift in Figure 3.1 can move sideways by its main engine, and it can independently raise or lower its fork by another motor. The total time it takes for the forklift to pick up an object 10m. away from a rack lying on the floor and place the object on a rack shelf 3m. above the floor is simply the maximum of the travel time and the raising time. A detailed formal discussion of metric spaces is located in Section 10.1.*

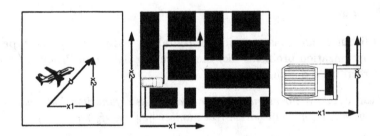

Fig. 3.1. Metric examples: $\|.\|_2$, $\|.\|_1$, $\|.\|_\infty$

Definition 3.1.4 *The length $\|x\|_2$ of a vector x in $\mathbb{R}^n$ is the positive square root of*

$$\|x\|^2 = \sum_{i=1}^{n} x_i^2.$$

Remark 3.1.5 $\|x\|_2^2$ *geometrically amounts to the Pythagoras formula applied (n-1) times.*

Definition 3.1.6 *The quantity* $x^T y$ *is called inner product of the vectors* x *and* y *in* $\mathbb{R}^n$

$$x^T y = \sum_{i=1}^{n} x_i y_i.$$

Proposition 3.1.7
$$x^T y = 0 \Leftrightarrow x \perp y.$$

Proof. ($\Leftarrow$) Pythagoras Formula: $\|x\|^2 + \|y\|^2 = \|x - y\|^2$,
$\|x - y\|^2 = \sum_{i=1}^{n} (x_i - y_i)^2 = \|x\|^2 + \|y\|^2 - 2x^T y$. The last two identities yield the conclusion, $x^T y = 0$.
($\Rightarrow$) $x^T y = 0 \Rightarrow \|x\|^2 + \|y\|^2 = \|x - y\|^2 \Rightarrow x \perp y$. $\square$

Theorem 3.1.8 (Schwartz Inequality)
$$\left| x^T y \right| \leq \|x\|_2 \|y\|_2, \quad x, y \in \mathbb{R}^n.$$

Proof. The following holds $\forall \alpha \in \mathbb{R}$:

$$0 \leq \|x + \alpha y\|_2^2 = x^T x + 2|\alpha| x^T y + \alpha^2 y^T y = \|x\|_2^2 + 2|\alpha| x^T y + \alpha^2 \|y\|_2^2, \quad (\maltese)$$

Case ($x \perp y$): In this case, we have $\Rightarrow x^T y = 0 \leq \|x\|_2 \|y\|_2$.
Case ($x \not\perp y$): Let us fix $\alpha = \frac{\|x\|_2^2}{x^T y}$. Then, $(\maltese)$ $0 \leq -\|x\|^2 + \frac{\|x\|_2^4 \|y\|_2^2}{(x^T y)^2}$. $\square$

3.1.2 Orthogonal Spaces

Definition 3.1.9 *Two subspaces* U *and* V *of the same space* $\mathbb{R}^n$ *are called orthogonal if* $\forall u \in U, \forall v \in V, \ u \perp v$.

Proposition 3.1.10 $\mathcal{N}(A)$ *and* $\mathcal{R}(A^T)$ *are orthogonal subspaces of* $\mathbb{R}^n$, $\mathcal{N}(A^T)$ *and* $\mathcal{R}(A)$ *are orthogonal subspaces of* $\mathbb{R}^m$.

Proof. Let $w \in \mathcal{N}(A)$ and $v \in \mathcal{R}(A^T)$ such that $Aw = \theta$, and $v = A^T x$ for some $x \in \mathbb{R}^n$. $w^T v = w^T (A^T x) = (w^T A^T) x = \theta^T x = 0$. $\square$

Definition 3.1.11 *Given a subspace* V *of* $\mathbb{R}^n$, *the space of all vectors orthogonal to* V *is called the orthogonal complement of* V, *denoted by* $V^\perp$.

Theorem 3.1.12 (Fundamental Theorem of Linear Algebra, Part 2)

$$\mathcal{N}(A) = (\mathcal{R}(A^T))^\perp, \quad \mathcal{R}(A^T) = (\mathcal{N}(A))^\perp,$$
$$\mathcal{N}(A^T) = (\mathcal{R}(A))^\perp, \quad \mathcal{R}(A) = (\mathcal{N}(A^T))^\perp.$$

Remark 3.1.13 *The following statements are equivalent.*

i. $W = V^\perp$.
ii. $V = W^\perp$.
iii. $W \perp V$ and $dimV + dimW = n$.

Proposition 3.1.14 *The following are true:*

i. $\mathcal{N}(AB) \supseteq \mathcal{N}(B)$.
ii. $\mathcal{R}(AB) \subseteq \mathcal{R}(A)$.
iii. $\mathcal{N}((AB)^T) \supseteq \mathcal{N}(A^T)$.
iv. $\mathcal{R}((AB)^T) \subseteq \mathcal{R}(B^T)$.

Proof. Consider the following:

i. $Bx = 0 \Rightarrow ABx = 0$. Thus, $\forall x \in \mathcal{N}(B)$, $x \in \mathcal{N}(AB)$.

ii. Let $b \ni ABx = b$ for some x, hence $\exists y = Bx \ni Ay = b$.

iii. Items (iii) and (iv) are similar, since $(AB)^T = B^T A^T$. □

Corollary 3.1.15
$$rank(AB) \leq rank(A),$$
$$rank(AB) \leq rank(B).$$

3.1.3 Angle between two vectors

See Figure 3.2 and below to prove the following proposition.

$$c = b - a \Rightarrow \cos c = \cos(b - a) = \cos b \cos a + \sin b \sin a$$

$$\cos c = \frac{u_1}{\|u\|} \frac{v_1}{\|v\|} + \frac{u_2}{\|u\|} \frac{v_2}{\|v\|} = \frac{u_1 v_1 + u_2 v_2}{\|u\| \|v\|}.$$

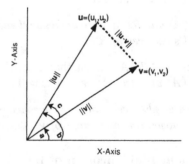

Fig. 3.2. Angle between vectors

Proposition 3.1.16 *The cosine of the angle between any two vectors u and v is*

$$\cos c = \frac{u^T v}{\|u\| \, \|v\|}.$$

Remark 3.1.17 *The law of cosines:*

$$\|u - v\|^2 = \|u\|^2 + \|v\|^2 - 2 \, \|u\| \, \|v\| \cos c.$$

3.1.4 Projection

Let $p = \bar{x}v$ where $\frac{\|p\|}{\|v\|} = \bar{x} \in \mathbb{R}$ is the scale factor. See Figure 3.3.

$$(u - p) \perp v \Leftrightarrow v^T(u - p) = 0 \Leftrightarrow \bar{x} = \frac{v^T u}{v^T v}.$$

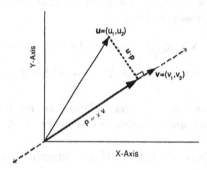

Fig. 3.3. Projection

Definition 3.1.18 *The projection p of the vector u onto the line spanned by the vector v is given by $p = \frac{u^T v}{v^T v} v$.*

The distance from the vector u to the line is (Schwartz inequality) therefore

$$\left\| u - \frac{v^T u}{v^T v} v \right\|^2 = u^T u - 2\frac{(v^T u)^2}{v^T v} + \left(\frac{v^T u}{v^T v}\right)^2 v^T v = \frac{(u^T u)(v^T v) - (v^T u)^2}{v^T v}.$$

3.1.5 Symmetric Matrices

Definition 3.1.19 *A square matrix A is called symmetric if $A^T = A$.*

Proposition 3.1.20 *Let $A \in \mathbb{R}^{m \times n}$, $rank(A) = r$. The product $A^T A$ is a symmetric matrix and $rank(A^T A) = r$.*

Proof. $(A^T A)^T = A^T (A^T)^T = A^T A.$
Claim: $\mathcal{N}(A) = \mathcal{N}(A^T A).$

i. $\mathcal{N}(A) \subseteq \mathcal{N}(A^T A)$: $x \in \mathcal{N}(A) \Rightarrow Ax = \theta \Rightarrow A^T Ax = A^T \theta = \theta \Rightarrow x \in \mathcal{N}(A^T A).$

ii. $\mathcal{N}(A^T A) \subseteq \mathcal{N}(A)$: $x \in \mathcal{N}(A^T A) \Rightarrow A^T Ax = \theta \Rightarrow x^T A^T Ax = \theta \Leftrightarrow \|Ax\|^2 = 0 \Leftrightarrow Ax = \theta, x \in \mathcal{N}(A).$ $\square$

Remark 3.1.21 $A^T A$ has n columns, so does A. Since $\mathcal{N}(A) = \mathcal{N}(A^T A)$, $dim\mathcal{N}(A) = n - r \Rightarrow dimR(A^T A) = n - (n - r) = r.$

Corollary 3.1.22 *If* $rank(A) = n \Rightarrow A^T A$ *is a square, symmetric, and invertible (non-singular) matrix.*

3.2 Projections and Least Squares Approximations

$Ax = b$ is solvable if $b \in R(A)$. If $b \notin R(A)$, then our problem is choose $\bar{x} \ni \|b - A\bar{x}\|$ is as small as possible.

$$A\bar{x} - b \perp R(A) \Leftrightarrow (Ay)^T (A\bar{x} - b) = 0 \Leftrightarrow$$

$$y^T [A^T A\bar{x} - A^T b] = 0 \ (y^T \neq \theta) \Rightarrow A^T A\bar{x} - A^T b = \theta \Rightarrow A^T A\bar{x} = A^T b.$$

Proposition 3.2.1 *The least squares solution to an inconsistent system* $Ax = b$ *of* m *equations and* n *unknowns satisfies* $A^T A\bar{x} = A^T b$ *(normal equations).*
If columns of A *are independent, then* $A^T A$ *is invertible, and the solution is*

$$\bar{x} = (A^T A)^{-1} A^T b.$$

The projection of b *onto the column space is therefore*

$$p = A\bar{x} = A(A^T A)^{-1} A^T b = Pb,$$

where the matrix $P = A(A^T A)^{-1} A^T$ *that describes this construction is known as projection matrix.*

Remark 3.2.2 $(I - P)$ *is another projection matrix which projects any vector* b *onto the orthogonal complement:* $(I - P)b = b - Pb.$

Proposition 3.2.3 *The projection matrix* $P = A(A^T A)^{-1} A^T$ *has two basic properties:*

a. *it is idempotent:* $P^2 = P.$
b. *it is symmetric:* $P^T = P.$

Conversely, any matrix with the above two properties represents a projection onto the column space of A.

Proof. The projection of a projection is itself.

$$P^2 = A[(A^T A)^{-1} A^T A](A^T A)^{-1} A^T = A(A^T A)^{-1} A^T = P.$$

We know that $(B^{-1})^T = (B^T)^{-1}$. Let $B = A^T A$.

$$P^T = (A^T)^T [(A^T A)^{-1}]^T A^T = A[A^T (A^T)^T]^{-1} A^T = A(A^T A)^{-1} A^T = P. \quad \square$$

3.2.1 Orthogonal bases

Definition 3.2.4 *A basis* $V = \{v_i\}_{i=1}^n$ *is called orthonormal if*

$$v_i^T v_j = \begin{cases} 0, \ i \neq j \ (ortagonality) \\ 1, \ i = j \ (normalization) \end{cases}$$

Example 3.2.5 $E = \{e_i\}_{i=1}^n$ *is an orthonormal basis for* $\mathbb{R}^n$, *whereas* $X = \{x_i\}_{i=1}^n$ *in Example 2.1.12 is not.*

Proposition 3.2.6 *If A is an m by n matrix whose columns are orthonormal (called an orthogonal matrix), then* $A^T A = I_n$.

$$P = AA^T = a_1 a_1^T + \cdots + a_n a_n^T \Rightarrow \bar{x} = A^T b$$

is the least squared solution for $Ax = b$.

Corollary 3.2.7 *An orthogonal matrix Q has the following properties:*

1. $Q^T Q = I = QQ^T$,
2. $Q^T = Q^{-1}$,
3. Q^T *is orthogonal.*

Example 3.2.8 *Suppose we project a point* $\alpha^T = (a, b, c)$ *into* $\mathbb{R}^2$ *plane. Clearly,* $p = (a, b, 0)$ *as it can be seen in Figure 3.4.*

$$e_1 e_1^T \alpha = \begin{bmatrix} a \\ 0 \\ 0 \end{bmatrix}, \ e_2 e_2^T \alpha = \begin{bmatrix} 0 \\ b \\ 0 \end{bmatrix}.$$

$$P = e_1 e_1^T + e_2 e_2^T = \begin{bmatrix} 1 & 0 & 0 \\ 0 & 1 & 0 \\ 0 & 0 & 0 \end{bmatrix}.$$

$$P\alpha = \begin{bmatrix} 1 & 0 & 0 \\ 0 & 1 & 0 \\ 0 & 0 & 0 \end{bmatrix} \begin{bmatrix} a \\ b \\ c \end{bmatrix} = \begin{bmatrix} a \\ b \\ 0 \end{bmatrix}.$$

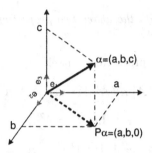

Fig. 3.4. Orthogonal projection

Remark 3.2.9 *When we find an orthogonal basis that spans the ground vector space and the coordinates of any vector with respect to this basis is on hand, the projection of this vector into a subspace spanned by any subset of the basis has coordinates 0 in the orthogonal complement and the same coordinates in the projected subspace. That is, the projection operation simply zeroes the positions other than the projected subspace like in the above example. One main aim of using orthogonal bases like $E = \{e_i\}_{i=1}^{n}$ for the Cartesian system, $\mathbb{R}^n$, is to have the advantage of simplifying projections, besides many other advantages like preserving lengths.*

Proposition 3.2.10 *Multiplication by an orthogonal Q preserves lengths*

$$\|Qx\| = \|x\|, \ \forall x;$$

and inner products

$$(Qx)^T(Qy) = x^T y, \ \forall x, y.$$

3.2.2 Gram-Schmidt Orthogonalization

Let us take two independent vectors a and b. We want to produce two perpendicular vectors v_1 and v_2:

$$v_1 = a, \ v_2 = b - p = b - \frac{v_1^T b}{v_1^T v_1} v_1 \Rightarrow v_1^T v_2 = 0 \Rightarrow v_1 \perp v_2.$$

If we have a third independent vector c, then

$$v_3 = c - \frac{v_1^T c}{v_1^T v_1} v_1 - \frac{v_2^T c}{v_2^T v_2} v_2 \Rightarrow v_3 \perp v_2, v_3 \perp v_1.$$

If we scale v_1, v_2, v_3, we will have orthonormal vectors:

$$q_1 = \frac{v_1}{\|v_1\|}, \ q_2 = \frac{v_2}{\|v_2\|}, \ q_3 = \frac{v_3}{\|v_3\|}.$$

Proposition 3.2.11 *Any set of independent vectors $a_1, a_2, \ldots, a_n$ can be converted into a set of orthogonal vectors $v_1, v_2, \ldots, v_n$ by the Gram-Schmidt process. First, $v_1 = a_1$, then each v_i is orthogonal to the preceding $v_1, v_2, \ldots, v_{i-1}$:*

$$v_i = a_i - \frac{v_1^T a_i}{v_1^T v_1} v_1 - \cdots - \frac{v_{i-1}^T a_i}{v_{i-1}^T v_{i-1}} v_{i-1}.$$

For every choice of i, the subspace spanned by original $a_1, a_2, \ldots, a_i$ is also spanned by $v_1, v_2, \ldots, v_i$. The final vectors

$$\left\{ q_i = \frac{v_i}{\|v_i\|} \right\}_{i=1}^{n}$$

are orthonormal.

Example 3.2.12 *Let* $a_1 = \begin{bmatrix} 1 \\ 0 \\ 1 \end{bmatrix}$, $a_2 = \begin{bmatrix} 1 \\ 1 \\ 0 \end{bmatrix}$, $a_3 = \begin{bmatrix} 0 \\ 1 \\ 1 \end{bmatrix}$.

$v_1 = a_1$, and

$$\frac{a_2^T v_1}{v_1^T v_1} = \tfrac{1}{2} \Rightarrow v_2 = a_2 - \tfrac{1}{2} v_1 = \begin{bmatrix} \frac{1}{2} \\ 1 \\ -\frac{1}{2} \end{bmatrix}.$$

$$\frac{a_3^T v_1}{v_1^T v_1} = \tfrac{1}{2}, \quad \frac{a_3^T v_2}{v_2^T v_2} = \tfrac{1}{3}, \Rightarrow v_3 = a_3 - \tfrac{1}{2} v_1 - \tfrac{1}{3} v_2 = \begin{bmatrix} -\frac{2}{3} \\ \frac{2}{3} \\ \frac{2}{3} \end{bmatrix}. \text{ Then,}$$

$$q_1 = \frac{v_1}{\|v_1\|} = \frac{1}{\sqrt{2}} \begin{bmatrix} 1 \\ 0 \\ 1 \end{bmatrix} = \begin{bmatrix} \frac{1}{\sqrt{2}} \\ 0 \\ \frac{1}{\sqrt{2}} \end{bmatrix}, \quad q_2 = \frac{v_2}{\|v_2\|} = \sqrt{\tfrac{2}{3}} \begin{bmatrix} \frac{1}{2} \\ 1 \\ -\frac{1}{2} \end{bmatrix} = \begin{bmatrix} \frac{1}{\sqrt{6}} \\ \frac{2}{\sqrt{6}} \\ -\frac{1}{\sqrt{6}} \end{bmatrix},$$

and $q_3 = \frac{v_3}{\|v_3\|} = \sqrt{\tfrac{9}{12}} \begin{bmatrix} -\frac{2}{3} \\ \frac{2}{3} \\ \frac{2}{3} \end{bmatrix} = \begin{bmatrix} -\frac{1}{\sqrt{3}} \\ \frac{1}{\sqrt{3}} \\ \frac{1}{\sqrt{3}} \end{bmatrix}.$

$$a_1 = v_1 = \sqrt{2} q_1$$
$$a_2 = \tfrac{1}{2} v_1 + v_2 = \sqrt{\tfrac{1}{2}} q_1 + \sqrt{\tfrac{3}{2}} q_2$$
$$a_3 = \tfrac{1}{2} v_1 + \tfrac{1}{3} v_2 + v_3 = \sqrt{\tfrac{1}{2}} q_1 + \sqrt{\tfrac{1}{6}} q_2 + \sqrt{\tfrac{4}{3}} q_3$$

$$\Leftrightarrow [a_1, a_2, a_3] = [q_1, q_2, q_3] \begin{bmatrix} \sqrt{2} & \sqrt{\frac{1}{2}} & \sqrt{\frac{1}{2}} \\ 0 & \sqrt{\frac{3}{2}} & \sqrt{\frac{1}{6}} \\ 0 & 0 & \sqrt{\frac{4}{3}} \end{bmatrix} \Leftrightarrow A = QR.$$

Proposition 3.2.13 *$A = QR$ where the columns of Q are orthonormal vectors, and R is upper-triangular with $\|v_i\|$ on the diagonal, therefore is invertible. If A is square, then so are Q and R.*

Definition 3.2.14 $A = QR$ *is known as Q–R decomposition.*

Remark 3.2.15 *If $A = QR$, then it is easy to solve $Ax = b$:*

$$\bar{x} = (A^T A)^{-1} A^T b = (R^T Q^T Q R)^{-1} R^T Q^T b = (R^T R)^{-1} R^T Q^T b = R^{-1} Q^T b.$$

$$R\bar{x} = Q^T b.$$

3.2.3 Pseudo (Moore-Penrose) Inverse

$$Ax = b \leftrightarrow A\bar{x} = p = Pb \leftrightarrow \bar{x} = (A^T A)^{-1} A^T b.$$

$A\bar{x} = p$ have only one solution $\Leftrightarrow$ The columns of A are linearly independent $\Leftrightarrow \mathcal{N}(A)$ contains only θ $\Leftrightarrow rank(A) = n$ $\Leftrightarrow A^T A$ is invertible.

Let $A^\dagger$ be *pseudo inverse* of A. If A is invertible, then $A^\dagger = A^{-1}$. Otherwise, $A^\dagger = (A^T A)^{-1} A^T$, if the above conditions hold. Then, $x = A^\dagger b$. Otherwise, the optimal solution is the solution of $A\bar{x} = p$ which is the one that has the minimum length.

Let $\bar{x}_0 \ni A\bar{x}_0 = p$, $\bar{x}_0 = \bar{x}_r + w$ where $\bar{x}_r \in \mathcal{R}(A^T)$ and $w \in \mathcal{N}(A)$. We have the following properties:

i. $A\bar{x}_r = A(\bar{x}_r + w) = A\bar{x}_0 = p$.
ii. $\forall \bar{x} \ni A\bar{x} = p$, $\bar{x} = \bar{x}_r + w$ with a variation in w part only, where $\bar{x}_r$ is fixed.
iii. $\|\bar{x}_r + w\|^2 = \|\bar{x}_r\|^2 + \|w\|^2$.

Proposition 3.2.16 *The optimal least squares solution to $Ax = b$ is $\bar{x}_r$ (or simply $\bar{x}$), which is determined by two conditions*

1. *$A\bar{x} = p$, where p is the projection of b onto the column space of A.*
2. *$\bar{x}$ lies in the row space of A.*

Then, $\bar{x} = A^\dagger b$.

Example 3.2.17 $A = \begin{bmatrix} 0 & 0 & 0 & 0 \\ 0 & \beta & 0 & 0 \\ 0 & 0 & \alpha & 0 \end{bmatrix}$ *where $\alpha > 0$, $\beta > 0$.*

Then, $\mathcal{R}(A) = \mathbb{R}^2$ and $p = Pb = (0, b_2, b_3)^T$.

$$A\bar{x} = p \Leftrightarrow \begin{bmatrix} 0 & 0 & 0 & 0 \\ 0 & \beta & 0 & 0 \\ 0 & 0 & \alpha & 0 \end{bmatrix} \begin{bmatrix} \bar{x}_1 \\ \bar{x}_2 \\ \bar{x}_3 \\ \bar{x}_4 \end{bmatrix} = \begin{bmatrix} 0 \\ b_2 \\ b_3 \end{bmatrix}$$

$$\bar{x}_2 = \frac{b_2}{\beta}, \quad \bar{x}_3 = \frac{b_3}{\alpha}, \quad \bar{x}_1 = \bar{x}_4 = 0, \text{ with the minimum length!}$$

$$\Rightarrow \bar{x} = \begin{bmatrix} 0 \\ \frac{b_2}{\beta} \\ \frac{b_3}{\alpha} \\ 0 \end{bmatrix} = A^\dagger b = \begin{bmatrix} 0 & 0 & 0 \\ 0 & \frac{1}{\beta} & 0 \\ 0 & 0 & \frac{1}{\alpha} \\ 0 & 0 & 0 \end{bmatrix} \begin{bmatrix} b_1 \\ b_2 \\ b_3 \end{bmatrix}. \text{ Thus, } A^\dagger = \begin{bmatrix} 0 & 0 & 0 \\ 0 & \frac{1}{\beta} & 0 \\ 0 & 0 & \frac{1}{\alpha} \\ 0 & 0 & 0 \end{bmatrix}.$$

3.2.4 Singular Value Decomposition

Definition 3.2.18 $A \in \mathbb{R}^{m \times n}$, $A = Q_1 \Sigma Q_2^T$ is known as singular value decomposition, where $Q_1 \in \mathbb{R}^{m \times m}$ orthogonal, $Q_2 \in \mathbb{R}^{m \times m}$ orthogonal, and Σ has a special diagonal form

$$\Sigma = \begin{bmatrix} \alpha & & & & & & \\ & \beta & & & & & \\ & & \ddots & & & & \\ & & & \gamma & & & \\ & & & & 0 & & \\ & & & & & \ddots & \\ & & & & & & 0 \end{bmatrix},$$

with the nonzero diagonal entries called singular values of A.

Proposition 3.2.19 $A^\dagger = Q_2 \Sigma^\dagger Q_1^T$ where $\Sigma^\dagger = \begin{bmatrix} \frac{1}{\alpha} & & & & & & \\ & \frac{1}{\beta} & & & & & \\ & & \ddots & & & & \\ & & & \frac{1}{\gamma} & & & \\ & & & & 0 & & \\ & & & & & \ddots & \\ & & & & & & 0 \end{bmatrix}.$

Proof. $\|Ax - b\| = \|Q_1 \Sigma Q_2^T x - b\| = \|\Sigma Q_2^T x - Q_1^T b\|.$
This is multiplied by Q_1^T $y = Q_2^T x = Q_2^{-1} x$ with $\|y\| = \|x\|$.

$$\min \|\Sigma y - Q_1^T b\| \rightarrow \bar{y} = \Sigma^\dagger Q_1^T b.$$

$$\Rightarrow \bar{x} = Q_2 \bar{y} = Q_2 \Sigma^\dagger Q_1^T b \Rightarrow A^\dagger = Q_2 \Sigma^\dagger Q_1^T. \quad \square$$

Remark 3.2.20 *A typical approach to the computation of the singular value decomposition is as follows. If the matrix has more rows than columns, a QR decomposition is first performed. The factor R is then reduced to a bidiagonal matrix. The desired singular values and vectors are then found by performing a bidiagonal QR iteration (see Remarks 6.2.3 and 6.2.8).*

3.3 Summary for $Ax = b$

Let us start with the simplest case which is illustrated in Figure 3.5. $A \in \mathbb{R}^{n \times n}$ is square, nonsingular (hence invertible), rank$(A) = n = r$. Thus, A represents a change-of-basis transformation from $\mathbb{R}^n$ onto itself. Since $n = r$, we have $\forall b \in \mathcal{R}(A) \equiv \mathbb{R}^n$. Therefore, there exists a unique solution $x = A^{-1}b$. If we have a decomposition of A ($PA = LU$, $A = QR$, $A = Q_1 \Sigma Q_2^T$), we follow an easy way to obtain the solution:

$(A = LU) \Rightarrow Lc = b, Ux = c$ using forward/backward substitutions as illustrated in the previous chapter;

$(A = QR) \Rightarrow Rx = Q^T b$ using backward substitution after multiplying the right hand side with Q^T;

$(A = Q_1 \Sigma Q_2^T) \Rightarrow x = Q_2 \Sigma^{-1} Q_1^T b$ using matrix multiplication operations after we take the inverse of the diagonal matrix Σ simply by inverting the diagonal elements.

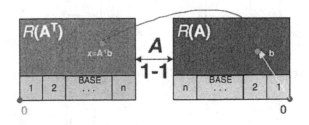

Fig. 3.5. Unique solution: $b \in \mathcal{R}(A)$, $A : n \times n$, and $r = n$

If $A \in \mathbb{R}^{m \times n}$ has full rank $r = m < n$, we choose any basis among the columns of $A = [B|N]$ to represent $\mathcal{R}(A) \equiv \mathbb{R}^m$ that contains b. In this case, we have a $p = n - m$ dimensional kernel $\mathcal{N}(A)$ whose elements, being the solutions to the homogeneous system $Ax = \theta$, extend the solution. Thus, we have infinitely many solutions $x_B = B^{-1}b - B^{-1}Nx_N$, given any basis B. One such solution is obtained by $x_N = \theta \Rightarrow x_B = B^{-1}b$ is called a *basic* solution. In this case, we may use decompositions of B ($B = LU$, $B = QR$, $B = Q_1 \Sigma Q_2^T$) to speed up the calculations.

If $A \in \mathbb{R}^{m \times n}$ has rank $r < m \leq n$ as given in Figure 3.6, we have $\dim(\mathcal{N}(A)) = p = n - r$, $\dim(\mathcal{N}(A^T)) = q = m - r$ and $\mathcal{R}(A) \equiv \mathcal{R}(A^T) \equiv \mathbb{R}^r$. The elementary row operations yield $A \to \begin{bmatrix} B|N \\ O_{q \times n} \end{bmatrix}$. There exists solution(s) only if $b \in \mathcal{R}(A)$. Assuming that we are lucky to have $b \in \mathcal{R}(A)$, and if $\hat{x}$ is a solution to the first r equations of $Ax = b$ (hence to $[B|N]x = b$), then $\hat{x} + \alpha \dot{x}$, $\forall \dot{x} \in \mathcal{N}(A) \setminus \{\theta\}$, $\forall \alpha \in \mathbb{R}$ is also a solution. Among all solutions $x_B = B^{-1}b - B^{-1}Nx_N$, $x_N = \theta \Rightarrow x_B = B^{-1}b$ is a basic solution. We may use decompositions of B to obtain x_B as well.

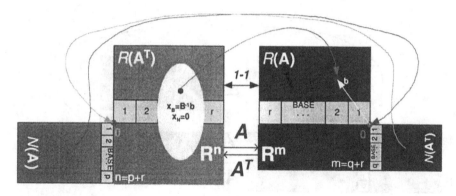

Fig. 3.6. Parametric solution: $b \in \mathcal{R}(A)$, $A : m \times n$, and $r = rank(A)$

What if $b \notin \mathcal{R}(A)$? We cannot find a solution. For instance, it is quite hard to fit a regression line passing through all observations. In this case, we are interested in the solutions, x, yielding the least squared error $\|b - Ax\|_2$. If $b \in \mathcal{N}(A^T)$, the projection of b over $\mathcal{R}(A)$ is the null vector θ. Therefore, $\mathcal{N}(A)$ is the collection of the solutions we seek.

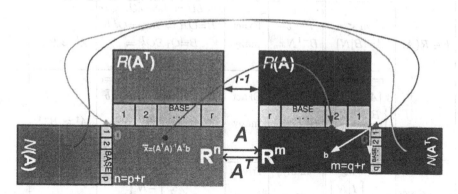

Fig. 3.7. Unique least squares solution: $(A^T A)$ is invertible and $A^\dagger = (A^T A)^{-1} A^T$

If b is contained totally in neither $\mathcal{R}(A)$ nor $\mathcal{N}(A^T)$, we are faced with the non-trivial least squared error minimization problem. If $A^T A$ is invertible, the unique solution is $\bar{x} = (A^T A)^{-1} A^T b$ as given in Figure 3.7. The regression line in Problem 3.2 is such a solution. We may use $A = QR$ or $A = Q_1 \Sigma Q_2^T$ decompositions to find this solution easily, in these ways: $R\bar{x} = Q^T b$ or $\bar{x} = Q_2 \Sigma^\dagger Q_1^T b$, respectively.

Otherwise, we have many $x \in \mathbb{R}^n$ leading to the least squared solution as in Figure 3.8. Among these solutions, we are interested in the solution with

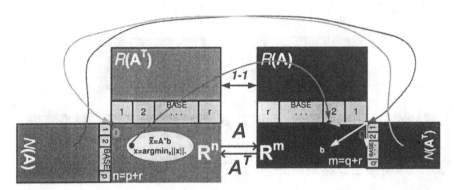

Fig. 3.8. Least norm squared solution: $(A^T A)$ is not invertible and $A^\dagger = Q_2 \Sigma^\dagger Q_1^T$

Table 3.1. How to solve $Ax = b$, where $A \in \mathbb{R}^{m \times n}$

Case	Subcase	Solution	Type	Special Forms	Inverse
$b \in \mathcal{R}(A)$	$r=n=m$	$x=A^{-1}b$	Exact unique	$A=LU \Rightarrow Lc=b, Ux=c$ $A=QR \Rightarrow Rx=Q^T b$ $A = Q_1 \Sigma Q_2^T \Rightarrow$ $x=Q_2 \Sigma^{-1} Q_1^T b$	$A^\dagger = A^{-1}$
	$r=m<n$ $A=[B\|N]$	$x_B=$ $B^{-1}b-$ $B^{-1}N x_n$	Exact many	$B=LU \Rightarrow Lc=b, U x_B=c$ $B=QR \Rightarrow R x_B=Q^T b$ $B=Q_1 \Sigma Q_2^T \Rightarrow$ $x_B=Q_2 \Sigma^{-1} Q_1^T b$	$A^\dagger \approx B^{-1}$
	$r=m<n$ $[A\|b] \to$ $\begin{bmatrix} B\|N\|\bar{b} \\ O \|\|O \end{bmatrix}$	$x_B=$ $B^{-1}\bar{b}-$ $B^{-1}N x_n$	Exact many	$B=LU \Rightarrow Lc=\bar{b}, U x_B=c$ $B=QR \Rightarrow R x_B=Q^T \bar{b}$ $B=Q_1 \Sigma Q_2^T \Rightarrow$ $x_B=Q_2 \Sigma^{-1} Q_1^T \bar{b}$	$A^\dagger \approx B^{-1}$
$b \in \mathcal{N}(A^T)$	$r<m$ $A \to \begin{bmatrix} I\|N \\ O \end{bmatrix}$	many $\forall \bar{x} \in$ $\mathcal{N}(A)$	Trivial Least Squares	$\bar{x}=\alpha^T \begin{bmatrix} -N \\ I \end{bmatrix},$ $\forall \alpha \in \mathbb{R}^{n-r}$	none
$b \notin \mathcal{R}(A)$ $b \notin \mathcal{N}(A)$	$(A^T A)$: invertible	$\bar{x}=A^\dagger b$	Unique Least Squares	$A=QR \Rightarrow R\bar{x}=Q^T b$ $A = Q_1 \Sigma Q_2^T \Rightarrow$ $\bar{x}=Q_2 \Sigma^{-1} Q_1^T b$	$A^\dagger =$ $(A^T A)^{-1} A^T$
	$(A^T A)$: not invertible	many $\bar{x}=A^\dagger b$ min.norm	Least Norm Squares	$A=Q_1 \Sigma Q_2^T \Rightarrow$ $\bar{x}=Q_2 \Sigma^\dagger Q_1^T b$	$A^\dagger =$ $Q_2 \Sigma^\dagger Q_1^T$

the smallest magnitude, in some engineering applications. We may use the singular value decomposition in this process.

The summary of the discussions about $Ax = b$ is listed in Table 3.1.

Problems

3.1. Q-R Decomposition

Find QR decomposition of $A = \begin{bmatrix} 1 & 2 & 0 & -1 \\ 1 & -1 & 3 & 2 \\ 1 & -1 & 3 & 2 \\ -1 & 1 & -3 & 1 \end{bmatrix}$.

3.2. Least Squares Approximation: Regression

Assume that you have sampled n pairs of data of the form (x,y). Find the regression line that minimizes the squared errors. Give an example for n=5.

3.3. Ax=b

Solve the following $Ax = b$ using the special decomposition forms.

(a) Let $A_1 = \begin{bmatrix} 1 & 3 & 2 \\ 2 & 1 & 3 \\ 3 & 2 & 1 \end{bmatrix}$ and $b_1 = \begin{bmatrix} 8 \\ 19 \\ 3 \end{bmatrix}$ using LU decomposition.

(b) $A_2 = \begin{bmatrix} 2 & 1 & 3 & 1 & 0 \\ 1 & 3 & 2 & 0 & 1 \\ 3 & 2 & 1 & 1 & 0 \end{bmatrix}$ and $b_2 = \begin{bmatrix} 8 \\ 19 \\ 3 \end{bmatrix}$ using LU decomposition. Find at least two solutions.

(c) $A_3 = \begin{bmatrix} 1 & 2 \\ 4 & 5 \\ 7 & 8 \\ 10 & 11 \end{bmatrix}$ and $b_3 = \begin{bmatrix} 2 \\ 5 \\ 6 \\ 8 \end{bmatrix}$ using QR decomposition.

(d) $A_4 = \begin{bmatrix} -1 & 0 & 0 & 1 \\ 1 & -1 & 0 & 0 \\ 0 & 1 & -1 & 0 \\ 0 & 0 & -1 & 1 \end{bmatrix}$ and $b_4 = \begin{bmatrix} 2 \\ 4 \\ 3 \\ 3 \end{bmatrix}$, using singular value decomposition.

Web material

http://abel.math.harvard.edu/~knill/math21b2002/10-orthogonal/
    orthogonal.pdf
http://astro.temple.edu/~dhill001/modern/1-sect6-2.pdf
http://avalon.math.neu.edu/~bridger/lschwart/lschwart.html
http://ccrma-www.stanford.edu/~jos/mdft/

Norm_Induced_Inner_Product.html
http://ccrma-www.stanford.edu/~jos/r320/Inner_Product.html
http://ccrma-www.stanford.edu/~jos/sines/
    Geometric_Signal_Theory.html
http://ccrma.stanford.edu/~jos/mdft/Inner_Product.html
http://cnx.org/content/m10561/latest/
http://elsa.berkeley.edu/~ruud/cet/excerpts/PartIOverview.pdf
http://en.wikipedia.org/wiki/Inner_product_space
http://en.wikipedia.org/wiki/Lp_space
http://en.wikipedia.org/wiki/Moore-Penrose_inverse
http://en.wikipedia.org/wiki/QR_decomposition
http://en.wikipedia.org/wiki/Singular_value_decomposition
http://engr.smu.edu/emis/8371/book/chap2/node8.html
http://eom.springer.de/P/p074290.htm
http://eom.springer.de/R/r130070.htm
http://epoch.uwaterloo.ca/~ponnu/syde312/algebra/page09.htm
http://epubs.siam.org/sam-bin/dbq/article/30478
http://genome-www.stanford.edu/SVD/
http://geosci.uchicago.edu/~gidon/geos31415/genLin/svd.pdf
http://info.wlu.ca/~wwwmath/faculty/vaughan/ma255/
    ma255orthogproj05.pdf
http://ingrid.ldeo.columbia.edu/dochelp/StatTutorial/SVD/
http://iria.pku.edu.cn/~jiangm/courses/IRIA/node119.html
http://isolatium.uhh.hawaii.edu/linear/lectures.htm
http://kwon3d.com/theory/jkinem/svd.html
http://library.lanl.gov/numerical/bookcpdf/c2-10.pdf
http://linneus20.ethz.ch:8080/2_2_1.html
http://lmb.informatik.uni-freiburg.de/people/dkats/
    DigitalImageProcessing/pseudoInvNew.pdf
http://mathnt.mat.jhu.edu/matlab/5-15.html
http://mathnt.mat.jhu.edu/matlab/5-6.html
http://maths.dur.ac.uk/~dma0wmo/teaching/1h-la/LAnotes/node18.html
http://mathworld.wolfram.com/InnerProductSpace.html
http://mathworld.wolfram.com/MatrixInverse.html
http://mathworld.wolfram.com/Moore-PenroseMatrixInverse.html
http://mathworld.wolfram.com/QRDecomposition.html
http://mathworld.wolfram.com/SchwarzsInequality.html
http://mathworld.wolfram.com/SingularValueDecomposition.html
http://mcraefamily.com/MathHelp/BasicNumberIneqCauchySchwarz.htm
http://mymathlib.webtrellis.net/matrices/vectorspaces.html
http://planetmath.org/encyclopedia/CauchySchwarzInequality.html
http://planetmath.org/encyclopedia/InnerProduct.html
http://planetmath.org/encyclopedia/InnerProductSpace.html
http://planetmath.org/encyclopedia/
    MoorePenroseGeneralizedInverse.html
http://planetmath.org/encyclopedia/NormedVectorSpace.html
http://planetmath.org/encyclopedia/OrthogonalityRelations.html
http://planetmath.org/encyclopedia/QRDecomposition.html
http://planetmath.org/encyclopedia/SingularValueDecomposition.html

http://psblade.ucdavis.edu/papers/ginv.pdf
http://public.lanl.gov/mewall/kluwer2002.html
http://rkb.home.cern.ch/rkb/AN16pp/node224.html
http://robotics.caltech.edu/~jwb/courses/ME115/handouts/pseudo.pdf
http://staff.science.uva.nl/~brandts/NW2/DOWNLOADS/hoofdstuk1.pdf
http://tutorial.math.lamar.edu/AllBrowsers/2318/
    InnerProductSpaces.asp
http://web.mit.edu/be.400/www/SVD/Singular_Value_Decomposition.htm
http://wks7.itlab.tamu.edu/Math640/notes7b.html
http://world.std.com/~sweetser/quaternions/quantum/bracket/
    bracket.html
http://www-ccrma.stanford.edu/~jos/mdft/Inner_Product.html
http://www.axler.net/Chapter6.pdf
http://www.ccmr.cornell.edu/~muchomas/8.04/1995/ps5/node17.html
http://www.cco.caltech.edu/~mihai/Ma8-Fall2004/Notes/Notes4/n4.pdf
http://www.cs.brown.edu/research/ai/dynamics/tutorial/Postscript/
    SingularValueDecomposition.ps
http://www.cs.hartford.edu/~bpollina/m220/html/7.1/
    7.1_InnerProducts.html
http://www.cs.rpi.edu/~flaherje/pdf/lin11.pdf
http://www.cs.unc.edu/~krishnas/eigen/node6.html
http://www.cs.ut.ee/~toomas_l/linalg/lin1/node10.html
http://www.csit.fsu.edu/~gallivan/courses/NLA2/set9.pdf
http://www.ctcms.nist.gov/~wcraig/variational/node2.html
http://www.ctcms.nist.gov/~wcraig/variational/node3.html
http://www.davidson.edu/math/will/svd/index.html
http://www.ee.ic.ac.uk/hp/staff/dmb/matrix/decomp.html
http://www.emis.de/journals/AM/99-1/cruells.ps
http://www.eurofreehost.com/ca/Cauchy-Schwartz_inequality.html
http://www.everything2.com/index.pl?node_id=53160
http://www.fiu.edu/~economic/wp2004/04-08.pdf
http://www.fmrib.ox.ac.uk/~tkincses/jc/SVD.pdf
http://www.fon.hum.uva.nl/praat/manual/
    generalized_singular_value_decomposition.html
http://www.free-download-soft.com/info/sdatimer.html
http://www.iro.umontreal.ca/~ducharme/svd/svd/index.html
http://www.library.cornell.edu/nr/bookcpdf/c2-6.pdf
http://www.mast.queensu.ca/~speicher/Section6.pdf
http://www.math.duke.edu/education/ccp/materials/linalg/leastsq/
    leas2.html
http://www.math.duke.edu/education/ccp/materials/linalg/orthog/
http://www.math.harvard.edu/~knill/teaching/math21b2002/
    10-orthogonal/orthogonal.pdf
http://www.math.ohio-state.edu/~gerlach/math/BVtypset/node6.html
http://www.math.psu.edu/~anovikov/math436/h-out/gram.pdf
http://www.math.sfu.ca/~ralfw/math252/week13.html
http://www.math.ucsd.edu/~gnagy/teaching/06-winter/Math20F/w9-F.pdf
http://www.math.umd.edu/~hck/461/s04/461s04m6.pdf
http://www.mathcs.duq.edu/larget/math496/qr.html

http://www.mathreference.com/top-ms,csi.html
http://www.maths.adelaide.edu.au/people/pscott/linear_algebra/lapf/
    32.html
http://www.maths.qmw.ac.uk/~sm/LAII/01ch5.pdf
http://www.mathworks.com/access/helpdesk/help/techdoc/math/
    mat_linalg25.html
http://www.matrixanalysis.com/Chapter5.pdf
http://www.mccormick.northwestern.edu/jrbirge/lec31_14nov2000.ppt
http://www.mccormick.northwestern.edu/jrbirge/lec33_17nov2000.ppt
http://www.nada.kth.se/kurser/kth/2D1220/Hsvd.pdf
http://www.nasc.snu.ac.kr/sheen/nla/html/node19.html
http://www.nationmaster.com/encyclopedia/Inner-product
http://www.netlib.org/lapack/lug/node53.html
http://www.public.asu.edu/~sergei/classes/mat242f99/LinAlg4.doc
http://www.reference.com/browse/wiki/Inner_product_space
http://www.sciencedaily.com/encyclopedia/lp_space
http://www.uwlax.edu/faculty/will/svd/
http://www.vias.org/tmdatanaleng/cc_matrix_pseudoinv.html
http://www.wooster.edu/math/linalg/LAFacts04.pdf
www.chu.edu.tw/~chlee/NA2003/NA2003-1.pdf
www.math.umn.edu/~olver/appl_/ort.pdf
www.math.uwo.ca/~aricha7/courses/283/week10.pdf
www.maths.lse.ac.uk/Personal/martin/fme5a.pdf
www.maths.qmw.ac.uk/~sm/LAII/01ch5.pdf

4

Eigen Values and Vectors

In this chapter, we will analyze determinant and its properties, definition of eigen values and vectors, different ways how to diagonalize square matrices and finally the complex case with Hermitian, unitary and normal matrices.

4.1 Determinants

4.1.1 Preliminaries

Proposition 4.1.1 $\det A \neq 0 \Rightarrow A$ is nonsingular.

Remark 4.1.2 Is $A - \lambda I$ (where λ is the vector of eigen values) invertible?

$$\det(A - \lambda I) =^? 0$$

where $\det(A - \lambda I)$ is a polynomial of degree n in λ, thus it has n roots.

Proposition 4.1.3 (Cramer's Rule) $Ax = b$ where A is nonsingular. Then, the solution for the jth unknown is

$$x_j = \frac{\det(A(j \leftarrow b))}{\det A},$$

where $A(j \leftarrow b)$ is the matrix obtained from A by interchanging column j with the right hand side b.

Proposition 4.1.4 $\det A = \pm$ [product of pivots].

Proposition 4.1.5 $|\det A| = Vol(P)$, where $P = conv\{\sum_{i=1}^{n} e_i a_i, e_i$ is the jth unit vector$\}$ is parallelepiped whose edges are from rows of A. See Figure 4.1.

Corollary 4.1.6 $|\det A| = \prod_{i=1}^{n} \|a_i\|$.

Definition 4.1.7 Let $\det A^{-1} = \frac{1}{\det A}$.

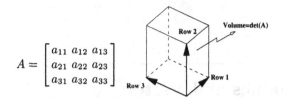

$$A = \begin{bmatrix} a_{11} & a_{12} & a_{13} \\ a_{21} & a_{22} & a_{23} \\ a_{31} & a_{32} & a_{33} \end{bmatrix}$$

Fig. 4.1. $|\det A| = Volume(P)$.

4.1.2 Properties

1. The determinant of I is 1.

   Example 4.1.8
   $$\begin{vmatrix} 1 & 0 \\ 0 & 1 \end{vmatrix} = 1.$$

2. The determinant is a linear function of any row, say the first row.

   Example 4.1.9
   $$\det \begin{bmatrix} a & b \\ c & d \end{bmatrix} = \begin{vmatrix} a & b \\ c & d \end{vmatrix} = ad - cb.$$

   $$\begin{vmatrix} ta & tb \\ c & d \end{vmatrix} = tad - tcd = t \begin{vmatrix} a & b \\ c & d \end{vmatrix}.$$

3. If A has a zero row, then $\det A = 0$.

   Example 4.1.10
   $$\begin{vmatrix} 0 & 0 \\ c & d \end{vmatrix} = 0.$$

4. The determinant changes sign when two rows are exchanged.

   Example 4.1.11
   $$\begin{vmatrix} c & d \\ a & b \end{vmatrix} = cb - ad = - \begin{vmatrix} a & b \\ c & d \end{vmatrix}.$$

5. The elementary row operations of subtracting a multiple of one row from another leaves the determinant unchanged.

   Example 4.1.12
   $$\begin{vmatrix} a - \alpha c & b - \alpha d \\ c & d \end{vmatrix} = (ad - \alpha cd) - (bc - \alpha cd) = ad - bc = \begin{vmatrix} a & b \\ c & d \end{vmatrix}.$$

6. If two rows are equal (singularity!), then $\det A = 0$.

   Example 4.1.13
   $$\begin{vmatrix} a & b \\ a & b \end{vmatrix} = 0.$$

7. $\det A^T = \det A$.

Example 4.1.14

$$\begin{vmatrix} a & c \\ b & d \end{vmatrix} = ad - cb = \begin{vmatrix} a & b \\ c & d \end{vmatrix}.$$

8. If A is triangular, then $\det A = \prod_{i=1}^{n} a_{ii}$ ($\det I = 1$).

Example 4.1.15

$$\begin{vmatrix} a & b \\ 0 & d \end{vmatrix} = ad, \quad \begin{vmatrix} a & 0 \\ c & d \end{vmatrix} = ad.$$

9. $A, B \in \mathbb{R}^{n \times n}$, nonsingular, $\det(AB) = (\det A)(\det B)$.

Example 4.1.16

$$\begin{vmatrix} a & b \\ c & d \end{vmatrix} \begin{vmatrix} e & f \\ g & h \end{vmatrix} = (ad - cb)(eh - gf) = adeh - adgf - cbeh + cbgf.$$

$$\begin{vmatrix} ae + bg & af + bh \\ ce + dg & cf + dh \end{vmatrix} = (ae + bg)(cf + dh) - (af + bh)(ce + dg)$$

$$= aecf + aedh + bgcf + bgdh - afce - afdg - bhce - bhdg$$

$$= adeh - adgf - cbeh + cbgf.$$

10. Let A be nonsingular, $A = P^{-1}LDU$. Then,

$$\det A = \det P^{-1} \det L \det D \det U = \pm(\text{product of pivots}).$$

The sign $\pm$ is the determinant of P^{-1} (or P) depending on whether the number of row exchanges is even or odd. We know $\det L = \det U = 1$ from property 7.

Example 4.1.17 *By one Gaussian elimination step, we have*

$$\begin{vmatrix} a & b \\ c & d \end{vmatrix} = \begin{vmatrix} 1 & 0 \\ \frac{a}{c} & 1 \end{vmatrix} \begin{vmatrix} a & 0 \\ 0 & \frac{ad-bc}{a} \end{vmatrix} \begin{vmatrix} 1 & \frac{b}{a} \\ 0 & 1 \end{vmatrix}, \text{ since } \begin{bmatrix} a & b \\ c & d \end{bmatrix} \rightarrow \begin{bmatrix} a & b \\ 0 & d - \frac{bc}{a} \end{bmatrix}. \text{ Thus,}$$

$$\begin{vmatrix} a & b \\ c & d \end{vmatrix} = ad - bc = \det D.$$

11. $\det A = a_{i1}A_{i1} + a_{i2}A_{i2} + \cdots + a_{in}A_{in}$ (property 1!) where A_{ij}'s are cofactors

$$A_{ij} = (-1)^{i+j} \det M_{ij}$$

where the minor M_{ij} is formed from A by deleting row i and column j.

Example 4.1.18

$$\begin{vmatrix} a_{11} & a_{12} & a_{13} \\ a_{21} & a_{22} & a_{23} \\ a_{31} & a_{32} & a_{33} \end{vmatrix} = \begin{vmatrix} a_{11} & & \\ & a_{22} & a_{23} \\ & a_{32} & a_{33} \end{vmatrix} + \begin{vmatrix} & a_{12} & \\ a_{21} & & a_{23} \\ a_{31} & & a_{33} \end{vmatrix} + \begin{vmatrix} & & a_{13} \\ a_{21} & a_{22} & \\ a_{31} & a_{32} & \end{vmatrix}$$

$$= a_{11}(a_{22}a_{33} - a_{23}a_{32}) + a_{12}(a_{23}a_{31} - a_{21}a_{33}) + a_{13}(a_{21}a_{32} - a_{22}a_{31})$$

$$= a_{11}a_{22}a_{33} + a_{12}a_{23}a_{31} + a_{13}a_{21}a_{32} - a_{11}a_{23}a_{32} - a_{12}a_{21}a_{33} - a_{13}a_{22}a_{31}.$$

4.2 Eigen Values and Eigen Vectors

Definition 4.2.1 *The number λ is an eigen value of A, with a corresponding nonzero eigen vector v such that $Av = \lambda v$.*

The last equation can be organized as $(\lambda I - A)v = \theta$. In order to have a non-trivial solution $v \neq \theta$, the corresponding null space (kernel) $\mathcal{N}(\lambda I - A)$ should contain vectors other than θ. Thus, the kernel has dimension larger than 0, which means we get at least one zero row in Gaussian elimination. Therefore, $(\lambda I - A)$ is singular. Hence, λ should be chosen such that $\det(\lambda I - A) = 0$. This equation is known as *characteristic equation* for A.

$$d(s) = \det(sI - A) = s^n + d_1 s^{n+1} + \cdots + d_n = 0.$$

Then, the eigen values are the roots.

$$d(s) = (s - \lambda_1)^{n_1}(s - \lambda_2)^{n_2} \ldots (s - \lambda_k)^{n_k} = \prod_{i=1}^{k}(s - \lambda_i)^{n_i}.$$

The sum of multiplicities should be equal to the dimension, i.e. $\sum_i n_i = n$. The sum of n-eigen values equals the sum of n-diagonal entries of A.

$$\lambda_1 + \cdots + \lambda_n = n_1 \lambda_1 + \cdots + n_k \lambda_k = a_{11} + \cdots + a_{nn}.$$

This sum is known as *trace* of A. Furthermore, the product of the n-eigen values equals the determinant of A.

$$\prod_{i=1}^{n} \lambda_i = \prod_{j=1}^{k} \lambda_i^{n_j} = \det A.$$

Remark 4.2.2 *If A is triangular, the eigen values $\lambda_1, \ldots, \lambda_n$ are the diagonal entries $a_{11}, \ldots, a_{nn}$.*

Example 4.2.3

$$A = \begin{bmatrix} \frac{1}{2} & 0 & 0 \\ \frac{1}{2} & 1 & 0 \\ 0 & \frac{1}{4} & \frac{3}{4} \end{bmatrix}.$$

$\det A = \frac{1}{2}(1)\frac{3}{4} = \frac{3}{8}$ *(property 8)*.

$$d(s) = \begin{vmatrix} s - \frac{1}{2} & 0 & 0 \\ -\frac{1}{2} & s - 1 & 0 \\ 0 & -\frac{1}{4} & s - \frac{3}{4} \end{vmatrix} = \left(s - \frac{1}{2}\right)(s - 1)\left(s - \frac{3}{4}\right).$$

So, $\lambda_1 = \frac{1}{2} = a_{11}$, $\lambda_2 = 1 = a_{22}$, $\lambda_3 = \frac{3}{4} = a_{33}$. *Finally,*

$$tr(A) = \frac{1}{2} + 1 + \frac{3}{4} = \frac{9}{4}.$$

4.3 Diagonal Form of a Matrix

Proposition 4.3.1 *Eigen vectors associated with distinct eigen values form a linearly independent set.*

Proof. Let $\lambda_i \leftrightarrow v_i$, $i = 1, \ldots, k$.
Consider $\sum_{i=1}^{n} \alpha_i v_i = \theta$. Multiply from the left by $\prod_{i=2}^{k}(A - \lambda_i I)$.
Since $(A - \lambda_i I) = \theta$, we obtain $(A - \lambda_i I)v_j = (\lambda_j - \lambda_i)v_j$, which yields

$$\alpha_1 (\lambda_1 - \lambda_2)(\lambda_1 - \lambda_3) \cdots (\lambda_1 - \lambda_k)v_1 = \theta.$$

$v_1 \neq \theta$, $\lambda_1 - \lambda_2 \neq 0$, $\ldots$, $\lambda_1 - \lambda_k \neq 0 \Rightarrow \alpha_1 = 0$. Then, we have $\sum_{i=2}^{n} \alpha_i v_i = \theta$.
Repeat by multiplying $\prod_{i=3}^{k}(A - \lambda_i I)$ to get $\alpha_2 = 0$, and so on.   □

4.3.1 All Distinct Eigen Values

$d(s) = \prod_{i=1}^{n}(s - \lambda_i)$. The n eigen vectors $v_1, \ldots, v_n$ form a linearly independent set. Choose them as a basis: $\{v_i\}_{i=1}^{n}$.

$$Av_1 = \lambda_1 v_1 + 0v_2 + \cdots + 0v_n$$

$$Av_2 = 0v_1 + \lambda_2 v_2 + \cdots + 0v_n$$

$$\vdots$$

$$Av_n = 0v_1 + 0v_2 + \cdots + \lambda_n v_n$$

Thus, A has representation $\Lambda = \begin{bmatrix} \lambda_1 & 0 & \cdots & 0 \\ 0 & \lambda_2 & \cdots & 0 \\ \vdots & \vdots & \ddots & \vdots \\ 0 & 0 & \cdots & \lambda_n \end{bmatrix}$.

Alternatively, let $S = [v_1 | v_2 | \cdots | v_n]$

$$AS = [Av_1 | Av_2 | \cdots | Av_n] = [\lambda_1 v_1 | \lambda_2 v_2 | \cdots | \lambda_n v_n]$$

$$AS = [v_1 | v_2 | \cdots | v_n] \begin{bmatrix} \lambda_1 & & & \\ & \lambda_2 & & \\ & & \ddots & \\ & & & \lambda_n \end{bmatrix} = S\Lambda.$$

Thus, $S^{-1}AS = \Lambda$ (Change of basis). Hence, we have proven the following theorem.

Theorem 4.3.2 *Suppose the n by n matrix A has n linearly independent eigen vectors. If these vectors are columns of a matrix S, then*

$$S^{-1}AS = \Lambda = \begin{bmatrix} \lambda_1 & & & \\ & \lambda_2 & & \\ & & \ddots & \\ & & & \lambda_n \end{bmatrix}.$$

Example 4.3.3 *From the previous example,*

$$A = \begin{bmatrix} \frac{1}{2} & 0 & 0 \\ \frac{1}{2} & 1 & 0 \\ 0 & \frac{1}{4} & \frac{3}{4} \end{bmatrix} \Rightarrow \lambda_1 = \frac{1}{2}, \ \lambda_2 = 1, \ \lambda_3 = \frac{3}{4}.$$

$$Ax = \lambda_1 x \Leftrightarrow \begin{bmatrix} \frac{1}{2}x_1 \\ \frac{1}{2}x_1 + x_2 \\ \frac{1}{4}x_2 + \frac{3}{4}x_3 \end{bmatrix} = \begin{bmatrix} \frac{1}{2}x_1 \\ \frac{1}{2}x_2 \\ \frac{1}{2}x_3 \end{bmatrix}.$$

$$\Leftrightarrow \left. \begin{matrix} \frac{1}{2}x_1 + \frac{1}{2}x_2 = 0 \Leftrightarrow x_1 + x_2 = 0. \\ \frac{1}{4}x_2 + \frac{1}{4}x_3 = 0 \Leftrightarrow x_2 + x_3 = 0. \end{matrix} \right\} \ \textit{Thus, } v_1 = \begin{bmatrix} 1 \\ -1 \\ 1 \end{bmatrix}.$$

$$Ax = \lambda_2 x \Leftrightarrow \begin{bmatrix} \frac{1}{2}x_1 \\ \frac{1}{2}x_1 + x_2 \\ \frac{1}{4}x_2 + \frac{3}{4}x_3 \end{bmatrix} = \begin{bmatrix} x_1 \\ x_2 \\ x_3 \end{bmatrix}.$$

$$\Leftrightarrow \left. \begin{matrix} x_1 = 0. \\ \frac{1}{4}x_2 - \frac{1}{4}x_3 = 0 \Leftrightarrow x_2 - x_3 = 0. \end{matrix} \right\} \ \textit{Thus, } v_2 = \begin{bmatrix} 0 \\ 1 \\ 1 \end{bmatrix}.$$

$$Ax = \lambda_3 x \Leftrightarrow \begin{bmatrix} \frac{1}{2}x_1 \\ \frac{1}{2}x_1 + x_2 \\ \frac{1}{4}x_2 + \frac{3}{4}x_3 \end{bmatrix} = \begin{bmatrix} \frac{3}{4}x_1 \\ \frac{3}{4}x_2 \\ \frac{3}{4}x_3 \end{bmatrix}.$$

$$\Leftrightarrow \left. \begin{matrix} x_1 = 0. \\ \frac{1}{2}x_1 - \frac{1}{4}x_2 = 0 \Rightarrow 2x_1 + x_2 = 0. \\ x_2 = 0. \end{matrix} \right\} \ \textit{Thus, } v_3 = \begin{bmatrix} 0 \\ 0 \\ 1 \end{bmatrix}.$$

*Therefore, * $S = \begin{bmatrix} 1 & 0 & 0 \\ -1 & 1 & 0 \\ 1 & 1 & 1 \end{bmatrix}.$

$$[S|I] = \begin{bmatrix} 1 & 0 & 0 & | & 1 & 0 & 0 \\ -1 & 1 & 0 & | & 0 & 1 & 0 \\ 1 & 1 & 1 & | & 0 & 0 & 1 \end{bmatrix} \to \begin{bmatrix} 1 & 0 & 0 & | & 1 & 0 & 0 \\ 0 & 1 & 0 & | & 1 & 1 & 0 \\ 0 & 1 & 1 & | & -1 & 0 & 1 \end{bmatrix} \to \begin{bmatrix} 1 & 0 & 0 & | & 1 & 0 & 0 \\ 0 & 1 & 0 & | & 1 & 1 & 0 \\ 0 & 0 & 1 & | & -2 & -1 & 1 \end{bmatrix} = [I|S^{-1}].$$

Then,

$$S^{-1}AS = \begin{bmatrix} 1 & 0 & 0 \\ 1 & 1 & 0 \\ -2 & -1 & 1 \end{bmatrix} \begin{bmatrix} \frac{1}{2} & 0 & 0 \\ \frac{1}{2} & 1 & 0 \\ 0 & \frac{1}{4} & \frac{3}{4} \end{bmatrix} \begin{bmatrix} 1 & 0 & 0 \\ -1 & 1 & 0 \\ 1 & 1 & 1 \end{bmatrix} = \begin{bmatrix} \frac{1}{2} & 0 & 0 \\ 0 & 1 & 0 \\ 0 & 0 & \frac{3}{4} \end{bmatrix} = \Lambda.$$

Remark 4.3.4 *Any matrix with distinct eigen values can be diagonalized. However, the diagonalization matrix S is not unique; hence neither is the basis $\{v\}_{i=1}^n$. If we multiply an eigen vector with a scalar, it will still remain an eigen vector. Not all matrices posses n linearly independent eigen vectors; therefore, some matrices are not dioganalizable.*

4.3.2 Repeated Eigen Values with Full Kernels

In this case, (recall that $d(s) = \prod_{i=1}^{k}(s - \lambda_i)^{n_i}$), we have $dim\mathcal{N}([A - \lambda_i I]) = n_i, \forall i$. Thus, there exists n_i linearly independent vectors in $\mathcal{N}([A - \lambda_i I])$, each of which is an eigen vector associated with λ_i, $\forall i$.

$$\lambda_1 \leftrightarrow v_{11}, v_{12}, \ldots, v_{1n_1}$$

$$\lambda_2 \leftrightarrow v_{21}, v_{22}, \ldots, v_{2n_2}$$

$$\vdots$$

$$\lambda_k \leftrightarrow v_{k1}, v_{k2}, \ldots, v_{kn_k}$$

$\bigcup_{i=1}^{n_i} \{v_{ij}\}_{j=1}^{n_i}$ is linearly independent (Exercise). Thus, we have obtained n linearly independent vectors, which constitute a basis. Consequently, we get

$$S^{-1}AS = \begin{bmatrix} \lambda_1 & & & & & & \\ & \ddots & & & & & \\ & & \lambda_1 & & & & \\ & & & \ddots & & & \\ & & & & \lambda_k & & \\ & & & & & \ddots & \\ & & & & & & \lambda_k \end{bmatrix}.$$

Example 4.3.5

$$A = \begin{bmatrix} 3 & 1 & -1 \\ 1 & 3 & -1 \\ 0 & 0 & 2 \end{bmatrix}.$$

$$d(s) = \det(sI - A) = \begin{vmatrix} s-3 & -1 & 1 \\ -1 & s-3 & 1 \\ 0 & 0 & s-2 \end{vmatrix} = 0$$

$$= (s-3)^2(s-2) - (s-2) = (s-2)[(s-3)^2 - 1]$$

$$= (s-2)(s-4)(s-2) = (s-2)^2(s-4).$$

$\Rightarrow \lambda_1 = 2, n_1 = 2$ and $\lambda_2 = 4, n_2 = 1$.

$$A - \lambda_1 I = \begin{bmatrix} 1 & 1 & -1 \\ 1 & 1 & -1 \\ 0 & 0 & 0 \end{bmatrix} \Rightarrow dim(\mathcal{N}([A - \lambda_1 I])) = 2.$$

$$v_{11} = (1, -1, 0)^T, v_{12} = (0, 1, 1)^T.$$

$$A - \lambda_2 I = \begin{bmatrix} -1 & 1 & -1 \\ 1 & -1 & -1 \\ 0 & 0 & -2 \end{bmatrix}.$$

$$v_2 = (1, 1, 0)^T.$$

$$S = \begin{bmatrix} 1 & 0 & 1 \\ -1 & 1 & 1 \\ 0 & 1 & 0 \end{bmatrix}, \quad S^{-1} = \begin{bmatrix} \frac{1}{2} & -\frac{1}{2} & \frac{1}{2} \\ 0 & 0 & 1 \\ \frac{1}{2} & \frac{1}{2} & -\frac{1}{2} \end{bmatrix}, \quad S^{-1}AS = \begin{bmatrix} 2 & 0 & 0 \\ 0 & 2 & 0 \\ 0 & 0 & 4 \end{bmatrix}.$$

4.3.3 Block Diagonal Form

In this case, we have

$$\exists i \ni n_i > 1, \ dim(\mathcal{N}[A - \lambda_i I]) < n_i$$

Definition 4.3.6 *The least degree monic (the polynomial with leading coefficient one) polynomial $m(s)$ that satisfies $m(A)=0$ is called the minimal polynomial of A.*

Proposition 4.3.7 *The following are correct for the minimal polynomial.*

i. $m(s)$ divides $d(s)$;
ii. $m(\lambda_i) = 0$, $\forall i = 1, 2, \ldots, k$;
iii. $m(s)$ is unique.

Example 4.3.8

$$A = \begin{bmatrix} c & 1 & 0 \\ 0 & c & 0 \\ 0 & 0 & c \end{bmatrix}, \ d(s) = \det(sI - A) = \begin{vmatrix} s-c & -1 & 0 \\ 0 & s-c & 0 \\ 0 & 0 & s-c \end{vmatrix} = (s-c)^3 = 0.$$

$\lambda_1 = c$, $n_1 = 3$. $m(s) =^? (s-c), (s-c)^2, (s-c)^3$:

$$[A - \lambda_1 I] = \begin{bmatrix} 0 & 1 & 0 \\ 0 & 0 & 0 \\ 0 & 0 & 0 \end{bmatrix} \neq O_3 \Rightarrow m(s) \neq (s-c).$$

$$[A - \lambda_1 I]^2 = \begin{bmatrix} 0 & 1 & 0 \\ 0 & 0 & 0 \\ 0 & 0 & 0 \end{bmatrix} \begin{bmatrix} 0 & 1 & 0 \\ 0 & 0 & 0 \\ 0 & 0 & 0 \end{bmatrix} = O_3 \Rightarrow m(s) = (s-c)^2.$$

Then, to find the eigen vectors

$$(A - cI)x = \theta \Leftrightarrow \begin{bmatrix} 0 & 1 & 0 \\ 0 & 0 & 0 \\ 0 & 0 & 0 \end{bmatrix} x = \theta \Rightarrow v_{11} = \begin{bmatrix} 1 \\ 0 \\ 0 \end{bmatrix}, \ v_{12} = \begin{bmatrix} 0 \\ 0 \\ 1 \end{bmatrix}.$$

Proposition 4.3.9

$$d(s) = \Pi_{i=1}^k (s - \lambda_i)^{n_i}, \ m(s) = \Pi_{i=1}^k (s - \lambda_i)^{m_i}, \ 1 \leq m_i \leq n_i, \ i = 1, 2, \ldots, k.$$

$$\mathcal{N}[(A - \lambda_i I)] \subsetneq \mathcal{N}[(A - \lambda_i I)^2] \subsetneq \cdots \subsetneq \mathcal{N}[(A - \lambda_i I)^{m_i}]$$
$$= \mathcal{N}[(A - \lambda_i I)^{m_i+1}] = \cdots = \mathcal{N}[(A - \lambda_i I)^{n_i}]$$

Proposition 4.3.10 $m(s) = \Pi_{i=1}^{k}(s - \lambda_i)^{m_i}$, then

$$\mathbb{C}^n = \mathcal{N}[(A - \lambda_1)^{m_1}] \oplus \cdots \oplus \mathcal{N}[(A - \lambda_k)^{m_k}],$$

where $\oplus$ is the direct sum of vector spaces.

Theorem 4.3.11 $d(s) = \Pi_{i=1}^{k}(s - \lambda_i)^{n_i}$, $m(s) = \Pi_{i=1}^{k}(s - \lambda_i)^{m_i}$.

i. $dim(\mathcal{N}[(A - \lambda_i)^{m_i}]) = n_i$;

ii. If columns of $n \times n_i$ matrices B_i form bases for $\mathcal{N}[(A - \lambda_i)^{m_i}]$ and $B = [B_1|\cdots|B_k]$, then B is nonsingular and

$$B^{-1}AB = \begin{bmatrix} \bar{A}_1 & & & \\ & \bar{A}_2 & & \\ & & \ddots & \\ & & & \bar{A}_k \end{bmatrix},$$

   where $\bar{A}_i$ are $n_i \times n_i$;

iii. Independent of the bases chosen for $\mathcal{N}[(A - \lambda_i)^{m_i}]$,

$$det(sI - \bar{A}_i) = (s - \lambda_i)^{n_i};$$

iv. Minimal polynomial of $\bar{A}_i$ is $(s - \lambda_i)^{m_i}$.

Example 4.3.12

$$A = \begin{bmatrix} 0 & 1 & 0 \\ 0 & 0 & 1 \\ 2 & -5 & 4 \end{bmatrix}, \ d(s) = \begin{vmatrix} s & -1 & 0 \\ 0 & s & -1 \\ -2 & 5 & s-4 \end{vmatrix} = (s-1)^2(s-2) = 0.$$

$\lambda_1 = 1, n_1 = 2; \lambda_2 = 2, n_2 = 1.$

$$[A - \lambda_1 I] = \begin{bmatrix} -1 & 1 & 0 \\ 0 & -1 & 1 \\ 2 & -5 & 3 \end{bmatrix}, \ dim(\mathcal{N}[(A - \lambda_1)]) = 1 < 2 = n_1(!)$$

$m_1 > 1 \Rightarrow m_1 = 2 \Rightarrow m(s) = (s - 1)^2(s - 2) = d(s).$

$$[A - \lambda_1 I]^2 = \begin{bmatrix} 1 & -2 & 1 \\ 2 & -4 & 2 \\ 4 & -8 & 4 \end{bmatrix}, \ dim(\mathcal{N}[(A - \lambda_1)^2]) = 2$$

$v_{11} = (1, 0, -1)^T, \ v_{12} = (0, 1, 2)^T, \ B_1 = \begin{bmatrix} 1 & 0 \\ 0 & 1 \\ -1 & 2 \end{bmatrix}.$

$$\lambda_2 = 2, \ [A - \lambda_2 I] = \begin{bmatrix} -2 & 1 & 0 \\ 0 & -2 & 1 \\ 2 & -5 & 2 \end{bmatrix}, \ dim(\mathcal{N}[(A - \lambda_2)]) = 1.$$

$v_2 = (1, 2, 4)^T$, $B_2 = \begin{bmatrix} 1 \\ 2 \\ 4 \end{bmatrix}$. *Therefore,*

$$B = \begin{bmatrix} 1 & 0 & 1 \\ 0 & 1 & 2 \\ -1 & 2 & 4 \end{bmatrix} \Rightarrow B^{-1} = \begin{bmatrix} 0 & 2 & -1 \\ -2 & 5 & -2 \\ 1 & -2 & 1 \end{bmatrix} \Rightarrow B^{-1}AB = \begin{bmatrix} 0 & 1 & \\ -1 & 2 & \\ & & 2 \end{bmatrix},$$

where $\bar{A}_1 = \begin{bmatrix} 0 & 1 \\ -1 & 2 \end{bmatrix}$ *and* $\bar{A}_2 = [2]$.

4.4 Powers of A

Example 4.4.1 (Compound Interest) *Let us take an example from engineering economy. Suppose you invest $ 500 for six years at 4 % in Citibank. Then,*

$$P_{k+1} = 1.04 P_k, \quad P_6 = (1.04)^6, \quad P_0 = (1.04)^6 500 = \$632.66.$$

Suppose, the time bucket is reduced to a month:

$$P_{k+1} = \left(1 + \frac{0.04}{12}\right) P_k, \quad P_{72} = \left(1 + \frac{0.04}{12}\right)^{72}, \quad P_0 = (1.00\bar{3})^{72} 500 = \$635.37.$$

What if we compound the interest daily?

$$P_{k+1} = \left(1 + \frac{0.04}{364}\right) P_k, \quad P_{6(364)+1.5} = \left(1 + \frac{0.04}{364}\right)^{2185.5}, \quad P_0 = \$635.72.$$

Thus, we have

$$\frac{P_{k+1} - P_k}{\Delta t} = 0.04 P_k \rightarrow \frac{dP}{dt} = 0.04 P \Rightarrow P(t) = e^{0.04t} P_0.$$

In the above simplest case, what we have is a difference/differential equation with one scalar variable. What if we have a matrix representing a set of difference/differential equation systems? What is e^{-At}?

Example 4.4.2 (Fibonacci Sequence)

$$F_{k+2} = F_{k+1} + F_k, \quad F_1 = 0, \quad F_2 = 1.$$

$$u_k = \begin{bmatrix} F_{k+1} \\ F_k \end{bmatrix}, \quad u_{k+1} = \begin{bmatrix} F_{k+2} \\ F_{k+1} \end{bmatrix} = \begin{bmatrix} 1 & 1 \\ 1 & 0 \end{bmatrix} \begin{bmatrix} F_{k+1} \\ F_k \end{bmatrix} = Au_k.$$

$$u_k = A^k u_0 = \begin{bmatrix} 1 & 1 \\ 1 & 0 \end{bmatrix}^k \begin{bmatrix} 1 \\ 0 \end{bmatrix}.$$

Hence, we sometimes need powers of a matrix!

4.4.1 Difference equations

Theorem 4.4.3 *If A can be diagonalized $(A = S\Lambda S^{-1})$, then*

$$u_k = A^k u_0 = (S\Lambda S^{-1})(S\Lambda S^{-1})\cdots(S\Lambda S^{-1})u_0 = S\Lambda^k S^{-1} u_0.$$

Remark 4.4.4

$$u_k = [v_1, \cdots, v_n] \begin{bmatrix} \lambda_1^k & & \\ & \ddots & \\ & & \lambda_n^k \end{bmatrix} S^{-1} u_0 = \alpha_1 \lambda_1^k v_1 + \cdots + \alpha_n \lambda_n^k v_n.$$

The general solution is a combination of special solutions $\lambda_i^k v_i$ and the coefficients α_i that match the initial condition u_0 are $\alpha_1 \lambda_1^0 v_1 + \cdots + \alpha_n \lambda_n^0 v_n = u_0$ or $S\alpha = u_0$ or $\alpha = S^{-1} u_0$. Thus, we have three different forms to the same equation.

Example 4.4.5 (Fibonacci Sequence, continued)

$$A = \begin{bmatrix} 1 & 1 \\ 1 & 0 \end{bmatrix}, \quad d(s) = \begin{vmatrix} s-1 & -1 \\ -1 & s \end{vmatrix} = s^2 - s - 1 = 0.$$

$$\lambda_1 = \frac{1+\sqrt{5}}{2}, \quad \lambda_2 = \frac{1-\sqrt{5}}{2};$$

$$A = S\Lambda S^{-1} = \begin{bmatrix} \lambda_1 & \lambda_2 \\ 1 & 1 \end{bmatrix} \begin{bmatrix} \lambda_1 & \\ & \lambda_2 \end{bmatrix} \begin{bmatrix} 1 & -\lambda_2 \\ -1 & \lambda_1 \end{bmatrix} \frac{1}{\lambda_1 - \lambda_2}.$$

$$\begin{bmatrix} F_{k+1} \\ F_k \end{bmatrix} = u_k = A^k u_0 = \begin{bmatrix} \lambda_1 & \lambda_2 \\ 1 & 1 \end{bmatrix} \begin{bmatrix} \lambda_1^k & \\ & \lambda_2^k \end{bmatrix} \begin{bmatrix} 1 \\ -1 \end{bmatrix} \frac{1}{\lambda_1 - \lambda_2}.$$

$$F_k = \frac{\lambda_1^k}{\lambda_1 - \lambda_2} - \frac{\lambda_2^k}{\lambda_1 - \lambda_2} = \frac{1}{\sqrt{5}} \left[\left(\frac{1+\sqrt{5}}{2} \right)^k - \left(\frac{1-\sqrt{5}}{2} \right)^k \right].$$

Since $\frac{1}{\sqrt{5}} \left(\frac{1-\sqrt{5}}{2} \right)^k < \frac{1}{2}$, $F_{1000} =$ the nearest integer to $\frac{1}{\sqrt{5}} \left(\frac{1+\sqrt{5}}{2} \right)^{1000}$.
Note that the ratio $\frac{F_{k+1}}{F_k} = \frac{1+\sqrt{5}}{2} \cong 1.618$ is known as the Golden Ratio, which represents the ratio of the lengths of the sides of the most elegant rectangle.

Example 4.4.6 (Markov Process) *Assume that the number of people leaving Istanbul annually is 5 % of its population, and the number of people entering is 1 % of Turkey's population outside Istanbul. Then,*

$$\begin{bmatrix} \#inside \\ \#outside \end{bmatrix} = \begin{bmatrix} y_1 \\ z_1 \end{bmatrix} = \begin{bmatrix} 0.95 & 0.01 \\ 0.05 & 0.99 \end{bmatrix} \begin{bmatrix} y_0 \\ z_0 \end{bmatrix}.$$

$$A = \begin{bmatrix} 0.95 & 0.01 \\ 0.05 & 0.99 \end{bmatrix}, \quad d(s) = \begin{vmatrix} s-0.95 & -0.01 \\ -0.05 & s-0.99 \end{vmatrix} = (s-1.0)(s-0.94).$$

$$\lambda_1 = 1.0, \ \lambda_2 = 0.94 \Rightarrow v_1 = \begin{bmatrix} \frac{1}{5} \\ 1 \end{bmatrix}, \ v_2 = \begin{bmatrix} 1 \\ -1 \end{bmatrix} \Rightarrow$$

$$A = S\Lambda S^{-1} = \begin{bmatrix} \frac{1}{5} & 1 \\ -1 & 1 \end{bmatrix} \begin{bmatrix} 1.00 & \\ & 0.94 \end{bmatrix} \begin{bmatrix} \frac{5}{6} & \frac{5}{6} \\ \frac{5}{6} & -\frac{1}{6} \end{bmatrix}.$$

$$\begin{bmatrix} y_k \\ z_k \end{bmatrix} = \begin{bmatrix} 0.95 & 0.01 \\ 0.05 & 0.99 \end{bmatrix}^k \begin{bmatrix} y_0 \\ z_0 \end{bmatrix} = \begin{bmatrix} \frac{1}{5} & 1 \\ -1 & 1 \end{bmatrix} \begin{bmatrix} 1.00^k & \\ & 0.94^k \end{bmatrix} \begin{bmatrix} \frac{5}{6} & \frac{5}{6} \\ \frac{5}{6} & -\frac{1}{6} \end{bmatrix} \begin{bmatrix} y_0 \\ z_0 \end{bmatrix}.$$

$$= \left(\frac{5}{6} y_0 + \frac{5}{6} z_0 \right) \begin{bmatrix} \frac{1}{5} \\ 1 \end{bmatrix} + \left(\frac{5}{6} y_0 - \frac{1}{6} z_0 \right) 0.94^k \begin{bmatrix} 1 \\ -1 \end{bmatrix}.$$

Since $0.94^k \to 0$ as $k \to \infty$,

$$\begin{bmatrix} y_\infty \\ z_\infty \end{bmatrix} = \left(\frac{5}{6} y_0 + \frac{5}{6} z_0 \right) \begin{bmatrix} \frac{1}{5} \\ 1 \end{bmatrix} = \begin{bmatrix} \frac{1}{6} & \frac{5}{6} \end{bmatrix} \begin{bmatrix} y_0 \\ z_0 \end{bmatrix}.$$

The steady-state probabilities are computed as in the classical way, $Au_\infty = 1 \cdot u_\infty$, corresponding to the eigen value of one. Thus, the steady-state vector is the eigen vector of A corresponding to $\lambda = 1$, after normalization to have legitimate probabilities (see Remark 4.3.4):

$$u_\infty = \alpha v_1 = \frac{5}{6} \begin{bmatrix} \frac{1}{5} \\ 1 \end{bmatrix} = \begin{bmatrix} \frac{1}{6} \\ \frac{5}{6} \end{bmatrix}.$$

4.4.2 Differential Equations

Example 4.4.7

$$\frac{du}{dt} = Au = \begin{bmatrix} 2 & 3 \\ 1 & 4 \end{bmatrix} u \Leftrightarrow u(t) = e^{At} u_0.$$

$$\lambda_1 = 5, \ v_1 = (1,1)^T, \ \lambda_2 = 1, \ v_2 = (-3,1)^T,$$

$$u(t) = \alpha_1 e^{\lambda_1 t} v_1 + \alpha_2 e^{\lambda_2 t} v_2 = \alpha_1 e^{5t} \begin{bmatrix} 1 \\ 1 \end{bmatrix} + \alpha_2 e^t \begin{bmatrix} -3 \\ 1 \end{bmatrix}.$$

$$u_0 = \alpha_1 \begin{bmatrix} 1 \\ 1 \end{bmatrix} + \alpha_2 \begin{bmatrix} -3 \\ 1 \end{bmatrix} = \begin{bmatrix} 1 & -3 \\ 1 & 1 \end{bmatrix} \begin{bmatrix} \alpha_1 \\ \alpha_2 \end{bmatrix}.$$

$$u(t) = \begin{bmatrix} 1 & -3 \\ 1 & 1 \end{bmatrix} \begin{bmatrix} e^{5t} & \\ & e^t \end{bmatrix} \begin{bmatrix} \alpha_1 \\ \alpha_2 \end{bmatrix} = S \begin{bmatrix} e^{5t} & \\ & e^t \end{bmatrix} S^{-1} u_0.$$

The power series expansion of the exponentiation of one scalar is

$$e^x = 1 + x + \frac{x^2}{2!} + + \frac{x^3}{3!} + \cdots$$

and if we generalize to the matrices

$$e^{At} = I + At + \frac{(At)^2}{2!} + \frac{(At)^3}{3!} + \cdots$$

If we take the derivative of both sides, we have

$$\frac{de^{At}}{dt} = I + A + \frac{A^2(2t)}{2!} + + \frac{A^3(3t^2)}{3!} + \cdots$$

$$= A\left[I + At + \frac{(At)^2}{2!} + \frac{(At)^3}{3!} + \cdots\right] = Ae^{At}.$$

If $A = SAS^{-1}$,

$$e^{At} = I + SAS^{-1} + \frac{SA^2S^{-1}t^2}{2!} + \frac{SA^3S^{-1}t^3}{3!} + \cdots$$

$$= S\left[I + At + \frac{(At)^2}{2!} + \frac{(At)^3}{3!} + \cdots\right]S^{-1} = Se^{At}S^{-1}.$$

Thus, we have the following theorem.

Theorem 4.4.8 *If A can be diagonalized as($A = SAS^{-1}$), then $\frac{du}{dt} = Au$ has the solution $u(t) = e^{At}u_0 = Se^{At}S^{-1}u_0$, or equivalently $u(t) = \alpha_1 e^{\lambda_1 t}v_1 + \cdots + \alpha_n e^{\lambda_n t}v_n$, where $\alpha = S^{-1}u_0$.*

4.5 The Complex case

In this section, we will investigate Hermitian and unitary matrices. The complex field $\mathbb{C}$ is defined over complex numbers (of the form $x + iy$ where $x, y \in \mathbb{R}$ and $i^2 = -1$) with the following operations:

$$(a + ib) + (c + id) = ((a + c) + i(b + d)) \quad (a + ib)(c + id) = ((ac - bd) + i(cb + ad)).$$

Definition 4.5.1 *The complex conjugate of $a + ib \in \mathbb{C}$ is $\overline{a + ib} = a - ib$. See Figure 4.2.*

Properties:

i. $\overline{(a + ib)(c + id)} = \overline{(a + ib)}\overline{(c + id)}$,

ii. $\overline{(a + ib) + (c + id)} = \overline{(a + ib)} + \overline{(c + id)}$,

iii. $(a + ib)\overline{a + ib} = a^2 + b^2 = r^2$ *where r is called modulus of $a + ib$.*
   We have $a = \sqrt{a^2 + b^2}\cos\theta$ and $b = \sqrt{a^2 + b^2}\sin\theta$ and

$$a + ib = \sqrt{a^2 + b^2}(\cos\theta + i\sin\theta) = re^{i\theta} \text{ (Polar Coordinates)},$$

where $re^{i\theta} = \cos\theta + i\sin\theta$.

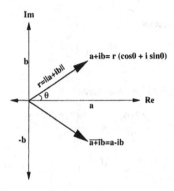

Fig. 4.2. Complex conjugate

Definition 4.5.2 $\overline{A}^T = A^H$ *with entries* $(A^H)_{ij} = (\overline{A})_{ij}$ *is known as conjugate transpose (Hermitian transpose).*

Properties:

i. $< x, y >= x^H y,\ x \perp y \Leftrightarrow x^H y = 0,$
ii. $\|x\| = (x^H x)^{\frac{1}{2}},$
iii. $(AB)^H = B^H A^H.$

Definition 4.5.3 *A is Hermitian if* $A^H = A.$

Properties:

i. $A^H = A,\ \forall x \in \mathbb{C}^n,\ x^H A x \in \mathbb{R}^n.$
ii. *Every eigen value of a Hermitian matrix is real.*
iii. *The eigen vectors of a Hermitian matrix, if they correspond to different eigen values, are orthogonal to each other.*
iv. *(Spectral Theorem)*
    $A = A^H$, *there exists a diagonalizing unitary (complex matrix of orthonormal vectors as columns) U such that*

$$U^{-1}AU = U^H AU = \Lambda.$$

Therefore, any Hermitian matrix can be decomposed into

$$A = U \Sigma U^H = \lambda_1 v_1 v_1^H + \cdots + \lambda_n v_n v_n^H.$$

Definition 4.5.4 *If* $B = M^{-1}AM$ *(change of variables), then A and B have the same eigen values with the same multiplicities, termed as A is similar to B.*

Properties:

i. $A \in \mathbb{C}^{m \times n}$, $\exists$ unitary $M = U \ni U^{-1}AU = T$ is upper-triangular. The eigen values of A must be shared by the similar matrix T and appear along the main diagonal.

ii. Any Hermitian matrix A can be diagonalized by a suitable U.

Definition 4.5.5 *The matrix N is called normal if $NN^H = N^H N$. Only for normal matrices, $T = U^{-1}NU = \Lambda$ where Λ is diagonal.*

Problems

4.1. Determinant
Prove property 11 in Section 4.1.2.

4.2. Jordan form

$$\text{Let } A = \begin{bmatrix} 1 & 1 & -1 & -1 & -1 \\ 2 & 1 & 1 & 2 & 1 \\ 0 & 1 & 1 & 0 & -1 \\ 1 & -1 & 1 & 3 & 1 \\ 2 & -2 & 2 & 2 & 4 \end{bmatrix}. \text{ Find } S \text{ such that } S^{-1}AS = \begin{bmatrix} 2 & 1 & & & \\ & 2 & & & \\ & & 2 & 1 & \\ & & & 2 & \\ & & & & 2 \end{bmatrix}.$$

Hint:
Choose $v_2 \in \mathcal{N}[(A - \lambda I)^2]$, $v_1 = [A - \lambda I]v_2$. Similarly, choose v_4 and v_3. Finally, choose $v_5 \in \mathcal{N}[(A - \lambda I)]$.

4.3. Using Jordan Decomposition

$$\text{Let } A = \begin{bmatrix} \frac{1}{10} & \frac{1}{10} & 0 \\ 0 & \frac{1}{10} & \frac{1}{10} \\ 0 & 0 & \frac{1}{10} \end{bmatrix}. \text{ Find } A^{10}.$$

4.4. Differential Equation System
Let the Blue (allied) forces be in a combat situation with the Red (enemy) forces. There are two Blue units (X_1, X_2) and two Red military units (Y_1, Y_2). At the start of the combat, the first Blue unit has 100 ($X_1^0 = 100$) combatants, the second Blue unit has 60 ($X_2^0 = 60$) combatants. The initial conditions for the Red force are $Y_1^0 = 40$ and $Y_2^0 = 30$. Since the start of the battle ($t = 0$), the number of surviving combatants (less than the initial values due to attrition) decrease monotonically and the values are denoted by X_1^t, X_2^t, Y_1^t, and Y_2^t.

The first Blue unit is subjected to directed fire from all the Red forces, with an attrition rate coefficient of 0.03 Blue 1 targets/Red 1 firer per unit time and 0.02 Blue 1 targets/Red 2 firer per unit time. The second Blue unit is also subjected to directed fire from all the Red forces, with an attrition rate coefficient of 0.04 Blue 2 targets/Red 1 firer per unit time and 0.01 Blue 2 targets/Red 2 firer per unit time. The first Red unit is under directed fire from both Blue units, with an attrition rate coefficient of 0.05 Red 1 targets/Blue 1 firer per unit time and 0.02 Red 1 targets/Blue 2 firer per unit time. The second Red unit is subjected to directed fire from only Blue 1, with an attrition rate coefficient of 0.03 Red 2 targets/Blue 1 firer per unit time.

(a) Write down the differential equation system to represent the combat dynamics.

(b) Find the closed form values as a function of time t for X_1^t, X_2^t, Y_1^t, Y_2^t.

(c) Calculate X_1^t, X_2^t, Y_1^t, Y_2^t, $t = 0, 1, 2, 3, 4, 5$.

Web material

http://149.170.199.144/multivar/eigen.htm
http://algebra.math.ust.hk/determinant/03_properties/lecture1.shtml
http://algebra.math.ust.hk/eigen/01_definition/lecture2.shtml
http://bass.gmu.edu/ececourses/ece521/lecturenote/chap1/node3.html
http://c2.com/cgi/wiki?EigenValue
http://ceee.rice.edu/Books/LA/eigen/
http://cepa.newschool.edu/het/essays/math/eigen.htm
http://cio.nist.gov/esd/emaildir/lists/opsftalk/msg00017.html
http://cnx.org/content/m2116/latest/
http://cnx.rice.edu/content/m10742/latest/
http://college.hmco.com/mathematics/larson/elementary_linear/4e/
    shared/downloads/c08s5.pdf
http://college.hmco.com/mathematics/larson/elementary_linear/5e/
    students/ch08-10/chap_8_5.pdf
http://ece.gmu.edu/ececourses/ece521/lecturenote/chap1/node3.html
http://en.wikipedia.org/wiki/Determinant
http://en.wikipedia.org/wiki/Eigenvalue
http://en.wikipedia.org/wiki/Hermitian_matrix
http://en.wikipedia.org/wiki/Jordan_normal_form
http://en.wikipedia.org/wiki/Skew-Hermitian_matrix
http://encyclopedia.laborlawtalk.com/Unitary_matrix
http://eom.springer.de/C/c023840.htm
http://eom.springer.de/E/e035150.htm
http://eom.springer.de/H/h047070.htm
http://eom.springer.de/J/j054340.htm
http://eom.springer.de/L/l059520.htm
http://everything2.com/index.pl?node=determinant
http://fourier.eng.hmc.edu/e161/lectures/algebra/node3.html
http://fourier.eng.hmc.edu/e161/lectures/algebra/node4.html
http://gershwin.ens.fr/vdaniel/Doc-Locale/Cours-Mirrored/
    Methodes-Maths/white/math/s3/s3spm/s3spm.html
http://home.iitk.ac.in/~arlal/book/nptel/mth102/node57.html
http://homepage.univie.ac.at/Franz.Vesely/cp0102/dx/node28.html
http://hyperphysics.phy-astr.gsu.edu/hbase/deter.html

```
http://kr.cs.ait.ac.th/~radok/math/mat/51.htm
http://kr.cs.ait.ac.th/~radok/math/mat3/m132.htm
http://kr.cs.ait.ac.th/~radok/math/mat3/m133.htm
http://kr.cs.ait.ac.th/~radok/math/mat3/m146.htm
http://kr.cs.ait.ac.th/~radok/math/mat7/step17.htm
http://linneus20.ethz.ch:8080/2_2_1.html
http://math.carleton.ca:16080/~daniel/teaching/114W01/117_EigVal.ps
http://math.fullerton.edu/mathews/n2003/JordanFormBib.html
http://mathworld.wolfram.com/Determinant.html
http://mathworld.wolfram.com/DeterminantExpansionbyMinors.html
http://mathworld.wolfram.com/Eigenvalue.html
http://mathworld.wolfram.com/Eigenvector.html
http://mathworld.wolfram.com/HermitianMatrix.html
http://mathworld.wolfram.com/JordanCanonicalForm.html
http://mathworld.wolfram.com/UnitaryMatrix.html
http://meru.rnet.missouri.edu/people/hai/research/jacobi.c
http://mpec.sc.mahidol.ac.th/radok/numer/STEP17.HTM
http://mysoftwear.com/go/0110/10406671133e894d172cd42.html
http://ocw.mit.edu/NR/rdonlyres/Electrical-Engineering-and-Computer-
    Science/6-241Fall2003/A685C9EE-6FF0-4E1A-81AC-04A8981C4FD9/0/
    rec5.pdf
http://oonumerics.org/MailArchives/oon-list/2000/06/0486.php
http://oonumerics.org/MailArchives/oon-list/2000/06/0499.php
http://orion.math.iastate.edu/hentzel/class.510/May.23
http://ourworld.compuserve.com/homepages/fcfung/mlaseven.htm
http://planetmath.org/encyclopedia/Determinant2.html
http://planetmath.org/encyclopedia/
    DeterminantIonTermsOfTracesOfPowers.html
http://planetmath.org/encyclopedia/Eigenvalue.html
http://planetmath.org/encyclopedia/JordanCanonicalForm.html
http://planetmath.org/encyclopedia/
    ProofOfJordanCanonicalFormTheorem.html
http://psroc.phys.ntu.edu.tw/cjp/v41/221.pdf
http://rakaposhi.eas.asu.edu/cse494/f02-hw1-qn1.txt
http://rkb.home.cern.ch/rkb/AN16pp/node68.html
http://schwehr.org/software/density/html/Eigs_8C.html
http://sherry.ifi.unizh.ch/mehrmann99structured.html
http://sumantsumant.blogspot.com/2004/12/one-of-beauty-of-matrix-
    operation-is.html
http://www-gap.dcs.st-and.ac.uk/~history/Search/historysearch.cgi?
    SUGGESTION=Determinant&CONTEXT=1
http://www-history.mcs.st-andrews.ac.uk/history/Biographies/
    Jordan.html
http://www-history.mcs.st-andrews.ac.uk/history/HistTopics/
    Matrices_and_determinants.html
http://www-math.mit.edu/18.013A/HTML/chapter04/section01.html#
    DeterminantVectorProducts
http://www.bath.ac.uk/mech-eng/units/xx10118/eigen.pdf
http://www.caam.rice.edu/software/ARPACK/UG/node46.html
```

http://www.cap-lore.com/MathPhys/Implicit/eigen.html
http://www.cs.berkeley.edu/~wkahan/MathH110/jordan.pdf
http://www.cs.ucf.edu/courses/cap6411/cot6505/Lecture-2.PDF
http://www.cs.ucf.edu/courses/cap6411/cot6505/spring03/Lecture-2.pdf
http://www.cs.uleth.ca/~holzmann/notes/eigen.pdf
http://www.cs.ut.ee/~toomas_l/linalg/lin1/node14.html
http://www.cs.ut.ee/~toomas_l/linalg/lin1/node16.html
http://www.cs.ut.ee/~toomas_l/linalg/lin2/node18.html
http://www.cs.ut.ee/~toomas_l/linalg/lin2/node20.html
http://www.cs.utk.edu/~dongarra/etemplates/
http://www.dpmms.cam.ac.uk/site2002/Teaching/IB/LinearAlgebra/
    jordan.pdf
http://www.ece.tamu.edu/~chmbrlnd/Courses/ELEN601/ELEN601-Chap7.pdf
http://www.ece.uah.edu/courses/ee448/appen4_2.pdf
http://www.ee.bilkent.edu.tr/~sezer/EEE501/Chapter8.pdf
http://www.ee.ic.ac.uk/hp/staff/www/matrix/decomp.html
http://www.emunix.emich.edu/~phoward/f03/416f3fh.pdf
http://www.freetrialsoft.com/free-download-1378.html
http://www.gold-software.com/MatrixTCL-review1378.htm
http://www.itl.nist.gov/div898/handbook/pmc/section5/pmc532.htm
http://www.mat.univie.ac.at/~kratt/artikel/detsurv.html
http://www.math.colostate.edu/~achter/369/help/jordan.pdf
http://www.math.ku.dk/ma/kurser/symbolskdynamik/konjug/node14.html
http://www.math.lsu.edu/~verrill/teaching/linearalgebra/linalg/
    linalg8.html
http://www.math.missouri.edu/courses/math4140/331eigenvalues.pdf
http://www.math.missouri.edu/~hema/331eigenvalues.pdf
http://www.math.poly.edu/courses/ma2012/Notes/Eigenvalues.pdf
http://www.math.sdu.edu.cn/mathency/math/u/u062.htm
http://www.math.tamu.edu/~dallen/m640_03c/lectures/chapter8.pdf
http://www.math.uah.edu/mathclub/talks/11-9-2001.html
http://www.math.ucdavis.edu/~daddel/linear_algebra_appl/
    Applications/Determinant/Determinant/Determinant.html
http://www.math.ucdavis.edu/~daddel/linear_algebra_appl/
    Applications/Determinant/Determinant/node3.html
http://www.math.ucdavis.edu/~daddel/Math22al_S02/LABS/LAB9/lab9_w00/
    node15.html
http://www.math.umd.edu/~hck/Normal.pdf
http://www.mathreference.com/la-det,eigen.html
http://www.mathreference.com/la-jf,canon.html
http://www.maths.gla.ac.uk/~tl/minimal.pdf
http://www.maths.lancs.ac.uk/~gilbert/m306c/node16.html
http://www.maths.liv.ac.uk/~vadim/M298/108.pdf
http://www.maths.lse.ac.uk/Personal/james/old_ma201/lect11.pdf
http://www.maths.mq.edu.au/~wchen/lnlafolder/la12.pdf
http://www.maths.surrey.ac.uk/interactivemaths/emmaspages/
    option3.html
http://www.mathwords.com/d/determinant.htm
http://www.mines.edu/~rtankele/cs348/LA%207.doc

http://www.nova.edu/~zhang/01CommAlgJordanForm.pdf
http://www.numbertheory.org/courses/MP274/realjord.pdf
http://www.numbertheory.org/courses/MP274/uniq.pdf
http://www.oonumerics.org/MailArchives/oon-list/2000/05/0481.php
http://www.oonumerics.org/oon/oon-list/archive/0502.html
http://www.perfectdownloads.com/audio-mp3/other/
    download-matrix-tcl.htm
http://www.ping.be/~ping1339/determ.htm
http://www.ppsw.rug.nl/~gladwin/eigsvd.html
http://www.reference.com/browse/wiki/Hermitian_matrix
http://www.reference.com/browse/wiki/Unitary_matrix
http://www.riskglossary.com/link/eigenvalue.htm
http://www.sosmath.com/matrix/determ0/determ0.html
http://www.sosmath.com/matrix/determ2/determ2.html
http://www.sosmath.com/matrix/inverse/inverse.html
http://www.stanford.edu/class/ee263/jcf.pdf
http://www.stanford.edu/class/ee263/jcf2.pdf
http://www.techsoftpl.com/matrix/doc/eigeń.htm
http://www.tversoft.com/computer/eigen.html
http://www.wikipedia.org/wiki/Determinant
http://www.wikipedia.org/wiki/Unitary_matrix
http://www.yotor.com/wiki/en/de/Determinant.htm
http://www.zdv.uni-tuebingen.de/static/hard/zrsinfo/x86_64/nag/
    mark20/NAGdoc/fl/html/indexes/kwic/determinant.html
http://www1.mengr.tamu.edu/aparlos/MEEN651/
    EigenvaluesEigenvectors.pdf
http://www2.maths.unsw.edu.au/ForStudents/courses/math2509/ch9.pdf

5

Positive Definiteness

Positive definite matrices are of both theoretical and computational importance in a wide variety of applications. They are used, for example, in optimization algorithms and in the construction of various linear regression models. As an initiation of our discussion in this chapter, we investigate first the properties for maxima, minima and saddle points when we have scalar functions with two variables. After introducing the quadratic forms, various tests for positive (semi) definiteness are presented.

5.1 Minima, Maxima, Saddle points

5.1.1 Scalar Functions

Let us remember the properties for maxima, minima and saddle points when we have scalar functions with two variables with the help the following examples.

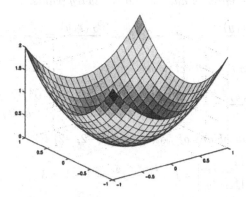

Fig. 5.1. Plot of $f(x, y) = x^2 + y^2$

Example 5.1.1 *Let $f(x,y) = x^2 + y^2$. Find the extreme points of $f(x,y)$:*

$$\frac{\partial f(x,y)}{\partial x} = 2x \doteq 0 \Rightarrow x = 0, \quad \frac{\partial f(x,y)}{\partial y} = 2y \doteq 0 \Rightarrow y = 0.$$

Since we have only one critical point, it is either the maximum or the minimum. We observe that $f(x,y)$ takes only nonnegative values. Thus, we see that the origin is the minimum point.

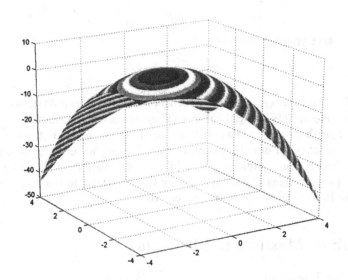

Fig. 5.2. Plot of $f(x,y) = xy - x^2 - y^2 - 2x - 2y + 4$

Example 5.1.2 *Find the extreme points of $f(x,y) = xy - x^2 - y^2 - 2x - 2y + 4$. The function is differentiable and has no boundary points.*

$$f_x = \frac{\partial f(x,y)}{\partial x} = y - 2x - 2, \; f_y = \frac{\partial f(x,y)}{\partial y} = x - 2y - 2.$$

Thus, $x = y = -2$ is the critical point.

$$f_{xx} = \frac{\partial^2 f(x,y)}{\partial x^2} = -2 = \frac{\partial^2 f(x,y)}{\partial y^2} = f_{yy}, \; f_{xy} = \frac{\partial^2 f(x,y)}{\partial x \partial y} = 1.$$

The discriminant (Jacobian) of f at $(a,b) = (-2,-2)$ is

$$\begin{vmatrix} f_{xx} & f_{xy} \\ f_{xy} & f_{yy} \end{vmatrix} = f_{xx}f_{yy} - f_{xy}^2 = 4 - 1 = 3.$$

Since $f_{xx} < 0$, $f_{xx}f_{yy} - f_{xy}^2 > 0 \Rightarrow f$ has a local maximum at $(-2,-2)$.

Theorem 5.1.3 *The extreme values for $f(x, y)$ can occur only at*

i. Boundary points of the domain of f.
ii. Critical points (interior points where $f_x = f_y = 0$, or points where f_x or f_y fails to exist).

If the first and second order partial derivatives of f are continuous throughout an open region containing a point (a, b) and $f_x(a, b) = f_y(a, b) = 0$, you may be able to classify (a, b) with the second derivative test:

i. $f_{xx} < 0$, $f_{xx}f_{yy} - f_{xy}^2 > 0$ at $(a, b) \Rightarrow$ local maximum;
ii. $f_{xx} > 0$, $f_{xx}f_{yy} - f_{xy}^2 > 0$ at $(a, b) \Rightarrow$ local minimum;
iii. $f_{xx}f_{yy} - f_{xy}^2 < 0$ at $(a, b) \Rightarrow$ saddle point;
iv. $f_{xx}f_{yy} - f_{xy}^2 = 0$ at $(a, b) \Rightarrow$ test is inconclusive (f is singular).

5.1.2 Quadratic forms

Definition 5.1.4 *The quadratic term $f(x, y) = ax^2 + 2bxy + cy^2$ is positive definite (negative definite) if and only if $a > 0$ ($a < 0$) and $ac - b^2 > 0$. f has a minimum (maximum) at $x = y = 0$ if and only if $f_{xx}(0,0) > 0$ ($f_{xx}(0,0) < 0$) and $f_{xx}(0,0)f_{yy}(0,0) > f_{xy}^2(0,0)$. If $f(0,0) = 0$, we term f as positive (negative) semi-definite provided the above conditions hold.*

Now, we are able to introduce matrices to the quadratic forms:

$$ax^2 + 2bxy + cy^2 = [x, y] \begin{bmatrix} a & b \\ b & c \end{bmatrix} \begin{bmatrix} x \\ y \end{bmatrix}.$$

Thus, for any symmetric A, the product $f = x^T A x$ is a pure quadratic form: it has a stationary point at the origin and no higher terms.

$$xA^T x = [x_1, x_2, \cdots, x_n] \begin{bmatrix} a_{11} & a_{12} & \cdots & a_{1n} \\ a_{21} & a_{22} & \cdots & a_{2n} \\ \vdots & \vdots & \ddots & \vdots \\ a_{n1} & a_{n2} & \cdots & a_{nn} \end{bmatrix} \begin{bmatrix} x_1 \\ x_2 \\ \vdots \\ x_n \end{bmatrix}$$

$$= a_{11}x_1^2 + a_{12}x_1x_2 + \cdots + a_{nn}x_n^2 = \sum_{i=1}^{n}\sum_{j=1}^{n} a_{ij}x_ix_j.$$

Definition 5.1.5 *If A is such that $a_{ij} = \frac{\partial^2 f}{\partial x_i \partial x_j}$ (hence symmetric), it is called the Hessian matrix. If A is positive definite ($x^T A x > 0$, $\forall x \neq 0$) and if f has a stationary point at the origin (all first derivatives at the origin are zero), then f has a minimum.*

Remark 5.1.6 *Let $f : \mathbb{R}^n \mapsto \mathbb{R}$ and $x^* \in \mathbb{R}^n$ be the local minimum, $\nabla f(x^*) = \theta$ and $\nabla^2 f(x^*)$ is positive definite. We are able to explore the neighborhood of x^* by means of $x^* + \Delta x$, where $\|\Delta x\|$ is sufficiently small (such that the second order Taylor's approximation is pretty good) and positive. Then,*

$$f(x^* + \Delta x) \cong f(x^*) + \Delta x^T \nabla f(x^*) + \frac{1}{2} \Delta x^T \nabla^2 f(x^*) \Delta x.$$

The second term is zero since x^* is a critical point and the third term is positive since the Hessian evaluated at x^* is positive definite. Thus, the left hand side is always strictly greater than the right hand side, indicating the local minimality of x^*.

5.2 Detecting Positive-Definiteness

Theorem 5.2.1 *A real symmetric matrix A is positive definite if and only if one of the following holds:*

i. $x^T A x > 0$, $\forall x \neq \theta$;
ii. All the eigen values of A satisfy $\lambda_i > 0$;
iii. All the submatrices A_k have positive determinants;
iv. All the pivots (without row exchanges) satisfy $d_i > 0$;
v. $\exists$ a nonsingular matrix $W \ni A = W^T W$ (called Cholesky Decomposition);

Proof. A is positive definite.

1. $(i) \Leftrightarrow (ii)$

   $(i) \Rightarrow (ii)$: Let x_i be the unit eigen vector corresponding to eigen value λ_i.
   $$A x_i = \lambda_i x_i \Leftrightarrow x_i^T A x_i = x_i^T \lambda_i x_i = \lambda_i.$$

   Then, $\lambda_i > 0$ since A is positive definite.

   $(i) \Leftarrow (ii)$: Since symmetric matrices have a full set of orthonormal eigen vectors
   (Exercise!).

   $$x = \sum \alpha_i x_i \Rightarrow A x = \sum \alpha_i A x_i \Rightarrow x^T A x = \left(\sum \alpha_i x_i^T \right) \left(\sum \alpha_i \lambda_i x_i \right).$$

   Because of orthonormality $x^T A x = \sum \alpha_i^2 \lambda_i > 0$.

2. $(i) \Leftrightarrow (iii) \Leftrightarrow (iv) \Leftrightarrow (v)$

   $(i) \Rightarrow (iii)$: $\det A = \lambda_1 \cdot \lambda_2 \cdots \lambda_n$, since $(i) \Leftrightarrow (ii)$.
   Claim: If A is positive definite, so is every A_k.
   Proof: If $x = \begin{bmatrix} x_k \\ 0 \end{bmatrix}$, then

$$x^T A x = [x_k, 0] \begin{bmatrix} A_k & * \\ * & * \end{bmatrix} \begin{bmatrix} x_k \\ 0 \end{bmatrix} = x_k^T A_k x_k > 0.$$

If we apply $(i) \Leftrightarrow (ii)$ for A_k (its eigen values are different, but all are positive), then its determinant is the product of its eigen values yielding a positive result.

$(iii) \Rightarrow (iv)$: *Claim:* If $A = LDU$, then the upper left corner satisfy $A_k = L_k D_k U_k$.

Proof: $A = \begin{bmatrix} L_k & 0 \\ B & C \end{bmatrix} \begin{bmatrix} D_k & 0 \\ 0 & E \end{bmatrix} \begin{bmatrix} U_k & F \\ 0 & G \end{bmatrix} = \begin{bmatrix} L_k D_k U_k & L_k D_k F \\ B D_k U_k & B D_k F + CEG \end{bmatrix}.$

$$\det A_k = \det L_k \det D_k \det U_k = \det D_k = d_1 \cdot d_2 \cdots d_k \Rightarrow$$

$d_k = \frac{\det A_k}{\det A_{k-1}}$ (Pivot=Ratio of determinants). If all determinants are positive, then all pivots are positive.

$(iv) \Rightarrow (v)$: In a Gaussian elimination of a symmetric matrix $U = L^T$, then $A = LDL^T$. One can take the square root of positive pivots $d_i > 0$. Then,

$$A = L\sqrt{D}\sqrt{D}L^T = W^T W.$$

$(v) \Rightarrow (i)$:

$$x^T A x = x^T W^T W x = \|Wx\|^2 \geq 0.$$

$Wx = \theta \Rightarrow x = \theta$ since W is nonsingular.
Therefore, $x^T A x > 0$, $\forall x \neq \theta$. $\square$

Remark 5.2.2 *The above theorem would be exactly the same in the complex case, for Hermitian matrices $A = A^H$.*

5.3 Semidefinite Matrices

Theorem 5.3.1 *A real symmetric matrix A is positive semidefinite if and only if one of the following holds:*

i. $x^T A x \geq 0$, $\forall x \neq \theta$;
ii. *All the eigen values of A satisfy $\lambda_i \geq 0$;*
iii. *All the submatrices A_k have nonnegative determinants;*
iv. *All the pivots (without row exchanges) satisfy $d_i \geq 0$;*
v. *$\exists$ a possibly singular matrix $W \ni A = W^T W$;*

Remark 5.3.2 $x^T A x \geq 0 \Leftrightarrow \lambda_i \geq 0$ *is important.*

$$A = Q\Lambda Q^T \Rightarrow x^T A x = x^T Q\Lambda Q^T x = y^T \Lambda y = \lambda_1 y_1^2 + \cdots + \lambda_n y_n^2,$$

and it is nonnegative when Λ_i's are nonnegative. If A has rank r, there are r nonzero eigen values and r perfect squares.

Remark 5.3.3 (Indefinite matrices) *Change of Variables:* $y = Cx$. *The quadratic form becomes* $y^T C^T A C y$. *Then, we have congruence transformation:* $A \mapsto C^T A C$ *for some nonsingular* C. *The matrix* $C^T A C$ *has the same number of positive (negative) eigen values of* A, *and the same number of zero eigen values. If we let* $A = I$, $C^T A C = C^T C$. *Thus, for any symmetric matrix* A, *the signs of pivots agree with the signs of eigen values.* Λ *and* D *have the same number of positive (negative) entries, and zero entries.*

5.4 Positive Definite Quadratic Forms

Proposition 5.4.1 *If* A *is symmetric positive definite, then*

$$P(x) = \frac{1}{2}x^T A x - x^T b$$

assumes its minimum at the point $Ax = b$.

Proof. Let $x \ni Ax = b$. Then, $\forall y \in \mathbb{R}^n$,

$$P(y) - P(x) = \left(\frac{1}{2}y^T A y - y^T b\right) - \left(\frac{1}{2}x^T A x - x^T b\right)$$

$$= \frac{1}{2}y^T A y - y^T A x + \frac{1}{2}x^T A x$$

$$= \frac{1}{2}(y - x)^T A(y - x)$$

$$\geq 0.$$

Hence, $\forall y \neq x$, $P(y) \geq P(x) \Rightarrow x$ is the minimum. $\square$

Theorem 5.4.2 (Rayleigh's principle) *Without loss of generality, we may assume that*

$$\lambda_1 \leq \lambda_2 \leq \cdots \leq \lambda_n.$$

The quotient, $R(x) = \frac{x^T A x}{x^T x}$, *is minimized by the first eigen vector* v_1 *and its minimum value is the smallest eigen value* λ_1:

$$R(v_1) = \frac{v_1^T A v_1}{v_1^T v_1} = \frac{v_1^T \lambda_1 v_1}{v_1^T v_1} = \lambda_1.$$

Remark 5.4.3 $\forall x$, $R(x)$ *is an upper bound for* λ_1.

Remark 5.4.4 *Rayleigh's principle is the basis for the principle component analysis, which has many engineering applications like factor analysis of the variance covariance matrix (symmetric) in multivariate data analysis.*

Corollary 5.4.5 *If* x *is orthogonal to the eigen vectors* $v_1, \ldots, v_{j-1}$, *then* $R(x)$ *will be minimized by the next eigen vector* v_j.

Remark 5.4.6 $\lambda_j = \min_{x \in \mathbb{R}^n} R(x)$ $\qquad\qquad\qquad$ $\lambda_j = \max_{x \in \mathbb{R}^n} R(x)$
$\qquad\qquad$ s.t. $\qquad\qquad\qquad\qquad\qquad\qquad\qquad$ s.t.

$$x^T v_1 = 0 \qquad\qquad\qquad\qquad x^T v_{j+1} = 0$$

$$\vdots \qquad\qquad\qquad\qquad\qquad\qquad \vdots$$

$$x^T v_{j-1} = 0 \qquad\qquad\qquad\qquad x^T v_n = 0$$

Problems

5.1. Prove the following theorem.

Theorem 5.4.7 (Rayleigh-Ritz) *Let* A *be symmetric,* $\lambda_1 \leq \lambda_2 \leq \cdots \leq \lambda_n$.

$$\lambda_1 = \min_{\|x\|=1} x^T A x, \ \lambda_n = \max_{\|x\|=1} x^T A x.$$

5.2. Use

$$A = \frac{1}{100} \begin{bmatrix} 2 & 1 & 0 \\ 1 & 2 & 1 \\ 0 & 1 & 1 \end{bmatrix}$$

to show Theorem 5.3.1.

5.3. Let

$$f(x_1, x_2) = \frac{1}{3}x_1^3 + \frac{1}{2}x_1^2 + 2x_1 x_2 + \frac{1}{2}x_2^2 - x_2 + 19.$$

Find the stationary and boundary points, then find the minimizer and the maximizer over $-4 \leq x_2 \leq 0 \leq x_1 \leq 3$.

Web material

```
http://bmbiris.bmb.uga.edu/wampler/8200/using-ff/sld027.htm
http://delta.cs.cinvestav.mx/~mcintosh/comun/contours/node8.html
http://delta.cs.cinvestav.mx/~mcintosh/oldweb/lcau/node98.html
http://dft.rutgers.edu/~etsiper/rrosc.html
http://econ.lse.ac.uk/courses/ec319/M/lecture5.pdf
http://employees.oneonta.edu/GoutziCJ/fall_2003/math276/maple/
    Lesson_141.html
http://en.wikipedia.org/wiki/Cholesky_decomposition
http://en.wikipedia.org/wiki/Maxima_and_minima
http://en.wikipedia.org/wiki/Positive-semidefinite_matrix
http://en.wikipedia.org/wiki/Quadratic_form
http://eom.springer.de/b/b016370.htm
http://eom.springer.de/C/c120160.htm
```

http://eom.springer.de/N/n130030.htm
http://eom.springer.de/q/q076080.htm
http://epubs.siam.org/sam-bin/dbq/article/38133
http://esperia.iesl.forth.gr/~amo/nr/bookfpdf/f2-9.pdf
http://gaia.ecs.csus.edu/~hellerm/EEE242/chapter%201/pd.htm
http://homepage.tinet.ie/~phabfys/maxim.htm
http://iridia.ulb.ac.be/~fvandenb/mythesis/node72.html
http://kr.cs.ait.ac.th/~radok/math/mat3/m131.htm
http://kr.cs.ait.ac.th/~radok/math/mat5/algebra62.htm
http://kr.cs.ait.ac.th/~radok/math/mat9/03c.htm
http://mat.gsia.cmu.edu/QUANT/NOTES/chap1/node8.html
http://mathworld.wolfram.com/CholeskyDecomposition.html
http://mathworld.wolfram.com/Maximum.html
http://mathworld.wolfram.com/PositiveDefiniteMatrix.html
http://mathworld.wolfram.com/PositiveSemidefiniteMatrix.html
http://mathworld.wolfram.com/QuadraticForm.html
http://mathworld.wolfram.com/topics/MaximaandMinima.html
http://modular.fas.harvard.edu/docs/magma/htmlhelp/text654.htm
http://ocw.mit.edu/NR/rdonlyres/Chemical-Engineering/10-34Fall-2005/
    695E79DF-11F7-4FB7-AD7E-FEDA74B9BFEF/0/lecturenotes142.pdf
http://omega.albany.edu:8008/calc3/extrema-dir/define-m2h.html
http://oregonstate.edu/instruct/mth254h/garity/Fall2005/Notes/
    10_15_8.pdf
http://people.hofstra.edu/faculty/Stefan_Waner/realworld/
    Calcsumm8.html
http://planetmath.org/encyclopedia/CholeskyDecomposition.html
http://planetmath.org/encyclopedia/
    DiagonalizationOfQuadraticForm.html
http://planetmath.org/encyclopedia/PositiveDefinite.html
http://planetmath.org/encyclopedia/QuadraticForm.html
http://pruffle.mit.edu/3.016/collected_lectures/node39.html
http://pruffle.mit.edu/3.016/Lecture_10_web/node2.html
http://random.mat.sbg.ac.at/~ste/diss/node25.html
http://rkb.home.cern.ch/rkb/AN16pp/node33.html
http://scienceandreason.blogspot.com/2006/03/quadratic-forms.html
http://sepwww.stanford.edu/sep/prof/gem/hlx/paper_html/node11.html
http://slpl.cse.nsysu.edu.tw/chiaping/la/chap6.pdf
http://taylorandfrancis.metapress.com/media/59dam5dwuj2xwl8rvvtk/
    contributions/d/3/y/y/d3yy93fbcqpvu69n.pdf
http://tutorial.math.lamar.edu/AllBrowsers/1314/
    ReducibleToQuadratic.asp
http://web.mit.edu/18.06/www/Video/video-fall-99.html
http://web.mit.edu/wwmath/vectorc/minmax/hessian.html
http://www-math.mit.edu/~djk/18_022/chapter04/section02.html
http://www.analyzemath.com/Equations/Quadratic_Form_Tutorial.html
http://www.answers.com/topic/quadratic-form
http://www.artsci.wustl.edu/~e503jn/files/math/DefiniteMatrics.pdf
http://www.astro.cf.ac.uk/undergrad/module/PX3104/tp1/node12.html
http://www.chass.utoronto.ca/~osborne/MathTutorial/QF2F.HTM

http://www.chass.utoronto.ca/~osborne/MathTutorial/QFF.HTM
http://www.chass.utoronto.ca/~osborne/MathTutorial/QFS.HTM
http://www.chass.utoronto.ca/~osborne/MathTutorial/QUF.HTM
http://www.cs.ut.ee/~toomas_l/linalg/lin2/node25.html
http://www.csie.ncu.edu.tw/~chia/Course/LinearAlgebra/sec8-2.pdf
http://www.ece.mcmaster.ca/~kiruba/3sk3/lecture7.pdf
http://www.ece.uwaterloo.ca/~ece104/TheBook/04LinearAlgebra/
    cholesky/
http://www.ee.ic.ac.uk/hp/staff/dmb/matrix/property.html
http://www.es.ucl.ac.uk/undergrad/geomaths/pdilink6.htm
http://www.iam.ubc.ca/~norris/research/quadapp.pdf
http://www.ics.mq.edu.au/~chris/math123/chap05.pdf
http://www.imada.sdu.dk/~swann/MM02/QuadraticForms.pdf
http://www.imsc.res.in/~kapil/crypto/notes/node37.html
http://www.matf.bg.ac.yu/r3nm/NumericalMethods/LAESolve/
    Cholesky.html
http://www.math.niu.edu/~rusin/known-math/99/posdef
http://www.math.oregonstate.edu/home/programs/undergrad/
    CalculusQuestStudyGuides/vcalc/min_max/min_max.html
http://www.math.rutgers.edu/courses/251/s01bumby/slide011.ps2.pdf
http://www.math.tamu.edu/~bollingr/Notes/bzb83.pdf
http://www.math.ucla.edu/~xinweiyu/164.1.05f/1102.pdf
http://www.math.uga.edu/~chadm/quadratic.pdf
http://www.math.uic.edu/math210/labs/lab5.html
http://www.math.uic.edu/~math210/newlabs/critpts/critpts.html
http://www.math.uiuc.edu/documenta/lsu/vol-lsu-eng.html
http://www.math.umn.edu/~nykamp/multivar/Fall2003/lecture26.pdf
http://www.math.vt.edu/people/javance1/Section_10.1.pdf
http://www.math.wm.edu/~hugo/compl3.html
http://www.mathematics.jhu.edu/matlab/8-4.html
http://www.mathreference.com/ca-mv,local.html
http://www.mathreference.com/la-qf,intro.html
http://www.maths.abdn.ac.uk/~igc/tch/ma2001/notes/node70.html
http://www.maths.abdn.ac.uk/~igc/tch/mx3503/notes/node79.html
http://www.maths.lse.ac.uk/Courses/MA207/fqmso.pdf
http://www.mpri.lsu.edu/textbook/Chapter2.htm
http://www.numericalmathematics.com/maxima_and_minima1.htm
http://www.psi.toronto.edu/matrix/special.html
http://www.quantlet.com/mdstat/scripts/mva/htmlbook/
    mvahtmlnode16.html
http://www.reference.com/browse/wiki/Maxima_and_minima
http://www.reference.com/browse/wiki/Quadratic_form
http://www.riskglossary.com/link/positive_definite_matrix.htm
http://www.sciencenews.org/articles/20060311/bob9.asp
http://www.stanford.edu/class/ee263/symm.pdf
http://www.ucl.ac.uk/Mathematics/geomath/level2/pdiff/pd7.html
http://www.ucl.ac.uk/Mathematics/geomath/level2/pdiff/pd8.html
http://www.vision.caltech.edu/mweber/research/CNS248/node22.html

6

Computational Aspects

For square matrices, we can measure the sensitivity of the solution of the linear algebraic system $Ax = b$ with respect to changes in vector b and in matrix A by using the notion of the condition number of matrix A. If the condition number is large, then the matrix is said to be ill-conditioned. Practically, such a matrix is almost singular, and the computation of its inverse or solution of a linear system of equations is prone to large numerical errors. In this chapter, we will investigate computational methods for solving $Ax = b$, and obtaining eigen values/vectors of A.

6.1 Solution of $Ax = b$

Let us investigate small changes in the right hand side of $Ax = b$ as if we are making a sensitivity analysis:

$$b \mapsto b + \Delta_b \Rightarrow x \mapsto x + \Delta_x$$

$$A(x + \Delta_x) = b + \Delta_b \Leftrightarrow A(\Delta_x) = \Delta_b.$$

Similarly, one can investigate the effect of perturbing the coefficient matrix A:

$$A \mapsto A + \Delta_A \Rightarrow x \mapsto x + \Delta_x$$

We will consider these cases with respect to the form of the coefficient matrix A in the following subsections.

6.1.1 Symmetric and positive definite

Let A be symmetric. Without loss of generality, we may assume that we ordered the nonnegative eigen values: $0 \le \lambda_1 \le \lambda_2 \le \cdots \le \lambda_n$. Since Δ_b is a vector itself, it could be represented in terms of the basis formed by the associated eigen vectors $v_1, v_2, \ldots, v_n$. Moreover, we can express Δ_b as a convex combination because its norm is sufficiently small.

$$\Delta_b = \sum_{i=1}^{n} \alpha_i v_i \text{ where } v_i \leftrightarrow \lambda_i, \ \sum \alpha_i = 1, \ \alpha_i \geq 0, \forall i.$$

If Δ_b is along v_1, i.e. $\Delta_b = \epsilon v_1$, then $\Delta_x = \frac{\Delta_b}{\lambda_1}$ since $\Delta_x = A^{-1}\Delta_b$. That is, the error of size $\|\Delta_b\|$ is amplified by the factor $\frac{1}{\lambda_1}$, which is just the largest eigen value of A^{-1}. On the other hand, if $b = v_n$, then $x = A^{-1}b = \frac{b}{\lambda_n}$, which makes the relative error

$$\frac{\|\Delta_x\|}{\|x\|} = \frac{\frac{\|\Delta_b\|}{\lambda_1}}{\frac{\|b\|}{\lambda_n}} = \frac{\lambda_n}{\lambda_1} \frac{\|\Delta_b\|}{\|b\|}$$

as much as possible.

Proposition 6.1.1 *For a positive definite matrix, the solution $x = A^{-1}b$ and the error $\Delta_x = A^{-1}\Delta_b$ always satisfy*

$$\|x\| \geq \frac{\|b\|}{\lambda_n} \text{ and } \|\Delta_x\| \leq \frac{\|\Delta_b\|}{\lambda_1}.$$

Therefore, the relative error is bounded by

$$\frac{\|\Delta_x\|}{\|x\|} \leq \frac{\lambda_n}{\lambda_1} \frac{\|\Delta_b\|}{\|b\|}.$$

Definition 6.1.2 *The quantity $c = \frac{\lambda_n}{\lambda_1} = \frac{\lambda_{max}}{\lambda_{min}}$ is known as condition number of A.*

Remark 6.1.3 *Notice that c is not affected by the size of a matrix. If $A = I$ or $A' = \frac{I}{10}$ then $c_A = 1 = c_{A'} = \frac{\lambda_{max}}{\lambda_{min}}$. However, $\det A = 1$, $\det A' = 10^{-n}$. Thus, determinant is a terrible measure of ill conditioning.*

Example 6.1.4

$$A = \begin{bmatrix} 2.00002 & 2 \\ 2 & 2.00002 \end{bmatrix} \Rightarrow \lambda_1 = 2 \times 10^{-5}, \ \lambda_2 = 4.00002 \Rightarrow c \approx 2 \times 10^5.$$

In particular,

$$b = b_1 = \begin{bmatrix} 2.00001 \\ 2.00001 \end{bmatrix} \Rightarrow x = x_1 = \begin{bmatrix} 0.5 \\ 0.5 \end{bmatrix} \text{ and } b_2 = \begin{bmatrix} 2.00002 \\ 2 \end{bmatrix} \Rightarrow x_2 = \begin{bmatrix} 1 \\ 0 \end{bmatrix}.$$

Then, we have

$$\|b\| = 2.00001\sqrt{2}, \ \Delta_b = b_2 - b_1 = 10^{-5} \begin{bmatrix} 1 \\ -1 \end{bmatrix} \Rightarrow \|\Delta_b\| = \sqrt{2} \times 10^{-5};$$

$$\|x\| = \frac{\sqrt{2}}{2}, \ \Delta_x = x_2 - x_1 = \frac{1}{2}\begin{bmatrix} 1 \\ -1 \end{bmatrix} \Rightarrow \|\Delta_x\| = \frac{\sqrt{2}}{2}$$

$$\Rightarrow \frac{\|\Delta_x\|}{\|x\|} = 1 \ and \ \frac{\|\Delta_b\|}{\|b\|} \approx 5 \times 10^{-6}.$$

The relative amplification in this particular instance, $\frac{\|\Delta_x\|}{\|x\|} \approx \frac{10^5}{2}\frac{\|\Delta_b\|}{\|b\|}$*, is approximately* $\frac{10^5}{2}$*, which is a lower bound for the condition number* $c \approx 2 \times 10^5$*.*

Remark 6.1.5 *As a rule of thumb (experimentally verified), a computer can loose* $\log c$ *decimal places to the round-off errors in Gaussian elimination.*

6.1.2 Symmetric and not positive definite

Let us now drop the positivity assumption while we keep A still symmetric. Then, nothing is changed except

$$c = \frac{|\lambda_{max}|}{|\lambda_{min}|}.$$

6.1.3 Asymmetric

In this case, the ratio of eigen values cannot represent the relative amplification.

Example 6.1.6 *Let the parameter* $\kappa \gg 0$ *be large enough.*

$$A = \begin{bmatrix} 1 & \kappa \\ 0 & 1 \end{bmatrix} \Leftrightarrow A^{-1} = \begin{bmatrix} 1 & -\kappa \\ 0 & 1 \end{bmatrix}, \ \lambda_1 = \lambda_2 = 1.$$

In particular,

$$b = b_1 = \begin{bmatrix} \kappa \\ 1 \end{bmatrix} \Rightarrow x = x_1 = \begin{bmatrix} 0 \\ 1 \end{bmatrix} \ and \ b_2 = \begin{bmatrix} \kappa \\ 0 \end{bmatrix} \Rightarrow x_2 = \begin{bmatrix} \kappa \\ 0 \end{bmatrix}.$$

Then, we have

$$\|b\| = \sqrt{1 + \kappa^2}, \ \Delta_b = b_2 - b_1 = \begin{bmatrix} 0 \\ -1 \end{bmatrix} \Rightarrow \|\Delta_b\| = 1;$$

$$\|x\| = 1, \ \Delta_x = x_2 - x_1 = \begin{bmatrix} \kappa \\ -1 \end{bmatrix} \Rightarrow \|\Delta_x\| = \sqrt{1 + \kappa^2}$$

$$\Rightarrow \frac{\|\Delta_x\|}{\|x\|} = \sqrt{1 + \kappa^2} \ and \ \frac{\|\Delta_b\|}{\|b\|} = \frac{1}{\sqrt{1 + \kappa^2}}.$$

The relative amplification in this particular instance is $1 + \kappa^2$*. Hence, we should have* $1 \ll 1 + \kappa^2 \leq c(A)$*. The condition number* $c(A)$ *is not just the ratio of eigen values, which is 1; but it should have a considerably larger value in this example, since* A *is not symmetric.*

Definition 6.1.7 *The norm of A is the number defined* $\|A\| = \max_{x \neq \theta} \frac{\|Ax\|}{\|x\|}$.

Remark 6.1.8 $\|A\|$ *bounds the "amplifying power" of the matrix.*

$$\|Ax\| \leq \|A\| \|x\|, \ \forall x;$$

and equality holds for at least one nonzero x. It measures the largest amount by which any vector (eigen vector or not) is amplified by matrix multiplication.

Proposition 6.1.9 *For a square nonsingular matrix, the solution* $x = A^{-1}b$ *and the error* $\Delta_x = A^{-1}\Delta_b$ *satisfy*

$$\frac{\|\Delta_x\|}{\|x\|} \leq \|A\| \|A^{-1}\| \frac{\|\Delta_b\|}{\|b\|}.$$

Proof. Since
$b = Ax \Rightarrow \|b\| \leq \|A\| \|x\|$ and
$\Delta_x = A^{-1}\Delta_b \Rightarrow \|\Delta_x =\| \leq \|A^{-1}\| \|\Delta_b\|$, we have

$$\|b\| \leq \|A\| \|x\| \text{ and } \|\Delta_x\| \leq \|A^{-1}\| \|\Delta_b\|. \ \square$$

Remark 6.1.10 *When A is symmetric,*

$$\|A\| = |\lambda_n|, \ \ \|A^{-1}\| = \frac{1}{|\lambda_1|} \Rightarrow c = \|A\| \|A^{-1}\| = \frac{|\lambda_n|}{|\lambda_1|}$$

and the relative error satisfies

$$\frac{\|\Delta_x\|}{\|x\|} \leq c \frac{\|\Delta_b\|}{\|b\|}.$$

Example 6.1.11 *Let us continue the previous example, where*

$$A = \begin{bmatrix} 1 & \kappa \\ 0 & 1 \end{bmatrix}, \ b = \begin{bmatrix} \kappa \\ 1 \end{bmatrix}, \ \Delta_b = \begin{bmatrix} 0 \\ -1 \end{bmatrix}.$$

Since we have

$$\kappa \leq \|A\| \leq \kappa + 1, \ and \ \kappa \leq \|A^{-1}\| \leq \kappa + 1,$$

then the relative amplification is approximately $\kappa^2 \approx \|A\| \|A^{-1}\|$.

Remark 6.1.12

$$\|A\|^2 = \max \frac{\|Ax\|^2}{\|x\|^2} = \max \frac{x^T A^T A x}{x^T x} : Rayleigh \ quotient!$$

Proposition 6.1.13 *The norm of A is the square root of the largest eigen value of $A^T A$. The vector that is amplified the most is the corresponding eigen vector of $A^T A$.*

$$\frac{x^T A^T A x}{x^T x} = \frac{x^T \lambda_{\max} x}{x^T x} = \lambda_{\max} = \|A\|.$$

Example 6.1.14 *Let us further continue the previous example:*

$$A = \begin{bmatrix} 1 & \kappa \\ 0 & 1 \end{bmatrix} \text{ and } A^{-1} = \begin{bmatrix} 1 & -\kappa \\ 0 & 1 \end{bmatrix}$$

$$A^T A = \begin{bmatrix} 1 & \kappa \\ \kappa & \kappa^2 + 1 \end{bmatrix} \Rightarrow \begin{vmatrix} s - 1 & \kappa \\ -\kappa & s - \kappa^2 - 1 \end{vmatrix} \doteq 0 \Rightarrow s^2 - (\kappa^2 + 2)s + 1 = 0$$

$$\Delta^2 = (\kappa^2 + 2)^2 - 4(1)1 = \kappa^2(\kappa^2 + 4) \Rightarrow$$

$$\lambda_{\max} = \frac{-(-\kappa^2 - 2) + \sqrt{\kappa^2(\kappa^2 + 4)}}{2(1)} \approx \kappa^2 \Rightarrow \|A\| = \sqrt{\lambda_{\max}} \approx \kappa.$$

Similarly, $\|A^{-1}\| = \sqrt{\lambda_{\max}[(A^{-1})^T A^{-1}]} \approx \kappa$. *Thus, the relative amplification is controlled by* $\|A\| \, \|A^{-1}\| \approx \kappa^2$.

Remark 6.1.15 *If A is symmetric, then $A^T A = A^2$ and $\|A\| = \max |\lambda_i|$.*

Let us consider now the changes in the coefficient matrix.

Proposition 6.1.16 *If we perturb A, then*

$$\frac{\|\Delta_x\|}{\|x + \Delta_x\|} \leq c \frac{\|\Delta_A\|}{\|A\|} \text{ where } c = \|A\| \, \|A^{-1}\|.$$

Proof.

$$\left. \begin{array}{l} Ax = b \\ (A + \Delta_A)(x + \Delta_x) = b \end{array} \right\} \Rightarrow$$

$$A\Delta_x + \Delta_A(x + \Delta_x) = 0 \Leftrightarrow \Delta_x = -A^{-1}(\Delta_A)(x + \Delta_x).$$

$$\|\Delta_x\| \leq \|A^{-1}\| \, \|\Delta_A\| \, \|x + \Delta_x\| \Leftrightarrow \frac{\|\Delta_x\|}{\|x + \Delta_x\|} \leq \|A^{-1}\| \, \|\Delta_A\| = c \frac{\|\Delta_A\|}{\|A\|}. \quad \square$$

Example 6.1.17

$$A = \begin{bmatrix} 1 & 10 & 100 \\ 10 & \frac{1}{10} & 1 \\ 1 & \frac{1}{10} & \frac{1}{100} \end{bmatrix}, \; b = \begin{bmatrix} 111 \\ \frac{111}{10} \\ \frac{111}{100} \end{bmatrix} \Rightarrow x = x_1 = \begin{bmatrix} 1 \\ 1 \\ 1 \end{bmatrix}.$$

$$\Rightarrow \|A\| = \sqrt{\lambda_{\max}[A^T A]} = \sqrt{\frac{131329}{13}} = 100.5099, \; \|x\| = \sqrt{3},$$

$$A^{-1} = \begin{bmatrix} -\frac{1}{999} & \frac{100}{999} & 0 \\ \frac{10}{10989} & -\frac{1010}{999} & \frac{1000}{99} \\ \frac{100}{10989} & \frac{100}{999} & -\frac{100}{99} \end{bmatrix}$$

$$\Rightarrow \|A^{-1}\| = \sqrt{\lambda_{\max}[(A^{-1})^T A]} = \sqrt{\frac{28831}{277}} = 10.2021.$$

$$\Delta_A = \begin{bmatrix} -1 & -10 & 0 \\ -10 & \frac{9}{10} & -1 \\ 0 & -\frac{1}{10} & -\frac{1}{100} \end{bmatrix} \Rightarrow A + \Delta_A = \begin{bmatrix} 0 & 0 & 100 \\ 0 & 1 & 0 \\ 1 & 0 & 0 \end{bmatrix} \Rightarrow x_2 = \begin{bmatrix} \frac{111}{100} \\ \frac{111}{10} \\ \frac{111}{100} \end{bmatrix}$$

$$\Rightarrow \Delta_x = x_2 - x_1 = \begin{bmatrix} \frac{11}{100} \\ \frac{101}{10} \\ \frac{11}{100} \end{bmatrix} \Rightarrow \|\Delta_x\| = \frac{\sqrt{1020342}}{100} = 10.1012, \text{ and}$$

$$\|\Delta_A\| = \sqrt{\lambda_{\max}[\Delta_A^T \Delta_A]} = \sqrt{\frac{14963}{146}} = 10.1236.$$

$$\frac{\|\Delta_x\|}{\|x + \Delta_x\|} = \frac{10.1012}{\sqrt{3}} \leq \frac{10.1236}{100.5099} c = c \frac{\|\Delta_A\|}{\|A\|}$$

$$\Rightarrow \frac{\frac{10.1012}{\sqrt{3}}}{\frac{10.1236}{100.5099}} = 57.9 \leq c = \|A\| \, \|A^{-1}\| = 100.5099(10.2021) = 1025.412.$$

The relative amplification in this instance is 57.9 whereas the theoretic upper bound is 1025.412.

Remark 6.1.18 *The following are the main guidelines in practise:*

1. *c and $\|A\|$ are never computed but estimated.*
2. *c explains why $A^T Ax = A^T b$ are so hard to solve in least squares problems: $c(A^T A) = [c(A)]^2$ where $c(.)$ is the condition number. The remedy is to use Gram-Schmidt or singular value decomposition, $A = Q_1 \Sigma Q_2^T$. The entries σ_i in Σ are singular values of A, and σ_i^2 are the eigen values of $A^T A$. Thus, $\|A\| = \sigma_{max}$. Recall that $\|Ax\| = \|Q_1 \Sigma Q_2^T x\| = \|\Sigma x\|$.*

6.2 Computation of eigen values

There is no best way to compute eigen values of a matrix. But there are some terrible ways. In this section, a method recommended for large-sparse matrices, the *power method,* will be introduced.

Let u_0 be initial guess. Then, $u_{k+1} = Au_k = A^{k+1}u_0$. Assume A has full set of eigen vectors $x_1, x_2, \ldots, x_n$, then $u_k = \alpha_1 \lambda_1^k x_1 + \cdots + \alpha_n \lambda_n^k x_n$. Assume further that $\lambda_1 \leq \lambda_2 \leq \cdots \leq \lambda_{n-1} < \lambda_n$; that is, the last eigen value is not repeated.

$$\frac{u_k}{\lambda_n^k} = \alpha_1 \left(\frac{\lambda_1}{\lambda_n}\right)^k x_1 + \cdots + \alpha_n \left(\frac{\lambda_{n-1}}{\lambda_n}\right)^k x_{n-1} + \alpha_n x_n.$$

The vectors u_k point more and more accurately towards the direction of x_n, and the convergence factor is $r = \frac{|\lambda_{n-1}|}{|\lambda_n|}$.

Example 6.2.1 (Markov Process, continued) *Recall Example 4.4.6:*

$$A = \begin{bmatrix} 0.95 & 0.01 \\ 0.05 & 0.99 \end{bmatrix} \Rightarrow \lambda_1 = 1 \leftrightarrow \begin{bmatrix} \frac{1}{5} \\ 1 \end{bmatrix} = v_1, \; \lambda_2 = 0.94$$

$$u_0 = \begin{bmatrix} 1 \\ 0 \end{bmatrix}, \; u_1 = \begin{bmatrix} 0.95 \\ 0.05 \end{bmatrix}, \; u_2 = \begin{bmatrix} 0.903 \\ 0.097 \end{bmatrix}, \; u_3 = \begin{bmatrix} 0.85882 \\ 0.14118 \end{bmatrix},$$

$$u_4 = \begin{bmatrix} 0.817291 \\ 0.182709 \end{bmatrix}, \; \cdots, \; u_{210} = \begin{bmatrix} 0.166667 \\ 0.833333 \end{bmatrix} \approx u_\infty = \begin{bmatrix} \frac{1}{6} \\ \frac{5}{6} \end{bmatrix} = \alpha v_1.$$

The convergence rate is quite low $r = 0.94 = \frac{0.94}{1} = \frac{|\lambda_2|}{|\lambda_1|}$. Since the power method is designed especially for large sparse matrices, it converges after 210 iterations if the significance level is six digits after the decimal point.

Remark 6.2.2 (How to increase r) *If $r \approx 1$, the convergence is slow. If $|\lambda_{n-1}| = |\lambda_n|$, no convergence at all. There are some methods to increase the convergence rate:*

i. *Block power method: Work with several vectors at once. Start with p orthonormal vectors, multiply by A, then apply Gram-Schmidt to orthogonalize again. Then, we have $r' = \frac{|\lambda_{n-p}|}{|\lambda_n|}$.*

ii. *Inverse power method: Operate with A^{-1} instead of A. $v_{k+1} = A^{-1}v_k \Rightarrow Av_{k+1} = v_k$ (save L and U!). The convergence rate is $r'' = \frac{|\lambda_1|}{|\lambda_2|}$, provided that $r'' < 1$. This method guarantees convergence to the smallest eigen vector.*

iii. *Shifted inverse power method: The best method. Let A be replaced by $A - \beta I$. All of the eigen values are shifted by β. Consequently, $r''' = \frac{|\lambda_1 - \beta|}{|\lambda_2 - \beta|}$. If we choose β as a good approximation to λ_1, the convergence will be accelerated.*

$$(A - \beta I)w_{k+1} = w_k = \frac{\alpha_1 x_1}{(\lambda_1 - \beta)^k} + \frac{\alpha_2 x_2}{(\lambda_2 - \beta)^k} + \cdots + \frac{\alpha_n x_n}{(\lambda_n - \beta)^k}.$$

If we know β, then we may use $A - \beta I = LU$ and solve $Ux_1 = (1, 1, \cdots, 1)^T$ by back substitution. We can choose $\beta = \beta_k$ at each step $\ni (A - \beta_k I)w_{k+1} = w_k$. If $A = A^T$, $\beta_k = R(u_k) = \frac{u_k^T A u_k}{u_k^T u_k}$, then we will get the cubic convergence.

Remark 6.2.3 (QR Algorithm) *Start with A_0. Factor it using the Gram-Schmidt process into $Q_0 R_0$, then reverse factors $A_1 = R_0 Q_0$. A_1 is similar to A_0: $Q_0^{-1} A_0 Q_0 = Q_0^{-1}(Q_0 R_0) Q_0 = A_1$. So, $A_k = Q_k R_k \Rightarrow A_{k+1} = R_k Q_k$. A_k approaches to a triangular form in which we can read the eigen values from the main diagonal. There are some modifications to speed up this procedure as well.*

Definition 6.2.4 *If a matrix is less than a triangular form, one nonzero diagonal below the main diagonal, it is called in Hessenberg form. Furthermore, if it is symmetric then it is said to be in tridiagonal form.*

Definition 6.2.5 *A Householder transformation (or an elementary reflector) is a matrix of the form*

$$H = I - 2\frac{vv^T}{\|c\|^2}.$$

Remark 6.2.6 *Often v is normalized to become a unit vector $u = \frac{v}{\|v\|}$, then $H = I - 2uu^T$. In either case, H is symmetric and orthogonal:*

$$H^T H = (I - 2uu^T)^T (I - 2uu^T) = I - 4uu^T + 4uu^T uu^T = I.$$

In the complex case, H is both Hermitian and unitary.

H is sometimes called elementary reflector since

Proposition 6.2.7 *Let $z = e_1 = (1, 0, \cdots, 0)^T$, and $\sigma = \|x\|$, and $v = x + \sigma z$. Then, $Hx = -\sigma z = (-\sigma, 0, \cdots, 0)^T$.*

Proof.

$$Hx = x - 2\frac{vv^T x}{\|v\|^2} = x - (x + \sigma z)\frac{2(x + \sigma z)^T x}{(x + \sigma z)^T (x + \sigma z)}$$

$$Hx = x - (x + \sigma z) = -\sigma z. \quad \square$$

Remark 6.2.8 *Assume that we are going to transform A into a tridiagonal or Hessenberg form $U^{-1} A U$. Let*

$$x = \begin{bmatrix} a_{21} \\ a_{31} \\ \vdots \\ a_{n1} \end{bmatrix}, \quad z = \begin{bmatrix} 1 \\ 0 \\ \vdots \\ 0 \end{bmatrix}, \quad Hx = \begin{bmatrix} -\sigma \\ 0 \\ \vdots \\ 0 \end{bmatrix}.$$

$$U_1 = \begin{bmatrix} 1 & 0\,0\,0\,0 \\ 0 & \\ 0 & H \\ 0 & \\ 0 & \end{bmatrix} = U_1^{-1}, \text{ and } U^{-1} A U_1 = \begin{bmatrix} a_{11} & * & * & * & * \\ -\sigma & * & * & * & * \\ 0 & * & * & * & * \\ 0 & * & * & * & * \\ 0 & * & * & * & * \end{bmatrix}.$$

The second stage is similar: x consists of the last $n - 2$ entries in the second column, z is the first unit coordinate vector of matching length, and H_2 is of order $n - 2$:

$$U_2 = \begin{bmatrix} 1\,0\,0\ 0\ 0 \\ 0\,1\,0\ 0\ 0 \\ 0\ 0 \\ 0\ 0 \quad H_2 \\ 0\ 0 \end{bmatrix} = U_2^{-1}, \text{ and } U_2^{-1}(U^{-1}AU_1)U_2 = \begin{bmatrix} *\,*\,*\,*\,* \\ *\,*\,*\,*\,* \\ 0\,*\,*\,*\,* \\ 0\,0\,*\,*\,* \\ 0\,0\,*\,*\,* \end{bmatrix}.$$

Following a similar approach, one may operate on the upper right corner of A simultaneously to generate a tridiagonal matrix at the end. This process is the main motivation of the QR algorithm.

Problems

6.1. Show that for orthogonal matrices $\|Q\| = c(Q) = 1$. Orthogonal matrices and their multipliers (αQ) are only perfect condition matrices.

6.2. Apply the QR algorithm for

$$A = \begin{bmatrix} 0.5000 & -1.1180 & 0 & 0 & 0 & 0 \\ -1.1180 & 91.2000 & -80.0697 & 0 & 0 & 0 \\ 0 & -80.0697 & 81.0789 & 4.1906 & 0 & 0 \\ 0 & 0 & 4.1906 & 2.5913 & 0.2242 & 0 \\ 0 & 0 & 0 & 0.2242 & 0.1257 & -0.0100 \\ 0 & 0 & 0 & 0 & -0.0100 & 0.0041 \end{bmatrix}.$$

6.3. Let $A(n) \in \mathbb{R}^{n \times n}$, $A(n) = (a_{ij})$, where $a_{ij} = \frac{1}{i+j-1}$.
(a) Take $A(2)$.

1. Let $b_I = \begin{bmatrix} 1.0 \\ 0.5 \end{bmatrix}$ and $b_{II} = \begin{bmatrix} 1.5 \\ 1.0 \end{bmatrix}$. Calculate the relative error.
2. Find a good upper bound for the relative error obtained after perturbing the right hand side.

3. Find the relative error of perturbing $A(2)$ by $\Delta_{A(2)} = \begin{bmatrix} 0 & -\frac{1}{2} \\ -\frac{1}{2} & \frac{2}{3} \end{bmatrix}$. Take

$b_I = \begin{bmatrix} 1.0 \\ 0.5 \end{bmatrix}$ as the right hand side.
4. Find a good upper bound for the relative error obtained after perturbing $A(2)$.

(b) Take $A(3)^T A(3)$ and find its condition number and compare with the condition number of $A(3)$.
(c) Take $A(4)$ and calculate its condition number after finding the eigen values using the QR algorithm.

Web material

```
http://202.41.85.103/manuals/planetmath/entries/65/
    MatrixConditionNumber/MatrixConditionNumber.html
http://bass.gmu.edu/ececourses/ece499/notes/note4.html
http://beige.ucs.indiana.edu/B673/node30.html
http://beige.ucs.indiana.edu/B673/node35.html
http://csdl.computer.org/comp/mags/cs/2000/01/c1038abs.htm
http://csdl2.computer.org/persagen/DLAbsToc.jsp?resourcePath=/dl/
    mags/cs/&toc=comp/mags/cs/2000/01/c1toc.xml&DOI=10.1109/
    5992.814656
http://efgh.com/math/invcond.htm
http://en.powerwissen.com/G1+DpIQ8h2QSmPsQTtNO8Q==
    _QR_algorithm.html
http://en.wikipedia.org/wiki/Condition_number
http://en.wikipedia.org/wiki/Matrix_norm
http://en.wikipedia.org/wiki/QR_algorithm
http://en.wikipedia.org/wiki/Tridiagonal_matrix
http://epubs.siam.org/sam-bin/dbq/article/23653
http://esperia.iesl.forth.gr/~amo/nr/bookfpdf/f11-5.pdf
http://fish.cims.nyu.edu/educational/num_meth_I_2005/lectures/
    lec_11_qr_algorithm.pdf
http://gosset.wharton.upenn.edu/~foster/teaching/540/
    class_s_plus_1/Notes/node1.html
http://mate.dm.uba.ar/~matiasg/papers/condi-arxiv.pdf
http://math.arizona.edu/~restrepo/475A/Notes/sourcea/node53.html
http://math.fullerton.edu/mathews/n2003/hessenberg/HessenbergBib/
    Links/HessenbergBib_lnk_2.html
http://math.fullerton.edu/mathews/n2003/qrmethod/QRMethodBib/Links/
    QRMethodBib_lnk_2.html
http://mathworld.wolfram.com/ConditionNumber.html
http://mpec.sc.mahidol.ac.th/numer/STEP16.HTM
http://olab.is.s.u-tokyo.ac.jp/~nishida/la7/sld009.htm
http://planetmath.org/encyclopedia/ConditionNumber.html
http://planetmath.org/encyclopedia/MatrixConditionNumber.html
http://w3.cs.huji.ac.il/course/2005/csip/condition.pdf
http://web.ics.purdue.edu/~nowack/geos657/lecture8-dir/lecture8.htm
http://www-math.mit.edu/~persson/18.335/lec14handout6pp.pdf
http://www-math.mit.edu/~persson/18.335/lec15handout6pp.pdf
http://www-math.mit.edu/~persson/18.335/lec16.pdf
http://www.absoluteastronomy.com/encyclopedia/q/qr/
    qr_algorithm1.htm
http://www.acm.caltech.edu/~mlatini/research/
    presentation-qr-feb04.pdf
http://www.acm.caltech.edu/~mlatini/research/qr_alg-feb04.pdf
http://www.caam.rice.edu/~timwar/MA375F03/Lecture22.ppt
http://www.cas.mcmaster.ca/~qiao/publications/nm-2005.pdf
http://www.cs.colorado.edu/~mcbryan/3656.04/mail/54.htm
http://www.cs.unc.edu/~krishnas/eigen/node4.html
```

```
http://www.cs.unc.edu/~krishnas/eigen/node6.html
http://www.cs.ut.ee/~toomas_l/linalg/lin1/node18.html
http://www.cs.utk.edu/~dongarra/etemplates/node95.html
http://www.csc.uvic.ca/~dolesky/csc449-540/5.5.pdf
http://www.ee.ucla.edu/~vandenbe/103/lineqsb.pdf
http://www.efgh.com/math/invcond.htm
http://www.ims.cuhk.edu.hk/~cis/2004.4/04.pdf
http://www.krellinst.org/UCES/archive/classes/CNA/dir1.7/
    uces1.7.html
http://www.library.cornell.edu/nr/bookcpdf/c11-3.pdf
http://www.library.cornell.edu/nr/bookcpdf/c11-6.pdf
http://www.ma.man.ac.uk/~higham/pap-le.html
http://www.ma.man.ac.uk/~nareports/narep447.pdf
http://www.math.vt.edu/people/renardym/class_home/nova/bifs/
    node52.html
http://www.math.wsu.edu/faculty/watkins/slides/qr03.pdf
http://www.maths.lancs.ac.uk/~gilbert/m306c/node22.html
http://www.maths.nuigalway.ie/MA385/nov14.pdf
http://www.mathworks.com/company/newsletters/news_notes/pdf/
    sum95cleve.pdf
http://www.nasc.snu.ac.kr/sheen/nla/html/node13.html
http://www.nasc.snu.ac.kr/sheen/nla/html/node23.html
http://www.netlib.org/scalapack/tutorial/tsld191.htm
http://www.physics.arizona.edu/~restrepo/475A/Notes/sourcea/
    node53.html
http://www.sci.wsu.edu/math/faculty/watkins/slides/qr03.pdf
http://www.ugrad.cs.ubc.ca/~cs402/handouts/handout12.pdf
http://www.ugrad.cs.ubc.ca/~cs402/handouts/handout26.pdf
http://www.ugrad.cs.ubc.ca/~cs402/handouts/handout28.pdf
http://www.uwlax.edu/faculty/will/svd/condition/index.html
http://www.uwlax.edu/faculty/will/svd/norm/index.html
http://www2.msstate.edu/~pearson/num-anal/num-anal-notes/
    qr-algorithm.pdf
http://www4.ncsu.edu/eos/users/w/white/www/white/dir1.7/
    sec1.7.6.html
```

7

Convex Sets

This chapter is compiled to present a brief summary of the most important concepts related to convex sets. Following the basic definitions, we will concentrate on supporting and separating hyperplanes, extreme points and polytopes.

7.1 Preliminaries

Definition 7.1.1 *A set X in $\mathbb{R}^n$ is said to be convex if*

$$\forall x_1, x_2 \in X \text{ and } \forall \alpha \in \mathbb{R}_+, 0 < \alpha < 1, \text{ the point } \alpha x_1 + (1 - \alpha)x_2 \in X.$$

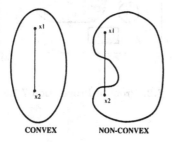

Fig. 7.1. Convexity

Remark 7.1.2 *Geometrically speaking, X is convex if for any points $x_1, x_2 \in X$, the line segment joining these two points is also in the set. This is illustrated in Figure 7.1.*

Definition 7.1.3 *A point $x \in X$ is an extreme point of the convex set X if and only if*

$$\not\exists\, x_1, x_2\; (x_1 \neq x_2) \in X \;\ni\; x = (1 - \alpha)x_1 + \alpha x_2,\; 0 < \alpha < 1.$$

Proposition 7.1.4 *Any extreme point is on boundary of the set.*

Proof. Let x_0 be any interior point of X. Then $\exists \epsilon > 0 \ni$ every point in this ϵ neighborhood of x_0 is in this set. Let $x_1 \neq x_0$ be a point in this ϵ neighborhood. Consider

$$x_2 = -x_1 + 2x_0,\; |x_2 - x_0| = |x_1 - x_0|$$

then x_2 is in ϵ neighborhood. Furthermore, $x_0 = \frac{1}{2}(x_1 + x_2)$; hence, x_0 is not an extreme point. □

Remark 7.1.5 *Not all boundary points of a convex set are necessarily extreme points. Some boundary points may lie between two other boundary points.*

Proposition 7.1.6 *Convex sets in $\mathbb{R}^n$ satisfy the following relations.*

i. *If X is a convex set and $\beta \in \mathbb{R}$, the set $\beta X = \{y : y = \beta x, x \in X\}$ is convex.*

ii. *If X and Y are convex sets, then the set $X + Y = \{z : z = x + y, x \in X, y \in Y\}$ is convex.*

iii. *The intersection of any collection of convex sets is convex.*

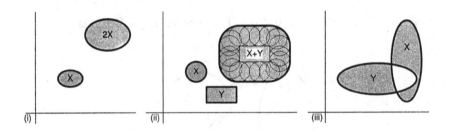

Fig. 7.2. Proof of Proposition 7.1.6

Proof. Obvious from Figure 7.2. □

Another important concept is to form the smallest convex set containing a given set.

Definition 7.1.7 *Let $S \subset \mathbb{R}^n$. The convex hull of S is the set which is the intersection of all convex sets containing S.*

Definition 7.1.8 *A cone C is a set such that if $x \in C$, then $\alpha x \in C, \forall \alpha \in \mathbb{R}_+$. A cone which is also convex is known as convex cone. See Figure 7.3.*

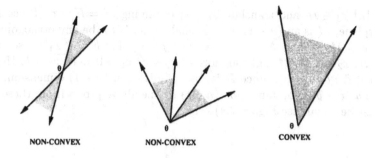

NON-CONVEX   NON-CONVEX   CONVEX

Fig. 7.3. Cones

7.2 Hyperplanes and Polytopes

The most important type of convex set (aside from single points) is the hyperplane.

Remark 7.2.1 *Hyperplanes dominate the entire theory of optimization; appearing in Lagrange multipliers, duality theory, gradient calculations, etc. The most natural definition for a hyperplane is the generalization of a plane in $\mathbb{R}^3$.*

Definition 7.2.2 *A set V in $\mathbb{R}^n$ is said to be linear variety, if, given any $x_1, x_2 \in V$, we have $\alpha x_1 + (1 - \alpha)x_2 \in V, \forall \alpha \in \mathbb{R}$.*

Remark 7.2.3 *The only difference between a linear variety and a convex set is that a linear variety is the entire line passing through any two points, rather than a simple line segment.*

Definition 7.2.4 *A hyperplane in $\mathbb{R}^n$ is an $(n-1)$-dimensional linear variety. It can be regarded as the largest linear variety in a space other than the entire space itself.*

Proposition 7.2.5 *Let $a \in \mathbb{R}^n, a \neq \theta$ and $b \in \mathbb{R}$. The set*

$$H = \{x \in \mathbb{R}^n : a^T x = b\}$$

is a hyperplane in $\mathbb{R}^n$.

Proof. Let $x_1 \in H$. Translate H by $-x_1$, we then obtain the set

$$M = H - x_1 = \{y \in \mathbb{R}^n : \exists x \in H \ni y = x - x_1\},$$

which is a linear subspace of $\mathbb{R}^n$. $M = \{y \in \mathbb{R}^n : a^T y = 0\}$ is also the set of all orthogonal vectors to $a \in \mathbb{R}^n$, which is clearly $(n-1)$ dimensional. $\square$

Proposition 7.2.6 *Let H be an hyperplane in $\mathbb{R}^n$. Then,*

$$\exists a \in \mathbb{R}^n \ni H = \{x \in \mathbb{R} : a^T x = b\}.$$

Proof. Let $x_1 \in H$, and translate by $-x_1$ obtaining $M = H - x_1$. Since H is a hyperplane, M is an $(n - 1)$-dimensional space. Let a be any orthogonal to M, i.e. $a \in M^{\perp}$. Thus, $M = \{y \in \mathbb{R}^n : a^T y = 0\}$. Let $b = a^T x_1$ we see that if $x_2 \in H$, $x_2 - x_1 \in M$ and therefore $a^T x_2 - a^T x_1 = 0 \Rightarrow a^T x_2 = b$. Hence, $H \subset \{x \in \mathbb{R} : a^T x = b\}$. Since H is, by definition, of $(n - 1)$ dimension, and $\{x \in \mathbb{R} : a^T x = b\}$ is of dimension $(n - 1)$ by the above proposition, these two sets must be equal (see Figure 7.4). $\square$

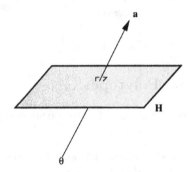

Fig. 7.4. Proof of Proposition 7.2.6

Definition 7.2.7 *Let $a \in \mathbb{R}^n$, $b \in \mathbb{R}$. Corresponding to the hyperplane $H = \{x : a^T x = b\}$, there are positive and negative closed half spaces:*

$$H_+ = \{x : a^T x \geq b\}, \; H_- = \{x : a^T x \leq b\}$$

and

$$\dot{H}_+ = \{x : a^T x > b\}, \; \dot{H}_- = \{x : a^T x < b\}.$$

Half spaces are convex sets and $H_+ \cup H_- = \mathbb{R}^n$.

Definition 7.2.8 *A set which can be expressed as the intersection of a finite number of closed half spaces is said to be a convex polyhedron.*

Convex polyhedra are the sets obtained as the family of solutions to a set of linear inequalities of the form

$$a_1^T x \leq b_1$$
$$a_2^T x \leq b_2$$
$$\vdots$$
$$a_m^T x \leq b_m$$

Since each individual entry defines a half space and the solution family is the intersection of these half spaces.

Definition 7.2.9 *A nonempty bounded polyhedron is called a polytope.*

7.3 Separating and Supporting Hyperplanes

Theorem 7.3.1 (Separating Hyperplane) *Let X be a convex set and y be a point exterior to the closure of X. Then, there exists a vector $a \in \mathbb{R}^n \ni a^T y < \inf_{x \in X} a^T x$. (Geometrically, a given point y outside X, a separating hyperplane can be passed through the point y that does not touch X. Refer to Figure 7.5)*

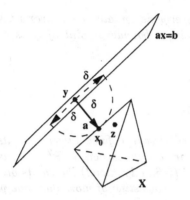

Fig. 7.5. Separating Hyperplane

Proof. Let $\delta = \inf_{x \in X} |x - y| > 0$ Then, there is an x_0 on the boundary of X such that $|x_0 - y| = \delta$. Let $z \in X$. Then,

$$\forall \alpha, \; 0 \leq \alpha \leq 1, \; x_0 + \alpha(z - x_0)$$

is the line segment between x_0 and z. Thus, by definition of x_0,

$$|x_0 + \alpha(z - x_0) - y|^2 \geq |x_0 - y|^2$$

$$\Leftrightarrow (x_0 - y)^T(x_0 - y) + 2\alpha(x_0 - y)^T(z - x_0) + \alpha^2(z - x_0)^T(z - x_0) \geq (x_0 - y)^T(x_0 - y)$$

$$\Leftrightarrow 2\alpha(x_0 - y)^T(z - x_0) + \alpha^2|z - x_0|^2 \geq 0$$

Let $\alpha \to 0^+$, then α^2 tends to 0 more rapidly than 2α. Thus,

$$(x_0 - y)^T(z - x_0) \geq 0 \Leftrightarrow (x_0 - y)^T z - (x_0 - y)^T x_0 \geq 0$$

$$\Leftrightarrow (x_0 - y)^T z \geq (x_0 - y)^T x_0 = (x_0 - y)^T y + (x_0 + y)^T(x_0 - y) = (x_0 - y)^T y + \delta^2$$

$$\Leftrightarrow (x_0 - y)^T y < (x_0 - y)^T x_0 \leq (x_0 - y)^T z, \; \forall z \in X \; (\text{Since } \delta > 0).$$

Let $a = (x_0 - y)$, then $a^T y < a^T x_0 = \inf_{z \in X} a^T z$. $\square$

Theorem 7.3.2 (Supporting Hyperplane) *Let X be a convex set, and let y be a boundary point of X. Then, there is a hyperplane containing y and containing X in one of its closed half spaces.*

Proof. Let $\{y_k\}$ be sequence of vectors, exterior to the closure of X, converging to y. Let $\{a_k\}$ be a sequence of corresponding vectors constructed according to the previous theorem, normalized so that $|a_k| = 1$, such that $a_k^T y_k < \inf_{x \in X}$. Since $\{a_k\}$ is a boundary sequence, it converges to a. For this vector, we have $a^T y = \lim a_k^T y_k \leq ax$. $\square$

Definition 7.3.3 *A hyperplane containing a convex set X in one of its closed half spaces and containing a boundary point of X is said to be supporting hyperplane of X.*

7.4 Extreme Points

Remark 7.4.1 *We have already defined extreme points. For example, the extreme points of a square are its corners in $\mathbb{R}^2$ whereas the extreme points of a circular disk are all (infinitely many!) the points on the boundary circle. Note that, a linear variety consisting of more than one point has no extreme points.*

Lemma 7.4.2 *Let X be a convex set, H be a supporting hyperplane of X and $T = X \cap H$. Every extreme point of T is an extreme point of X.*

Proof. Suppose $x_0 \in T$ is not an extreme point of X. Then,

$$x_0 = \alpha x_1 + (1 - \alpha)x_2 \text{ for some } x_1, x_2 \in X, \, 0 < \alpha < 1.$$

Let $H = \{x : a^T x = c\}$ with X contained in its closed positive half space. Then, $a^T x_1 \geq c$, $a^T x_2 \geq c$. However, since $x_0 \in H$,

$$c = a^T x_0 = \alpha a^T x_1 + (1 - \alpha)a^T x_2.$$

Thus, $x_1, x_2 \in H$. Hence, $x_1, x_2 \in T$ and x_0 is not an extreme point of T. $\square$

Theorem 7.4.3 *A closed bounded convex set in $\mathbb{R}^n$ is equal to the closed convex hull of its extreme points.*

Proof. This proof is by induction on n.
For $n = 1$, the statement is true for a line segment:

$$[a, b] = \{x \in \mathbb{R} : x = \alpha + (1 - \alpha)b, 0 \leq \alpha \leq 1\}.$$

Suppose that the theorem is true for $(n - 1)$. Let X be a closed bounded convex set in $\mathbb{R}^n$, and let K be the convex hull of the extreme points of X.

We will show that $X = K$.

Assume that $\exists y \in X \ni y \notin K$. Then, by Theorem 7.3.1, there is a hyperplane separating y and K;

$$\exists a \neq 0 \ni a^T y < \inf_{x \in K} a^T x$$

Let $x_0 = \inf_{x \in X}(a^T x)$. x_0 is finite and $\exists x_0 \in X \ni a^T x_0 = b_0$ (because by Weierstrass' Theorem: The continuous function $a^T x$ achieve its minimum over any closed bounded set).

Hence, the hyperplane $H = \{x : a^T x = b_0\}$ is a supporting hyperplane to X. Since $b_0 \leq a^T y \leq \inf_{x \in K} a^T x$, H is disjoint from K. Let $T = H \cap X$. Then, T is a bounded closed convex set of H, which can be regarded as a space in $\mathbb{R}^{n-1}$. $T \neq \emptyset$, since $x_0 \in T$. Hence, by induction hypothesis, T contains extreme points; and by the previous Lemma, these are the extreme points of X. Thus, we have found extreme points of X not in K, Contradiction. Therefore, $X \subseteq K$, and hence $X = K$ (since $K \subseteq X$, i.e. K is closed and bounded). $\square$

Remark 7.4.4 *Let us investigate the implications of this theorem for convex polytopes. A convex polytope is a bounded polyhedron. Being the intersection of closed halfspaces, a convex polytope is closed. Thus, any convex polyhedron is the closed convex hull of its extreme points. It can be shown that any polytope has at most a finite number of extreme points, and hence a convex polytope is equal to the convex hull of a finite number of points. The converse can also be established, yielding the following two equivalent characterizations.*

Theorem 7.4.5 *A convex polytope can be described either as a bounded intersection of a finite number of closed half spaces, or as the convex hull of a finite number of points.*

Problems

7.1. Characterize (draw, give an example, list extreme points and half spaces) the following polytopes:

a) zero dimensional polytopes.

b) one dimensional polytopes.

c) two dimensional polytopes.

7.2. d-simplex

d-simplex is the convex hull of any $d + 1$ independent points in $\mathbb{R}^n$ $(n \geq d)$. Standard $d - simplex$ with $d + 1$ vertices in $\mathbb{R}^{d+1}$ is

$$\Delta_d = \{x \in \mathbb{R}^{d+1} : \sum_{i=1}^{d+1} x_i = 1; x_i \geq 0, i = 1, \ldots, d + 1\}.$$

Characterize Δ_2 in $\mathbb{R}^3$.

7.3. Cube and Octahedron

Characterize cubes and octahedrons with the help of three dimensional cube C_3, and octahedron C_3^Δ.

7.4. Pyramid

Let $P_{n+1}=\mathrm{conv}(C_n, x_0)$ be a (n+1)–dimensional pyramid, where $x_0 \notin C_n$. Draw

$$P_3 = conv(C_2 : \alpha = 1, (1/2, 1/2, 1)^T)$$

and write down all describing inequalities.

7.5. Tetrahedron

The vertices of a *tetrahedron* of side length $\sqrt{2}$ can be given by a particularly simple form when the vertices are taken as corners of the unit cube. Such a tetrahedron inside a cube of side length 1 has side length $\sqrt{2}$ with vertices $(0,0,0)^T$, $(0,1,1)^T$, $(1,0,1)^T$, $(1,1,0)^T$. Draw and find a set of describing inequalities. Is it possible to express P_{n+1} as a union / intersection / direct sum of a cone and a polytope?

7.6. Dodecahedron

Find the vertices of a *dodecahedron* (see Figure 7.6) of side length $a = \sqrt{5} - 1$.

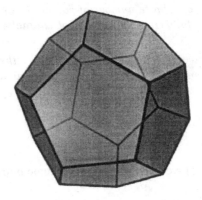

Fig. 7.6. A dodecahedron

Web material

http://cepa.newschool.edu/het/essays/math/convex.htm
http://cm.bell-labs.com/who/clarkson/cis677/lecture/6/index.html
http://cm.bell-labs.com/who/clarkson/cis677/lecture/8/

```
http://dimax.rutgers.edu/~sjaslar/
http://dogfeathers.com/java/hyperslice.html
http://en.wikipedia.org/wiki/Polytope
http://en.wikipedia.org/wiki/Wikipedia:WikiProject_Mathematics/
    PlanetMath_Exchange/52-XX_Convex_and_discrete_geometry
http://eom.springer.de/c/c026340.htm
http://grace.speakeasy.net/~dattorro/EDMAbstract.pdf
http://grace.speakeasy.net/~dattorro/Meboo.html
http://learningtheory.org/colt2004/colt04_boyd.pdf
http://math.sfsu.edu/beck/teach/870/lecture5.pdf
http://mathworld.wolfram.com/Convex.html
http://mathworld.wolfram.com/Polytope.html
http://mizar.uwb.edu.pl/JFM/pdf/convex3.pdf
http://ocw.mit.edu/NR/rdonlyres/Electrical-Engineering-and-Computer-
    Science/6-253Spring2004/14DD65AE-0A43-4353-AE09-7B107CC4AAD7/0/
    lec_11.pdf
http://ocw.mit.edu/NR/rdonlyres/Electrical-Engineering-and-Computer-
    Science/6-253Spring2004/81D31E98-C26B-4375-B089-FB5FAE4E99CF/0/
    lec_7.pdf
http://ocw.mit.edu/NR/rdonlyres/Electrical-Engineering-and-Computer-
    Science/6-253Spring2004/96203668-B98C-4F3C-A65D-4646F942EF71/0/
    lec_3.pdf
http://staff.polito.it/giuseppe.calafiore/cvx-opt/secure/
    02_cvx-sets_gc.pdf
http://www-math.mit.edu/~vempala/18.433/L4.pdf
http://www-personal.umich.edu/~mepelman/teaching/IOE611/Handouts/
    611Sets.pdf
http://www.cas.mcmaster.ca/~cs4te3/notes/convexopt.pdf
http://www.cas.mcmaster.ca/~deza/CombOptim_Ch7.ppt
http://www.cis.upenn.edu/~cis610/polytope.pdf
http://www.cs.cmu.edu/afs/cs/academic/class/16741-s06/www/
    Lecture13.pdf
http://www.cs.wustl.edu/~pless/506/12.html
http://www.cse.unsw.edu.au/~lambert/java/3d/ConvexHull.html
http://www.eecs.berkeley.edu/~wainwrig/ee227a/Scribe/
    lecture12_final_verB.pdf
http://www.eleves.ens.fr/home/trung/supporting_hyperplane.html
http://www.geom.uiuc.edu/graphics/pix/Special_Topics/
    Computational_Geometry/cone.html
http://www.geom.uiuc.edu/graphics/pix/Special_Topics/
    Computational_Geometry/half.html
http://www.hss.caltech.edu/~kcb/Ec101/index.shtml#Notes
http://www.ics.uci.edu/~eppstein/junkyard/polytope.html
http://www.irisa.fr/polylib/DOC/node16.html
http://www.isye.gatech.edu/~spyros/LP/node15.html
http://www.jstor.org/view/00029939/di970732/97p0127h/0
http://www.mafox.com/articles/Polytope
http://www.math.rutgers.edu/pub/sontag/pla.txt
http://www.maths.lse.ac.uk/Personal/martin/fme9a.pdf
```

http://www.mizar.org/JFM/Vol15/convex3.html
http://www.ms.uky.edu/~sills/webprelim/sec013.html
http://www.mtholyoke.edu/~jsidman/wolbachPres.pdf
http://www.princeton.edu/~chiangm/ele53912.pdf
http://www.stanford.edu/class/ee364/lectures/sets.pdf
http://www.stanford.edu/class/ee364/reviews/review1.pdf
http://www.stanford.edu/class/msande310/lecture03.pdf
http://www.stanford.edu/~dattorro/mybook.html
http://www.stat.psu.edu/~jiali/course/stat597e/notes2/percept.pdf
http://www.uni-bayreuth.de/departments/wirtschaftsmathematik/rambau/
    Diss/diss_MASTER/node35.html
http://www.wisdom.weizmann.ac.il/~feige/lp/lecture2.ps
http://www2.isye.gatech.edu/~spyros/LP/node15.html
http://www2.sjsu.edu/faculty/watkins/convex.htm

8

Linear Programming

A Linear Programming problem, or LP, is a problem of optimizing a given linear objective function over some polyhedron. We will present the forms of LPs in this chapter. Consequently, we will focus on the simplex method of G. B. Dantzig, which is the algorithm most commonly used to solve LPs; in practice it runs in polynomial time, but the worst-case running time is exponential. Following the various variants of the simplex method, the duality theory will be introduced. We will concentrate on the study of duality as a means of gaining insight into the LP solution. Finally, the series of Farkas' Lemmas, the most important theorems of alternatives, will be stated.

8.1 The Simplex Method

This section is about linear programming: optimization of a linear objective function subject to finite number (m) of linear constraints with n unknown and nonnegative decision variables.

Example 8.1.1 *The following is an LP:*

$$Min \quad z = 2x + 3y$$
$$s.t.$$
$$2x + y \geq 6$$
$$x + 2y \geq 6$$
$$x, y \geq 0.$$

Standard Form:

$$Min \quad z = c^T x$$
$$s.t.$$
$$Ax \geq b$$
$$x \geq \theta$$

Canonical form:

$$Min \ z = c^T x + \theta^T y$$

s.t.

$$Ax - y = b$$
$$x, y \geq \theta$$

$\Leftrightarrow$

$$Min \ z = [c^T | \theta^T] \begin{bmatrix} x \\ y \end{bmatrix}$$

s.t.

$$[A| - I] \begin{bmatrix} x \\ y \end{bmatrix} = b$$

$$\begin{bmatrix} x \\ y \end{bmatrix} \geq \theta.$$

Example 8.1.2

$$Min \ z = 2x + 3y$$

s.t.

$$2x + y \geq 6$$
$$x + 2y \geq 6$$
$$x, y \geq 0.$$

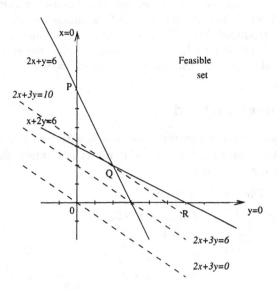

Fig. 8.1. The feasible solution region in Example 8.1.2

See Figure 8.1.

$$A = \begin{bmatrix} 1 & 2 & -1 & 0 \\ 2 & 1 & 0 & -1 \end{bmatrix}, b = \begin{bmatrix} 6 \\ 6 \end{bmatrix}, c = \begin{bmatrix} 2 \\ 3 \\ 0 \\ 0 \end{bmatrix}.$$

Definition 8.1.3 *The extreme points of the feasible set are exactly the basic feasible solutions of $Ax = b$. A solution is basic when n of its $m+n$ components are zero, and is feasible when it satisfies $x \geq \theta$. Phase I of the simplex method finds one basic feasible solution, and Phase II moves step by step to the optimal one.*

If we are already at a basic feasible solution x, and for convenience we reorder its components so that the n zeros correspond to free variables.

$$x = \begin{bmatrix} x_B \\ x_N = \theta \end{bmatrix}, A = [B, N], c^T = (c_B^T, c_N^T)$$

$$Min\ z = (c_B^T, c_N^T) \begin{bmatrix} x_B \\ x_N = \theta \end{bmatrix}$$

s.t.

$$[B|N] \begin{bmatrix} x_B \\ x_N = \theta \end{bmatrix} = b$$

$$\begin{bmatrix} x_B \\ x_N = \theta \end{bmatrix} \geq \theta.$$

$\Leftrightarrow$

$$Min\ z = c_B^T x_B + c_N^T x_N$$

s.t.

$$B x_B + N x_N = b$$

$$x_B, x_N \geq \theta$$

Let us take the constraints

$$B x_B + N x_N = b \Leftrightarrow B x_b = b - N x_N \Leftrightarrow x_B = B^{-1}[b - N x_N] = B^{-1}b - B^{-1}N x_N.$$

Now plug x_B in the objective function

$$z = c_B^T x_B + c_N^T x_N = c_B^T[B^{-1}b - B^{-1}N x_N] + c_N^T x_N$$

$$= c_B^T B^{-1}b + (c_N^T - c_B B^{-1}N)x_N.$$

If we let $x_N = \theta$, then $x_B = B^{-1}b \geq \theta \Rightarrow z = c_B^T B^{-1}b$.

Proposition 8.1.4 (Optimality Condition) *If the vector $(c_N^T - c_B^T B^{-1}N)$ is nonnegative, then no reduction in z can be achieved. The current extreme point $(x_B = B^{-1}b, x_N = \theta)$ is optimal and the minimum objective function value is $c_B B^{-1}b$.*

Assume that the optimality condition fails, the usual greedy strategy is to choose the most negative component of $c_N - c_B B^{-1}N$, known as Dantzig's rule. Thus, we have determined which component will move from free to basic, called as *entering variable* x_e. We have to decide which basic component is to become free, called as *leaving variable*, x_l. Let N^e be the column of N corresponding to x_e. $x_B = B^{-1}b - B^{-1}N^e x_e$. If we increase x_e from 0, some entries of x_B may begin to decrease, and we reach a a neighboring extreme point when a component of x_B reaches 0. It is the component corresponding to x_l. At this extreme point, we have reached a new x which is both feasible and basic: it is feasible because $x \geq 0$, it is basic since we again have n zero components. x_e is gone from zero to α, replaces x_l which is dropped to zero. The other components of x_B might have changed their values, but remain positive.

Proposition 8.1.5 (Min Ratio) *Suppose* $u = N^e$, *then the value of* x_e *will be:*

$$\alpha = \min_{x_j:basic} \frac{(B^{-1}b)_j}{(B^{-1}u)_j} = \frac{(B^{-1}b)_l}{(B^{-1}u)_l}$$

and the objective function will decrease to $c_B^T B^{-1}b - \alpha B^{-1}u$.

Remark 8.1.6 (Unboundedness) *The minimum is taken only over positive components of* $B^{-1}u$, *since negative entries will increase* x_B *and zero entries keeps* x_B *as their previous values. If there are no positive components, then the next extreme point is infinitely far away, then the cost can be reduced forever;* $z = -\infty$! *In this case we term the optimization problem as unbounded.*

Remark 8.1.7 (Degeneracy) *Suppose that more than* n *of the variables are zero or two different components if the minimum ratio formula give the same minimum ratio. We can choose either one of them to be made free, but the other will still be in the basis at zero level. Thus, the new extreme point will have* $(n + 1)$ *zero components. Geometrically, there is an extra supporting plane at the extreme point. In degeneracy, there is the possibility of cycling forever around the same set of extreme points without moving toward* x^*, *the optimal solution. In general, one may assume nondegeneracy hypothesis* $(x_B = B^{-1}b > 0)$.

Example 8.1.8 *Assume that we are at the extreme point* $\mathcal{P}$ *in Figure 8.1, corresponding to the following basic feasible solution:*

$$x = \begin{bmatrix} x_B \\ x_N \end{bmatrix} = \begin{bmatrix} 6 \\ 6 \\ \hline 0 \\ 0 \end{bmatrix} = \begin{bmatrix} z_1 \\ y \\ x \\ z_2 \end{bmatrix},$$

$$A = [B|N] = \begin{bmatrix} z_1 & y & x & z_2 \\ -1 & 2 & 1 & 0 \\ 0 & 1 & 2 & -1 \end{bmatrix}, \quad c^T = (c_B^T|c_N^T) = \begin{pmatrix} z_1 & y & x & z_2 \\ 0 & 3 & 2 & 0 \end{pmatrix}.$$

$$c_N^T - c_B^T B^{-1}N = [2\,0] - [0\,3] \begin{bmatrix} -1 & 2 \\ 0 & 1 \end{bmatrix}^{-1} \begin{bmatrix} 1 & 0 \\ 2 & -1 \end{bmatrix}.$$

$$\begin{bmatrix} -1 & 2 & 1 & 0 \\ 0 & 1 & 0 & 1 \end{bmatrix} \rightarrow \begin{bmatrix} 1 & -2 & -1 & 0 \\ 0 & 1 & 0 & 1 \end{bmatrix} \rightarrow \begin{bmatrix} 1 & 0 & -1 & 2 \\ 0 & 1 & 0 & 1 \end{bmatrix} \Rightarrow B^{-1} = \begin{bmatrix} -1 & 2 \\ 0 & 1 \end{bmatrix}.$$

$$c_N^T - c_B^T B^{-1} N = [\,2\;0\,] - [\,0\;3\,] \begin{bmatrix} -1 & 2 \\ 0 & 1 \end{bmatrix} \begin{bmatrix} 1 & 0 \\ 2 & -1 \end{bmatrix}$$

$$c_N^T - c_B^T B^{-1} N = [\,2\;0\,] - [\,0\;3\,] \begin{bmatrix} 3 & -2 \\ 2 & -1 \end{bmatrix} = \begin{pmatrix} x & z_2 \\ -4 & 3 \end{pmatrix}.$$

Since the first component is negative, P is not optimal; x should enter the basis, i.e.

$$x_e = x, N^e = \begin{bmatrix} 1 \\ 2 \end{bmatrix} \Rightarrow B^{-1} N^e = \begin{bmatrix} 3 \\ 2 \end{bmatrix}, B^{-1} b = \begin{bmatrix} 6 \\ 6 \end{bmatrix} = \begin{bmatrix} z_1 \\ y \end{bmatrix},$$

$$x_B = B^{-1} b - B^{-1} N^e x_e = \begin{bmatrix} z_1 \\ y \end{bmatrix} = \begin{bmatrix} 6 \\ 6 \end{bmatrix} - \begin{bmatrix} 3 \\ 2 \end{bmatrix} x \geq \begin{bmatrix} 0 \\ 0 \end{bmatrix}.$$

$$\Rightarrow \alpha = Min\{\tfrac{6}{3} = 2, \tfrac{6}{2} = 3\} = 2. \text{ Thus, } x_l = z_1, x_e = 2, y = 6 - 2\alpha = 2.$$

$$x = \begin{bmatrix} x_B \\ x_N \end{bmatrix} = \begin{bmatrix} 2 \\ 2 \\ 0 \\ 0 \end{bmatrix} = \begin{bmatrix} x \\ y \\ z_1 \\ z_2 \end{bmatrix}, \; A = [\,B|N\,] = \begin{bmatrix} 1 & 2 & -1 & 0 \\ 2 & 1 & 0 & -1 \end{bmatrix}.$$

$$c^T = [\,c_B^T | c_N^T\,] = [\,2\;3|0\;0\,], \; B = \begin{bmatrix} 1 & 2 \\ 2 & 1 \end{bmatrix},$$

$$[\,B|I\,] = \begin{bmatrix} 1 & 2 & 1 & 0 \\ 2 & 1 & 0 & 1 \end{bmatrix} \rightarrow \begin{bmatrix} 1 & 2 & 1 & 0 \\ 0 & -3 & -2 & 1 \end{bmatrix} \rightarrow \begin{bmatrix} 1 & 0 & -\tfrac{1}{3} & \tfrac{2}{3} \\ 0 & 1 & \tfrac{2}{3} & -\tfrac{1}{3} \end{bmatrix} = [\,I|B^{-1}\,].$$

$$c_N^T - c_B^T B^{-1} N = [\,0\;0\,] - [\,2\;3\,] \begin{bmatrix} -\tfrac{1}{3} & \tfrac{2}{3} \\ \tfrac{2}{3} & -\tfrac{1}{3} \end{bmatrix} \begin{bmatrix} -1 & 0 \\ 0 & -1 \end{bmatrix} =$$

$$[\,0\;0\,] - [\,2\;3\,] \begin{bmatrix} \tfrac{1}{3} & -\tfrac{2}{3} \\ -\tfrac{2}{3} & -\tfrac{1}{3} \end{bmatrix} = [\,\tfrac{4}{3}\;\tfrac{1}{3}\,] > 0.$$

Thus, extreme point Q in Figure 8.1 is optimal, $c_B^T B^{-1} b = 10$ is the optimal value of the objective function.

8.2 Simplex Tableau

We have achieved a transition from the geometry of the simplex method to algebra so far. In this section, we are going to analyze a simplex step which can be organized in different ways.

The Gauss-Jordan method gives rise to the *simplex tableau*.

$$[A\|b] = [B|N\|b] \longrightarrow [I|B^{-1}N\|B^{-1}b].$$

Adding the cost row

$$\left[\begin{array}{c|c||c} I & B^{-1}N & B^{-1}b \\ \hline c_B^T & c_N^T & 0 \end{array}\right] \longrightarrow \left[\begin{array}{c|c||c} I & B^{-1}N & B^{-1}b \\ \hline 0 & c_N^T - c_B^T B^{-1}N & -c_B^T B^{-1}b \end{array}\right].$$

The last result is the complete tableau. It contains the solution $B^{-1}b$, the crucial vector $c_N{}^T - c_B{}^T B^{-1}N$ and the current objective function value $c_B{}^T B^{-1}b$ with a superfluous minus sign indicating that our problem is minimization. The simplex tableau also contains reduced coefficient matrix $B^{-1}N$ that is used in the minimum ratio. After determining the entering variable x_e, we examine the positive entries in the corresponding column of $B^{-1}N$, $(v = B^{-1}u = B^{-1}N^e)$ and α is determined by taking the ratio of $\frac{(B^{-1}b)_j}{(B^{-1}N^e)_j}$ for all positive v_j's.

If the smallest ratio occurs in l^{th} component, then the lth column of B should be replaced by u. The lth element of $(B^{-1}N^e)_l = v_l$ is distinguished as pivot element.

It is not necessary to return the starting tableau, exchange two columns and start again. Instead we can continue with the current tableau. Without loss of generality, we may assume that the first row corresponds to the leaving variable, that is the pivot element is v_1.

$$\begin{bmatrix} 1 \vdots 0 \cdots 0 & * \cdots * & v_1 & * \cdots * & (B^{-1}b)_1 \\ \vdots 0 \vdots & & \vdots v_2 \vdots & & \\ \vdots \cdots \vdots & & \vdots \cdot \cdot \vdots & & \\ \vdots \cdot \vdots \; I & B^{-1}N \vdots & \vdots \; \vdots & \vdots B^{-1}N & B^{-1}b \\ \vdots \cdot \vdots & & \vdots \cdot \cdot \vdots & & \\ \vdots 0 \vdots & & \vdots v_m \vdots & & \\ \hline 0 \vdots 0 \cdots 0 & * \cdots * \vdots c_e - c_B^T v \vdots * \cdots * & & & -c_B^T B^{-1}b \end{bmatrix}$$

The first step in the pivot operation is to divide the leaving variable's row by the pivot element to create 1 in the pivot entry. Then, we have

$$\begin{bmatrix} \vdots & \frac{1}{v_1} & \vdots 0 \cdots 0 & *\cdots* & \vdots & 1 & \vdots * \cdots * & \Big\Vert & \alpha \\ & \vdots 0 \vdots & & & \vdots v_2 \vdots & & & \Big\Vert \\ & \vdots \cdots \vdots & & & \vdots \cdots \vdots & & & \Big\Vert \\ & \vdots \vdots & I & B^{-1}N \vdots & \vdots & \vdots & \vdots B^{-1}N & \Big\Vert B^{-1}b \\ & \vdots \vdots & & & \vdots \cdots \vdots & & & \Big\Vert \\ & \vdots 0 \vdots & & & \vdots v_m \vdots & & & \Big\Vert \\ \hline \vdots & 0 \vdots 0 \cdots 0 & & *\cdots* \vdots c_e - c_B^T v \vdots *\cdots* & & & & \Big\Vert & -c_B^T B^{-1}b \end{bmatrix}.$$

For all the rows except the objective function row, do the following operation. For row i, multiply v_1*(the updated first row) and subtract from row i. For the objective function row, multiply the first row by $(c_e - c_B^T v)$ and subtract from the objective function row.

What we have at the end is another simplex tableau.

$$\begin{bmatrix} \vdots & \frac{1}{v_1} & \vdots 0 \cdots 0 & *\cdots* \vdots 1 \vdots *\cdots* & \Big\Vert & \alpha \\ & \vdots \frac{-v_2}{v_1} \vdots & & \vdots 0 \vdots & \Big\Vert \\ & \vdots \cdot \vdots & & \vdots \cdots \vdots & \Big\Vert & + \\ & \vdots \vdots & \vdots \cdot \vdots I & * \vdots \vdots * & \Big\Vert & \cdot \\ & \vdots \vdots & & \vdots \cdots \vdots & \Big\Vert & + \\ & \vdots \frac{-v_m}{v_1} \vdots & & \vdots 0 \vdots & \Big\Vert \\ \hline & \vdots \frac{-c_e - c_B^T v}{v_1} \vdots 0 \cdots 0 & *\cdots* \vdots 0 \vdots *\cdots* & \Big\Vert & -c_B^T B^{-1}b - \alpha(c_e - c_B^T v) \end{bmatrix}.$$

Example 8.2.1 *The starting tableau at point P is*

$$\left[\begin{array}{c|c} A & b \\ \hline c^T & 0 \end{array}\right] = \left[\begin{array}{c|c|c} B & N & b \\ \hline c_B^T & c_N^T & 0 \end{array}\right] = \left[\begin{array}{cc|cc|c} -1 & 2 & 1 & 0 & 6 \\ 0 & 1 & 2 & -1 & 6 \\ \hline 0 & 3 & 2 & 0 & 0 \end{array}\right]$$

The final tableau after Gauss-Jordan iterations is

$$\left[\begin{array}{c|cc|cc|c} & z_1 & y & x & z_2 & RHS \\ \hline z_1 & 1 & 0 & 3 & -2 & 6 \\ y & 0 & 1 & 2 & -1 & 6 \\ \hline z & 0 & 0 & -4 & 3 & -18 \end{array}\right] = \left[\begin{array}{c|c|c} I & B^{-1}N & B^{-1}b \\ \hline 0 & c_N^T - c_B^T B^{-1}N & -cost \end{array}\right]$$

Since the reduced cost for x is $-4 < 0$, x should enter the basis. The minimum ratio $\alpha = Min\{\frac{6}{2}, \frac{6}{3}\} = 2$ due to z_1, thus z_1 should leave the basis.

$$
\begin{bmatrix} 1 & 0 & 3 & -2 & \| & 6 \\ 0 & 1 & 2 & -1 & \| & 6 \\ \hline 0 & 0 & -4 & 3 & \| & -18 \end{bmatrix} \longrightarrow
\begin{bmatrix} & | & z_1 & y & | & x & z_2 & \| & RHS \\ \hline x & | & \frac{1}{3} & 0 & | & 1 & -\frac{2}{3} & \| & 2 \\ y & | & -\frac{2}{3} & 1 & | & 0 & \frac{1}{3} & \| & 2 \\ \hline -z & | & \frac{4}{3} & 0 & | & 0 & \frac{1}{3} & \| & -10 \end{bmatrix}
$$

Thus, $x^* = 2 = y^* \Rightarrow z^* = 10$.

Remark 8.2.2 *All the pivot operation can be handled by multiplying the inverse of the following elementary matrix.*

$$
E =
\begin{bmatrix}
1 & & \vdots v_1 \vdots 0 & \ldots & 0 \\
. & 0 & \vdots . \vdots . & & . \\
. & & \vdots . \vdots . & & . \\
0 & . & \vdots . \vdots . & & . \\
& & 1 \vdots v_l \vdots 0 & \ldots & 0 \\
0 & \ldots & 0 \vdots . \vdots 1 & & \\
. & & \vdots . \vdots & 1 & 0 \\
. & & \vdots . \vdots & & . \\
. & & \vdots . \vdots & 0 & . \\
0 & \ldots & 0 \vdots v_m \vdots & & 1
\end{bmatrix}
\Leftrightarrow E^{-1} =
\begin{bmatrix}
1 & & \vdots \frac{-v_1}{v_l} \vdots & & \\
. & & \vdots . \vdots & & \\
. & & \vdots . \vdots & 0 & \\
. & & . \vdots . \vdots & & \\
& & \vdots \frac{1}{v_l} \vdots & & \\
& & \vdots & \vdots 1 & \\
& 0 & \vdots & \vdots . & \\
& & \vdots & \vdots & \\
& & \vdots . \vdots & & \\
& & \vdots \frac{-v_m}{v_l} \vdots & & 1
\end{bmatrix}
$$

Thus, the pivot operation is

$$
[I | B^{-1}N \| B^{-1}b] \longrightarrow [E^{-1}I | E^{-1}B^{-1}N \| E^{-1}B^{-1}b].
$$

New basis is BE (B except the lth column is replaced by $u = N^e$) and basis inverse is $(BE)^{-1} = E^{-1}B^{-1}$. This is called product form of the inverse. *Thus, if we store E^{-1}'s then we can implement the simplex method on a simplex tableau.*

8.3 Revised Simplex Method

Let us investigate what calculations are really necessary in the simplex method. Each iteration exchanges a column of N with a column of B, and one has to decide which columns to choose, beginning with a basis matrix B and the current solution $x_B = B^{-1}b$.

S1. Compute row vector $\lambda = c_B^T B^{-1}$ and then $c_N^T - \lambda N$.

S2. If $c_N^T - \lambda N \geq \theta$, stop; the current solution is optimal. Otherwise, if the most negative component is e^{th} component, choose e^{th} column of N to enter the basis. Denote it by u.

S3. Compute $v = B^{-1}u$.

S4. Calculate ratios of $B^{-1}b$ to $v = B^{-1}u$, admitting only positive components of v. If there are no positive components, the minimal cost is $-\infty$; if the smallest ratio occurs at component l, then l^{th} column of current B will be replaced with u.

S5. Update B (or B^{-1}) and the solution is $x_B = B^{-1}b$. Return to S1.

Remark 8.3.1 *We need to compute* $\lambda = c_B^{-1}B^{-1}, v = B^{-1}u$, *and* $x_B = B^{-1}b$. *Thus, the most popular way is to work only on* B^{-1}. *With the help of previous remark, we can update* B^{-1}*'s by premultiplying* E^{-1}*'s.*

The excessive computing (multiplying with E^{-1}'s) could be avoided by directly reinverting the current B at a time and deleting the current E^{-1}'s that contain the history.

Remark 8.3.2 *The alternative way of computing* λ, v *and* x_B *is* $\lambda B = c_B^T, Bv = u$, *and* $Bx_B = b$. *Then, the standard decompositions (B = QR or PB = LU) lead directly to these solutions.*

Remark 8.3.3 *How many simplex iterations do we have to take?*
There are at most $\binom{n}{m}$ *extreme points. In the worst case, the simplex method may travel almost all of the vertices. Thus, the complexity of the simplex method is exponential. However, experience supports the following average behavior. The simplex method travels about m extreme points, which means an operation count of about* m^2n, *which is comparable to ordinary elimination to solve* $Ax = b$, *and that is the reason of its success.*

8.4 Duality Theory

The standard *primal* problem is: Minimize $c^T x$ subject to $Ax \geq b$ and $x \geq \theta$. The *dual* problem starts from the same A, b, and c and reverses everything: Maximize $y^T b$ subject to $A^T y \leq c$ and $y \geq \theta$.

There is a complete symmetry between the two. The dual of the dual is the primal problem. Both problems are solved at once. However, one must recognize that the feasible sets of the two problems are completely different. The primal polyhedron is a subset of $\mathbb{R}^n$, marked out by matrix A and the right hand side b. The dual polyhedron is a subset of $\mathbb{R}^m$, determined by A^T and the cost vector c.

The whole theory of linear programming hinges on the relation between them.

Theorem 8.4.1 (Duality Theorem) *If either the primal problem or the dual has an optimal vector, then so does the other, and their values are the* *same: The minimum of $c^T x$ equals the maximum of $y^T b$. Otherwise, if optimal vectors do not exist, either both feasible sets are empty or else one is empty and the other problem is unbounded.*

Theorem 8.4.2 (Weak Duality) *If x and y are feasible vectors in the minimum and maximum problems, then $y^T b \leq c^T x$.*

Proof. Since they are feasible, $Ax \geq b$ and $A^T y \leq c$ ($\Leftrightarrow y^T A \leq c^T$). They should be nonnegative as well: $x \geq \theta$, $y \geq \theta$. Therefore, we can take inner products without ruining the inequalities: $y^T Ax \geq y^T b$ and $y^T Ax \leq c^T x$. Thus, $y^T b \leq c^T x$ since left-hand-sides are identical. □

Corollary 8.4.3 *If the vectors x and y are feasible, and if $c^T x = y^T b$, then these vectors must be optimal.*

Proof. No feasible y can make $y^T b$ larger than $c^T x$. Since our particular y achieves this value it should be optimal. Similarly, x should be optimal. □

Theorem 8.4.4 (Complementary Slackness) *Suppose the feasible vectors x and y satisfy the following complementary slackness conditions:*
if $(Ax)_i > b_i$, then $y_i = 0$ and if $(A^T y)_j < c_j$, then $x_j = 0$.
Then, x and y are optimal. Conversely, optimal vectors must satisfy complementary slackness.

Proof. At optimality we have

$$y^T b = y^T (Ax) = (y^T A)x = c^T x.$$

If $y \geq 0$ and $Ax \geq b \Rightarrow y^T b \leq y^T (Ax)$. When $y^T b = y^T (Ax)$ holds, if $b_i < (Ax)_i$, the corresponding factor y_i should be zero. The same is true for $y^T Ax \leq c^T x$. If $c_j > (A^T y)_j$ then $x_j = 0$ to have $y^T Ax = c^T x$. Thus, complementary slackness guarantees (and is guaranteed by) optimality. □

Proof (Strong Duality). We have to show that $y^T b = c^T x$ is really possible.

$$\text{Max } c^T x, \ Ax \geq b, \ x \geq \theta$$

$$\Leftrightarrow \text{Max } [c^T | \theta^T] \begin{bmatrix} x \\ z \end{bmatrix}, \ [A|-I] \begin{bmatrix} x \\ z \end{bmatrix} = b, \ \begin{bmatrix} x \\ z \end{bmatrix} \geq \theta.$$

$$[A|-I] \rightarrow [B|N], \ \begin{bmatrix} x \\ z \end{bmatrix} \rightarrow \begin{bmatrix} x_B \\ x_N \end{bmatrix} = \begin{bmatrix} B^{-1}b \\ 0 \end{bmatrix}, \ [c^T | \theta^T] \rightarrow [c_B^T | c_N^T].$$

Optimality condition: $N^T (B^T)^{-1} c_B \leq c_N$.

Since we have finite number of extreme points, the optimality condition is eventually met. At that moment, the minimum cost is $c^T x = c_B^T B^{-1} x_B$.

$$\text{Max } b^T y \text{ subject to } \begin{bmatrix} A^T \\ -I \end{bmatrix} y \leq \begin{bmatrix} c \\ 0 \end{bmatrix} \rightarrow \begin{bmatrix} B^T \\ N^T \end{bmatrix} y \leq \begin{bmatrix} c_B \\ c_N \end{bmatrix} \Leftrightarrow B^T y = c_B$$

$$\Leftrightarrow y^T B = c_B^T \Leftrightarrow y^T = c_B^T B^{-1} \Leftrightarrow y^T b = c_B^T B^{-1} b = c^T x!$$

Furthermore, this choice of y is optimal, and the strong duality theorem has been proven. This is a constructive proof, x^* and y^* were actually computed, which is convenient since we know that the simplex method finds the optimal values. □

8.5 Farkas' Lemma

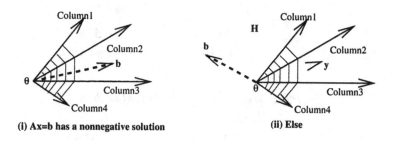

(i) Ax=b has a nonnegative solution

(ii) Else

Fig. 8.2. Farkas' Lemma

By the fundamental theorem of Linear Algebra,

$$\text{either } b \in \mathcal{R}(A) \text{ or } \exists y \in \mathcal{N}(A^T) \ni y \not\perp b,$$

that is, there is a component of b in the left null space. Here, we immediately have the following theorem of alternatives.

Proposition 8.5.1 *Either $Ax = b$ has a solution, or else there is a $y \ni$ $A^T y = \theta, y^T b \neq 0$.*

If $b \in \text{Cone}(a^1, a^2, a^3, \dots)$ then $Ax = b$ is solvable. If $b \notin \text{Cone}(\text{columns}$ of A), then there is a separating hyperplane which goes through the origin defined by y that has b on the negative side. The inner product of y and b is negative ($y^T b < 0$) since they make a wide angle ($> 90°$) whereas the inner product of y and every column of A is positive ($A^T y \geq \theta$). Thus, we have the following theorem.

Proposition 8.5.2 *Either $Ax = b, x \geq \theta$ has a solution, or else there is a y such that $A^T y \geq \theta, y^T b < 0$.*

Corollary 8.5.3 *Either $Ax \geq b, x \geq \theta$ has a solution, or else there is a y such that $A^T y \geq \theta, y^T b < 0, y \leq \theta$.*

Proof. $Ax \geq b, x \geq \theta \rightarrow Ax - Iz = b, z \geq \theta$.

Either $\begin{bmatrix} A & -I \end{bmatrix} \begin{bmatrix} x \\ z \end{bmatrix}$ has a nonnegative solution or $\exists y \ni \begin{bmatrix} A^T \\ -I \end{bmatrix} y \geq \theta, y^T b < 0$.

$\Rightarrow A^T y \geq \theta, y^T b < 0, y \leq \theta$. $\square$

Remark 8.5.4 *The propositions in this section can also be shown using the primal dual pair of linear programming problems: If the dual is unbounded, the primal is infeasible.*

1. *Either $Ax = b$ has a solution, or else there is a $y \ni A^T y = \theta, y^T b \neq 0$:*

$$(P1): \quad Max \ \theta^T x \qquad\qquad (D1): \quad Min \ b^T y$$
$$s.t. \qquad\qquad\qquad\qquad\qquad s.t.$$
$$Ax = b \qquad\qquad\qquad\qquad A^T y = \theta$$
$$x : URE \qquad\qquad\qquad\qquad y : URE$$

$$(P2): \quad Min \ \theta^T x \qquad\qquad (D2): \quad Max \ b^T y$$
$$s.t. \qquad\qquad\qquad\qquad\qquad s.t.$$
$$Ax = b \qquad\qquad\qquad\qquad A^T y = \theta$$
$$x : URE \qquad\qquad\qquad\qquad y : URE$$

Either P1 (or P2) is feasible, or D1 (or D2) is unbounded. For D1 (D2) to be unbounded, we must have $b^T y < 0$ ($b^T y > 0$). Thus, either $Ax = b$ or $\exists y \ni A^T y = \theta, y^T b \neq 0$.

2. *Either $Ax = b, x \geq \theta$ has a solution, or else there is a y such that $A^T y \geq \theta, y^T b < 0$:*

$$(P3): \quad Max \ \theta^T x \qquad\qquad (D3): \quad Min \ b^T y$$
$$s.t. \qquad\qquad\qquad\qquad\qquad s.t.$$
$$Ax = b \qquad\qquad\qquad\qquad A^T y \geq \theta$$
$$x \geq \theta \qquad\qquad\qquad\qquad\quad y : URE$$

Either P3 is feasible, or D3 is unbounded. For D3 to be unbounded, we must have $b^T y < 0$. Thus, either $Ax = b, x \geq \theta$ has a solution, or else $\exists y \ni A^T y \geq \theta, y^T b < 0$.

3. *Either $Ax \geq b, x \geq \theta$ has a solution, or else there is a y such that $A^T y \geq \theta, y^T b < 0, y \leq \theta$:*

$$(P4): \quad Max \ \theta^T x \qquad\qquad (D4): \quad Min \ b^T y$$
$$s.t. \qquad\qquad\qquad\qquad\qquad s.t.$$
$$Ax \geq b \qquad\qquad\qquad\qquad A^T y \geq \theta$$
$$x \geq \theta \qquad\qquad\qquad\qquad\quad y \leq \theta$$

Either P4 is feasible, or D4 is unbounded. For D4 to be unbounded, we must have $b^T y < 0$. Thus, either $Ax \geq b, x \geq \theta$ has a solution, or else $\exists y \ni A^T y \geq \theta, y^T b < 0, y \leq \theta$.

Problems

8.1. (P):

$$Max \ z = x_1 + 2x_2 + 2x_3$$
$$s.t.$$
$$2x_1 + x_2 \leq 8$$
$$x_3 \leq 10$$
$$x_2 \geq 2$$
$$x_1, x_2, x_3 \geq 0.$$

Let the slack/surplus variables be s_1, s_2, s_3.
a)Draw the polytope defined by the constraints in $\mathbb{R}^3$, identify its extreme points and the minimum set of supporting hyperplanes.
b) Solve (P) using

1. matrix form,
2. simplex tableau,
3. revised simplex with product form of the inverse,
4. revised simplex with $B = LU$ decomposition,
5. revised simplex with $B = QR$ decomposition.

c) Write the dual problem, draw its polytope.

8.2. Let $P = \{(x_1, x_2, x_3) \geq 0$ and

$$2x_1 - x_2 - x_3 \geq 3$$
$$x_1 - x_2 + x_3 \geq 2$$
$$x_1 - 2x_2 + 2x_3 \geq 4\}.$$

Let s_1, s_2, s_3 be the corresponding slack/surplus variables.
a) Find all the extreme points of P.
b) Find the extreme rays of P (if any).
c) Considering the extreme rays of P (if any) check whether we have a finite solution $x \in P$ if we maximize

1. $x_1 + x_2 + x_3$,
2. $-2x_1 - x_2 - 3x_3$,
3. $-x_1 - 2x_2 + 2x_3$.

d) Let $x_1 = 6, x_2 = 1, x_3 = \frac{1}{2}$. Express this solution with the convex combination of extreme points plus the canonical combination of extreme rays (if any) of P.

e) Let the problem be

$$\min x_1 + 2x_2 + 2x_3 \text{ subject to } (x_1, x_2, x_3) \in P.$$

1. Solve.
2. What if we reduce the right hand side of (1) by 3 and (3) by 1.
3. Consider the solution found above. What if we add a new constraint

$$2x_1 + 5x_2 + x_3 \leq 3.$$

8.3. Upper bounded simplex

Modify the simplex algorithm without treating the bounds as specific constraints but modifying the optimality, entering and leaving variable selection conditions to solve the following LP problem:

$$\max 2x_1 + 3x_2 + x_3 + 4x_4$$

s.t.

$$x_1 + 2x_2 + 3x_3 + 5x_4 \leq 30 \quad (1)$$

$$x_1 + x_2 \qquad\qquad \leq 13 \quad (2)$$

$$3x_3 + x_4 \leq 20 \quad (3)$$

$$1 \leq x_1 \leq 6, \;\; 0 \leq x_2 \leq 10, \;\; 3 \leq x_3 \leq 9, \;\; 0 \leq x_4 \leq 5$$

a) Start with the initial basis as $\{s_1, s_2, s_3\}$ where s_1, s_2, s_3 are the corresponding slack variables at their lower bounds. Use Bland's (lexicographically ordering) rule in determining the entering variables. Find the optimal solution.

b) Take the dual after expressing the nonzero lower/upper bounds as specific constraints. Find the optimal dual values by considering only the optimal primal solution.

8.4. Decomposition

Let $a \in A$ be an arc of a network $N = (V, A)$, where $\|V\| = n$, $\|A\| = m$. Given a node $i \in V$, let $T(i)$ be the set of arcs entering to i and $H(i)$ be the set of arcs leaving from i. Let there be $k = 1, \ldots, K$ commodities to be distributed; c_{ka} denotes the unit cost of sending a commodity through an arc, u_{ka} denotes the corresponding arc capacity, d_{ki} denotes the supply/demand at node i, and U_a is the total carrying capacity of arc a.

a) Let x_{ka} be the decision variable representing the flow of commodity k across arc a. Give the classical Node-Arc formulation of the minimum cost multi-commodity flow problem, where commodities share capacity. Discuss the size of the formulation.

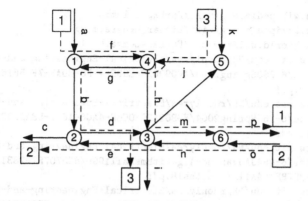

Fig. 8.3. Starting bfs solution for our multi-commodity flow instance

b) Let $\mathcal{P}^k$ be the set of paths from source node s_k to sink node t_k for commodity k. For $P \in \mathcal{P}$, let f_P: flow on path P (decision variable),

$$I_{aP} \doteq \begin{cases} 1, & \text{if } a \text{ is in } P \\ 0, & \text{otherwise} \end{cases}$$

C_{kP}: unit cost of flow $\doteq \sum_a I_{aP} c_{ka}$

D_k: demand for the circulation

μ_P: upper bound on flow $\doteq \min \{u_{ka} : I_{aP} = 1\}$

Give the Path-Cycle formulation, relate to the Node-Arc formulation, and discuss the size.

c) Take the path cycle formulation. Let w_a be the dual variable of the capacity constraint and π_k the dual variable of the demand constraint. What will be the reduced cost of path P? What will be the reduced cost of path P at the optimality? Write down a subproblem (column generation) that seeks a path with lower cost to displace the current flow. Discuss the properties.

d) Solve the example instance using column generation starting from the solution given in Figure 8.3. Let us fix all capacities at 10 and all positive supplies/demands at 10 with unit carrying costs.

e) Sketch briefly the row generation, which is equivalent to the Dantzig-Wolfe/Bender's decompositions' viewpoint.

Web material

http://agecon2.tamu.edu/people/faculty/mccarl-bruce/mccspr/new04.pdf
http://archives.math.utk.edu/topics/linearProg.html
http://catt.bus.okstate.edu/itorms/volumes/vol1/papers/murphy/
http://cepa.newschool.edu/het/essays/math/convex.htm#minkowski
http://cgm.cs.mcgill.ca/~beezer/Publications/avis-kaluzny.pdf
http://cis.poly.edu/rvslyke/simplex.pdf
http://citeseer.ist.psu.edu/62898.html

http://en.wikipedia.org/wiki/Farkas's_lemma
http://en.wikipedia.org/wiki/Linear_programming
http://mathworld.wolfram.com/FarkassLemma.html
http://ocw.mit.edu/NR/rdonlyres/Electrical-Engineering-and-Computer-
    Science/6-253Spring2004/A609002F-E9DF-47F0-8DAF-7F05F16920F7/0/
    lec_19.pdf
http://ocw.mit.edu/NR/rdonlyres/Electrical-Engineering-and-Computer-
    Science/6-253Spring2004/E9B02139-0C6E-4AC6-A8B7-5BA22D281DBF/0/
    lec_12.pdf
http://ocw.mit.edu/NR/rdonlyres/Electrical-Engineering-and-Computer-
    Science/6-854JAdvanced-AlgorithmsFall1999/8C3707F7-2831-4984-
    83FB-BD7BF754A11E/0/notes18.pdf
http://ocw.mit.edu/NR/rdonlyres/Electrical-Engineering-and-Computer-
    Science/6-854JFall2001/FB552487-8E11-4D14-A064-B521724DCE65/0/
    notes_lp.pdf
http://ocw.mit.edu/NR/rdonlyres/Mathematics/18-310Fall-2004/32478C79-
    6843-4775-B925-068489AD0774/0/liner_prog_3_dua.pdf
http://ocw.mit.edu/NR/rdonlyres/Mathematics/18-310Fall-2004/ACD5267C-
    0B38-4DDF-97AC-C4B32E20B4EE/0/linear_prog_ii.pdf
http://ocw.mit.edu/OcwWeb/Sloan-School-of-Management/15-066JSystem-
    Optimization-and-Analysis-for-ManufacturingSummer2003/
    LectureNotes/index.htm
http://opim.wharton.upenn.edu/~guignard/321/handouts/duality_OK.pdf
http://planetmath.org/encyclopedia/FarkasLemma.html
http://planetmath.org/encyclopedia/LinearProgrammingProblem.html
http://shannon.math.gatech.edu/~bourbaki/2602/lp/lp.pdf
http://web.mit.edu/15.053/www/AMP-Appendix-B.pdf
http://www-math.mit.edu/18.310/28.pdf
http://www-personal.umich.edu/~mepelman/teaching/IOE610/lecture5.pdf
http://www.comp.leeds.ac.uk/or21/OVERHEADS/sect5.pdf
http://www.cs.berkeley.edu/~vazirani/s99cs170/notes/linear3.pdf
http://www.cs.helsinki.fi/u/gionis/farkas.pdf
http://www.cs.nyu.edu/cs/faculty/overton/g22_lp/encyc/
    article_web.html
http://www.cs.toronto.edu/~avner/teaching/2411/index.html
http://www.cs.toronto.edu/~avner/teaching/2411/ln/lecture6.pdf
http://www.cs.uiuc.edu/class/fa05/cs473g/lectures/17-lp.pdf
http://www.cs.uiuc.edu/class/fa05/cs473g/lectures/18-simplex.pdf
http://www.cs.uleth.ca/~holzmann/notes/lpdual.pdf
http://www.cs.wisc.edu/~swright/525/handouts/dualexample.pdf
http://www.cse.ucsd.edu/~dasgupta/mcgrawhill/chap7.pdf
http://www.e-optimization.com/directory/trailblazers/hoffman/
    linear_programming.cfm
http://www.eecs.harvard.edu/~parkes/cs286r/spring02/lectures/
    class8.pdf
http://www.hss.caltech.edu/~kcb/Notes/LP.pdf
http://www.ici.ro/camo/books/rbb.htm
http://www.ie.boun.edu.tr/course_pages/ie501/Ch81.pdf
http://www.imada.sdu.dk/~jbj/DM85/lec4b.pdf

http://www.math.chalmers.se/Math/Grundutb/CTH/tma947/0506/
    lecture9.pdf
http://www.math.kth.se/optsyst/research/5B5749/13.pdf
http://www.math.mtu.edu/~msgocken/ma5630spring2003/lectures/ineq/
    ineq/node8.html
http://www.math.mun.ca/~sharene/cs3753_F05/BookPartII.pdf
http://www.math.niu.edu/~rusin/known-math/index/90-XX.html
http://www.math.washington.edu/~burke/crs/408f/notes/lpnotes/
http://www.maths.lse.ac.uk/Courses/MA208/notes6.pdf
http://www.me.utexas.edu/~jensen/ORMM/frontpage/tours/tour_lp.html
http://www.me.utexas.edu/~jensen/ORMM/supplements/methods/lpmethod/
    S3_dual.pdf
http://www.mosek.com/homepages/e.d.andersen/papers/linopt.ps
http://www.mpi-sb.mpg.de/~mehlhorn/Optimization/Linprog.ps
http://www.optimization-online.org/DB_FILE/2003/04/646.pdf
http://www.optimization-online.org/DB_HTML/2004/10/969.html
http://www.personal.psu.edu/tmc7/tmclinks.html
http://www.princeton.edu/~rvdb/542/lectures.html
http://www.princeton.edu/~rvdb/LPbook/onlinebook.pdf
http://www.scs.leeds.ac.uk/or21/OVERHEADS/sect5.pdf#
    search='linear%20programming%20duality%20theory'
http://www.stanford.edu/class/msande310/lecture03.pdf
http://www.stanford.edu/class/msande310/lecture06.pdf
http://www.stanford.edu/class/msande314/lecture02.pdf
http://www.stats.ox.ac.uk/~yu/
http://www.statslab.cam.ac.uk/~rrw1/opt/index98.html
http://www.tutor.ms.unimelb.edu.au/duality/duality.html
http://www.twocw.net/mit/NR/rdonlyres/Electrical-Engineering-and-
    Computer-Science/6-854JAdvanced-AlgorithmsFall1999/
    8C3707F7-2831-4984-83FB-BD7BF754A11E/0/notes18.pdf
http://www.utdallas.edu/~chandra/documents/6310.htm
http://www.wisdom.weizmann.ac.il/~feige/algs04.html
http://www.wisdom.weizmann.ac.il/~feige/lp02.html
http://www2.imm.dtu.dk/courses/02711/DualLP.pdf
http://www2.maths.unsw.edu.au/applied/reports/1999/amr99_2.pdf

Number Systems

In this chapter, we will review the basic concepts in real analysis: order relations, ordered sets and fields, construction and properties of the real and the complex fields, and finally the theory of countable and uncountable sets together with the cardinal numbers. The known sets of numbers that we will use in this chapter are

- $\mathbb{N}$: Natural
- $\mathbb{Z}$: Integer
- $\mathbb{Q}$: Rational
- $\mathbb{R}$: Real
- $\mathbb{C}$: Complex

9.1 Ordered Sets

Definition 9.1.1 *Let S be a set. An order on S is a relation $\prec$ such that*

i) If x, y are any two elements of S, then one and only one of the following is true:
$$x \prec y, x = y, y \prec x.$$
ii) If $x, y, z \in S$ and $x \prec y$ and $y \prec z$, then $x \prec z$.
$x \prec y \not\succ y \prec x.$
$x \preceq y$ *means* $x \prec y$ *or* $x = y$ *without specifying one.*

Example 9.1.2 $S = \mathbb{Q}$ *has an order; define* $x \prec y$ *if* $y - x$ *is positive.*

Definition 9.1.3 *An ordered set is a set S on which there is an order.*

Definition 9.1.4 *Let S be an ordered set and $\emptyset \neq E \subset S$. E is*

- *bounded above if $\exists b \in S \ni \forall x \in E, x \preceq b$ where b is an upper bound of E.*
- *bounded below if $\exists a \in S \ni \forall x \in E, a \preceq x$ where a is a lower bound of E.*

- *bounded if E is both bounded above and below.*

Example 9.1.5 $A = \{p \in \mathbb{Q} : p \succ 0, p^2 \prec 2\}$ *is*

- *bounded above, $b = 3/2, 2, \ldots$ are upper bounds.*
- *bounded below, $a = 0, -1/2, \ldots$ are lower bounds.*

Definition 9.1.6 *Let S be an ordered set and $\emptyset \neq E \subset S$ be bounded above. Suppose $\exists b \in S \ni$:*

1. *b is an upper bound of E.*
2. *if $b' \in S$ and $b' \prec b$ then b' is not an upper bound of E. Equivalently, if b'' is any upper bound of E if $b'' \succ b$.*

Then, b is called least upper bound (lub) or supremum (sup) of E and denoted by

$$b = \sup E = lub\, E.$$

Greatest lower bound (glb) or infimum (inf) of E is defined analogously.

Example 9.1.7 $S = \mathbb{Q}$, $E = \{p \in \mathbb{Q} : p \succ 0, \ p^2 \prec 2\}$ $\inf E = 0$, *but E has no supremum in $S = \mathbb{Q}$. Suppose $p_0 = \sup E$ exists in $\mathbb{Q}$. Then, either $p_0 \in E$ or $p_0 \notin E$.*
If $p_0 \in E$, $\exists q \in E \ni p_0 \prec q$ because E has no largest element; therefore, p is not an upper bound of E.
If $p_0 \notin E$, then $p_0 \succ 0$ because it is an upper bound and $p_0^2 \succeq 2$ because $p_0 \notin E$. Then, either $p_0^2 = 2$ (not true because $p_0 \in \mathbb{Q}$) or $p_0^2 \succ 2$ (true), then $p_0 \in B = \{p \in \mathbb{Q} : p \succ 0, \ p^2 \succ 2\}$. Then, $\exists q_0 \in B \ni q_0 \prec p_0$ (*) because B has no smallest element. $\forall p \in E$, $p^2 \prec 2 \prec q_0^2 \Rightarrow q_0$ is an upper bound of E. Moreover, $p_0 \prec q_0$ because lub Contradiction to (*).

Definition 9.1.8 *Let S be an ordered set. We say that S has the least upper bound property if every nonempty subset of S which is bounded above has lub in S.*

Example 9.1.9 $S = \mathbb{Q}$ *does not have lub-property.*

Theorem 9.1.10 *Let S be an ordered set with lub-property. Then, every nonempty subset of S which is bounded below has inf in S.*

Proof. Let $B \neq \emptyset, B \subset S$ be bounded below, L be the set of all lower bounds of B. Then, $L \neq \emptyset$ (because B is bounded below), $y \in B$ be arbitrary, then for any $x \in L$ we have $x \preceq y$. So, y is an upper bound of L; i.e. all elements of B are upper bounds of $L \Rightarrow L$ is bounded above. $\alpha = \sup L$, $\alpha \in S$ (because S has lub property).
Claim (i): $\alpha = \inf B$
Proof (i): Show α is lower bound of B; i.e. show $\forall x \in B$, $\alpha \preceq x$. Assume that it is not true; i.e. $\exists x_0 \in B \ni \alpha \succ x_0$. Then, x_0 is not an upper bound of α (because $\alpha = \sup L$) $\Rightarrow x_0 \notin B$ (because all elements of B are upper bounds of L). Contradiction! ($x_0 \in B$). Therefore, α is a lower bound of B.

Claim (ii): α is the greatest of the lower bounds.
Proof (ii): Show if $\alpha \prec \beta$, $\beta \in S \Rightarrow \beta$ is not a lower bound of B.
$\beta \notin L$ (because $\alpha \prec \beta$); i.e. β is not a lower bound of B.
Therefore, $\alpha = \inf B$. $\square$

9.2 Fields

Let us repeat Definition 2.1.1 for the sake of completeness.

Definition 9.2.1 *A field is a set $F \neq \emptyset$ with two operations, addition(+) and multiplication(.), which satisfy the following axioms:*
(A) Addition Axioms:
(A1) $\forall x, y \in F$, $x + y \in F$ (closed under +)
(A2) $\forall x, y \in F$, $x + y = y + x$ (commutative)
(A3) $\forall x, y, z \in F$, $(x + y) + z = x + (y + z)$ (associative)
(A4) $\exists 0 \in F \ni \forall x \in F$ $x + 0 = x$ (existence of ZERO element)
(A5) $\forall x \in F, \exists$ an element $-x \in F \ni x + (-x) = 0$ (existence of INVERSE element)

(M) Multiplication Axioms:
(M1) $\forall x, y \in F$, $x \cdot y \in F$ (closed under $\cdot$)
(M2) $\forall x, y \in F$, $x \cdot y = y \cdot x$ (commutative)
(M3) $\forall x, y, z \in F$, $(x \cdot y) \cdot z = x \cdot (y \cdot z)$ (associative)
(M4) $\exists 1 \neq 0 \ni \forall x \in F$, $1 \cdot x = x$ (existence of UNIT element)
(M5) $\forall x \neq 0 \exists$ an element $\frac{1}{x} \in F \ni x\frac{1}{x} = 1$ (existence of INVERSE element)

(D) Distributive Law:
$\forall x, y, z \in F$, $x \cdot (y + z) = xy + xz$

Notation :

$$x + (-y) = x - y; \ x\left(\frac{1}{y}\right) = \frac{x}{y}; \ x + (y + z) = (x + y) + z$$

$$x \cdot x = x^2; \ x + x = 2x; \ x(yz) = xyz, \ \cdots$$

Example 9.2.2 $F = \mathbb{Q}$ *with usual $+$ and $\cdot$ is a field.*

Example 9.2.3 *Let $F = \{a, b, c\}$ where $a \neq b$, $a \neq c$, $b \neq c$.*
Define

+	a	b	c		$\cdot$	a	b	c
a	a	b	c		a	a	a	a
b	b	c	a		b	a	b	c
c	c	a	b		c	a	c	b

F *is a field with $0 = a$, $1 = b$.*

Proposition 9.2.4 *In a field F, the following properties hold:*

(a) $x + y = x + z \Rightarrow y = z$ *(cancelation law for addition)*.
(b) $x + y = x \Rightarrow y = 0$.
(c) $x + y = 0 \Rightarrow y = -x$.
(d) $-(-x) = x$.
(e) $x \neq 0$ *and* $xy = xz \Rightarrow y = z$ *(cancelation law for multiplication)*.
(f) $x \neq 0$ *and* $xy = x \Rightarrow y = 1$.
(g) $x \neq 0$ *and* $xy = 1 \Rightarrow y = \frac{1}{x}$.
(h) $x \neq 0$, $\frac{1}{(1/x)} = x$.
(i) $\forall x \in F$, $0x = 0$.
(j) $x \neq 0$ *and* $y \neq 0$, *then* $xy \neq 0$ *(no zero divisors)*.
(k) $\forall x, y \in F$, $(-x)(-y) = xy$.

Definition 9.2.5 *Let F be an ordered set and a field of F is an ordered field if*

i) $x, y, z \in F$ *and* $x \prec y \Rightarrow x + z \prec y + z$,
ii) $x \succ 0$, $y \succ 0 \Rightarrow xy \succ 0$.

If $x \succ 0$, call x as <u>positive</u>, *If $x \prec 0$, call x as* <u>negative</u>.

Example 9.2.6 $S = \mathbb{Q}$ *is an ordered field.*

Proposition 9.2.7 *Let F be an ordered field. Then,*

(a) $x \succ 0 \Leftrightarrow -x \prec 0$.
(b) $x \succ 0$ *and* $y \prec z \Rightarrow xy \prec xz$.
(c) $x \prec 0$ *and* $y \prec z \Rightarrow xy \succ xz$.
(d) $x \neq 0 \Rightarrow x^2 \succ 0$. *In particular* $1 \succ 0$.
(e) $0 \prec x \prec y \Rightarrow 0 \prec \frac{1}{y} \prec \frac{1}{x}$.

Proof. F is an ordered field.

(a) Assume $x \succ 0 \Rightarrow x + (-x) \succ 0 + (-x) \Rightarrow 0 \succ -x$.

$$-x \prec 0 \Rightarrow -x + x \prec 0 + x \Rightarrow 0 \prec x.$$

(b) Let $x \succ 0$ and $y \prec z \Rightarrow 0 \prec z - y \Rightarrow 0 \prec x(z - y) = xz - xy \Rightarrow xy \prec xz$.
(c) $x \prec 0$ and $y \prec z \Rightarrow -x \succ 0$ and $z - y \succ 0 \Rightarrow -x(z - y) \succ 0 \Rightarrow x(z - y) \prec 0 \Rightarrow xz \prec xy$.
(d) $x \neq 0 \Rightarrow x \succ 0 \Rightarrow (y = x$ in (b)) $x^2 \succ 0$ or
$x \prec 0 \Rightarrow -x \succ 0$ $(y = -x) \Rightarrow (-x)(-x) = x^2 \succ 0$.
(e) Let $x \succ 0$. Show $\frac{1}{x} \succ 0$. If not, $\frac{1}{x} \preceq 0 \Rightarrow (x \succ 0)$, $x\frac{1}{x} = 1 \preceq 0$,
Contradiction!
Assume $0 \prec x \prec y \Rightarrow \frac{1}{y} \succ 0$, $\frac{1}{x} \succ 0$, therefore (by (b))

$$\left. \begin{array}{c} \frac{1}{x}\frac{1}{y} \succ 0 \\ x \prec y \end{array} \right\} \Rightarrow \frac{1}{y} \prec \frac{1}{x}. \qquad \square$$

Remark 9.2.8 $\mathbb{C}$ *with usual* $+$ *and* $\cdot$ *is a field. But it is not an ordered field. If* $x = i$ *then* $i^2 = -1 \succ 0$, *hence property (d) does not hold.*

Definition 9.2.9 *Let* F *(with* $+, \cdot$*) and* F' *(with* $\oplus, \odot$*) be two fields. We say* F *is a subfield of* F' *if* $F \subset F'$ *and two operations* $\oplus$ *and* $\odot$ *when restricted to* F *are* $+$ *and* $\cdot$, *respectively. That is, if* $x, y \in F \Rightarrow x \oplus y = x + y$, $x \odot y = x \cdot y$. *Then, we have* $0_F = 0_{F'}$, *and* $1_F = 1_{F'}$.
Moreover, if F *(with* $\prec$*) and* F' *with (with* $\prec'$*) are ordered fields, then we say* F *is an ordered subfield of* F' *if* F *is a subfield of* F' *and for* $\forall x \in F$ *with* $0_F \prec x \Rightarrow 0_{F'} \prec' x$.

9.3 The Real Field

Theorem 9.3.1 (Existence & Uniqueness) *There is an ordered field* $\mathbb{R}$ *with lub property* $\ni \mathbb{Q}$ *is an ordered subfield of* $\mathbb{R}$. *Moreover if* $\mathbb{R}'$ *is another such ordered field, then* $\mathbb{R}$ *and* $\mathbb{R}'$ *are "isomorphic":* $\exists$ *a function* $\phi : \mathbb{R} \mapsto \mathbb{R}' \ni$

i) ϕ *is 1-1 and onto,*
ii) $\forall x, y \in \mathbb{R}$, $\phi(x + y) = \phi(x) + \phi(y)$ *and* $\phi(xy) = \phi(x)\phi(y)$,
iii) $\forall x, \in \mathbb{R}$ *with* $x \succ 0$, *we have* $\phi(x) \succ 0$).

Theorem 9.3.2 (ARCHIMEDEAN PROPERTY)

$$x, y \in \mathbb{R} \text{ and } x \succ 0 \Rightarrow \exists n \in N \text{ (depending on } x \text{ and } y) \ni nx \succ y.$$

Proof. Suppose $\exists x, y \in \mathbb{R}$ with $x \succ 0$ for which claim is not true. Then, $\forall n \in N$, $nx \preceq y$. Let $A = \{nx : n \in N\}$. A is bounded above (by y). $\alpha = \sup A \in \mathbb{R}$, since $\mathbb{R}$ has lub property. $x \succ 0 \Rightarrow \alpha - x \prec \alpha$, so $\alpha - x$ is not an upper bound for A.
Therefore, $\exists m \in N \ni (\alpha - x) \prec mx \Rightarrow \alpha \prec (m + 1)x$. Contradiction $(\alpha = \sup A)$. $\square$

Theorem 9.3.3 ($\mathbb{Q}$ *is dense in* $\mathbb{R}$)

$$\forall x, y \in \mathbb{R} \text{ with } x \prec y, \exists p \in \mathbb{Q} \ni x \prec p \prec y.$$

Proof. $x, y \in \mathbb{R}$, $x \prec y \Rightarrow y - x \succ 0$
(By Theorem 9.3.2) $\exists n \in N \ni n(y - x) \succ 1 \Rightarrow ny \succ 1 + nx$.

$\exists m_1 \in N \ni m_1 \succ nx \leftarrow (y = nx, x = 1)$ in Theorem 9.3.2.

Let $A = \{m \in Z : nx \prec m\}$. $A \neq \emptyset$, because $m_1 \in A$. A is bounded below. So A has a smallest element m_0, then $nx \prec m_0 \Rightarrow (m_0 - 1) \preceq nx$.
If not, $nx \prec m_0 - 1$, but m_0 is the smallest element: Contradiction.
$\Rightarrow (m_0 - 1) \preceq nx \preceq m_0 \Rightarrow nx \prec m_0 \preceq nx + 1 \prec ny \Rightarrow x \prec \frac{m_0}{n} \prec y$. Let $p = \frac{m_0}{n} \in \mathbb{Q}$. $\square$

Theorem 9.3.4

$$\forall x \in \mathbb{R},\ x \succeq 0,\ \forall n \in \mathbb{N}\ \exists\ a\ \underline{unique}\ y \in \mathbb{R},\ y \succ 0 \ni y^n = x.$$

Proof. [*Existence*]:

Given $x \succ 0$, $n \in N$. Let $E = \{t \in \mathbb{R} : t \succ 0$ and $t^n \prec x\}$.

Claim 1: $E \neq \emptyset$

Let $t = \frac{x}{x+1} \succ 0$, $t \prec 1$, $t \prec x$; $0 \prec t \prec 1 \Rightarrow t^n \prec t$

$(0 \prec t \prec 1 \Rightarrow 0 \prec t^2 \prec t \prec 1 \Rightarrow \ldots \Rightarrow 0 \prec t^n \prec t \prec 1)$.

Also we have, $t \prec x \Rightarrow t^n \prec x$; therefore, $t = \frac{x}{x+1} \in E$.

Claim 2: E is bounded above

If $1 + x$ is an upper bound of E.

If not, $\exists t \in E \ni t \succ 1 + x$. In particular, $t \succ 1$ (because $x \succ 0$) $\Rightarrow$
$t^n \succ t \succ 1 + x \succ x$; therefore, $t \notin E$: Contradiction!

$y = \sup E \in \mathbb{R}$ because $\mathbb{R}$ has lub property.

$y \succ 0$, because $(x \succ 0)$.

Claim 3: $y^n = x$

If not, then either $y^n \prec x$ or $x \prec y^n$.

We know the following:

Let $0 \prec a \prec b$. Then, $b^n - a^n = (b-a)(b^{n-1} + b^{n-2}a + \cdots + a^{n-1}) \Rightarrow$

$(*): b^n - a^n \prec (b-a)nb^{n-1}$.

i) $y^n \prec x \Rightarrow \frac{x-y^n}{n(y+1)^{n-1}} \succ 0$. Find $n \in \mathbb{R} \ni 0 \prec h \prec 1$ and $0 \prec \frac{hn(y+1)^{n-1}}{hn(y+h)^n}$.

   $(*): (y+h)^n - (y)^n \prec hn(y+h)^n \prec hn(y+1)^{n-1} \prec x - y^n$

   Therefore, $(y+h)^n \prec x \Rightarrow y+h \in E$. But $y+h \succ y \Rightarrow y$ is not an upper
   bound of E, Contradiction!

ii) $x \prec y^n$. Let $k = \frac{y^n - x}{ny^{n-1}} \succ 0$ and $x \prec y$ [because $y^n - x \prec ny^{n-1}$].

   Claim: $y - k$ is an upper bound of E.

   Suppose not, $\exists t \in E \ni t \succ y - k \succ 0$.

   Then, $t^n \succ (y-k)^n \Rightarrow -t^n \prec -(y-k)^n \Rightarrow y^n - t^n \prec y^n - (y-k)^n$

   $(*): y^n - (y-k)^n \prec kny^{n-1} = y^n - x \Rightarrow y^n - t^n \prec y^n - x \Rightarrow t^n \succ x \Rightarrow t \notin E$,
   Contradiction!

   Therefore, $y - k$ is an upper bound of E.

   However, y is lub of E, Contradiction!

[*Uniqueness*]:

Suppose $y \succ 0$, $y' \succ 0$ are two positive roots $\ni y \neq y'$ and $y^n = x = (y')^n$.
Without loss of generality, we may assume that , $y' \succ y \succ 0$, (because $y \neq y'$) $\Rightarrow y^n \prec (y')^n$, Contradiction! Thus, y is unique. $\square$

Definition 9.3.5 *Real numbers which are not rational are called irrational numbers.*

Example 9.3.6 $\sqrt{2}$ *is an irrational number.*

Corollary 9.3.7 *Let $a \succ 0$, $b \succ 0$ and $n \in N$. Then, $(ab)^{1/n} = a^{1/n}b^{1/n}$.*

Proof. Let $\alpha = a^{1/n}$, $\beta = b^{1/n} \Rightarrow \alpha^n = a$, $\beta^n = b \Rightarrow (\alpha\beta)^n = \alpha^n \beta^n = ab \succ 0$ and n^{th} root is unique $\Rightarrow (ab)^{1/n} = \alpha\beta$. $\quad \Box$

Definition 9.3.8 (Extended real numbers) $\mathbb{R} \cup \{+\infty, -\infty\}$ $\ni$ *preserve the order in* $\mathbb{R}$ *and* $\forall x \in \mathbb{R}$, $-\infty \prec x \prec \infty$. $\mathbb{R} \cup \{+\infty, -\infty\}$ *is an ordered set and every non-empty subset has supremum/infimum in* $\mathbb{R} \cup \{+\infty, -\infty\}$.

In $\mathbb{R} \cup \{+\infty, -\infty\}$, we make the following conventions:

i) For $x \in \mathbb{R}$, $x + \infty = +\infty$, $x - \infty = -\infty$,
ii) If $x \prec 0$, we have $x \cdot (+\infty) = -\infty$, $x \cdot (-\infty) = +\infty$,
iii) $0 \cdot (+\infty)$, $0 \cdot (-\infty)$ are undefined.

9.4 The Complex Field

Let $\mathbb{C}$ be the set of all ordered pairs (a, b) of real numbers. We say

$$(a, b) = (c, d) \text{ if and only if } a = c \text{ and } b = d.$$

Let $x = (a, b)$, $y = (c, d)$. Define

$$x + y = (a + c, b + d), \quad xy = [ac - bd, ad + bc].$$

Under these operations $\mathbb{C}$ is a field with $(0, 0)$ being the zero element, and $(1, 0)$ being the multiplicative unit.
Define $\phi : \mathbb{R} \mapsto \mathbb{C}$ by $\phi(a) = (a, 0)$, then ϕ is 1-1.

$$\phi(a + b) = (a + b, 0) = (a, 0) + (b, 0) = \phi(a) + \phi(b).$$
$$\phi(ab) = (ab, 0) = (a, 0)(b, 0) = \phi(a)\phi(b).$$

Therefore, $\mathbb{R}$ can be identified by means of ϕ with a subset of $\mathbb{C}$ in such a way that addition and multiplication are preserved. This identification gives us the real field as a subfield of the complex field.
Let $i = (0, 1) \Rightarrow i^2 = (0, 1)(0, 1) = (-1, 0) = \phi(-1)$, i.e. i^2 corresponds to the real -1.

Let us introduce some notation.
$\phi(a) = (a, 0) = a \Rightarrow i^2 = \phi(-1) = -1$, also if $(a, b) \in \mathbb{C}$, $a + ib = (a, b)$.
Hence,

$$\mathbb{C} = \{a + ib : a, b \in \mathbb{R}\}.$$

If $z = a + ib \in \mathbb{C}$, we define $\bar{z} = a - ib$ (conjugate of z),

$$a = \text{Re}(z) = \frac{z + \bar{z}}{2}, \quad b = \text{Im}(z) = \frac{z - \bar{z}}{2i}.$$

If $z, w \in \mathbb{C} \Rightarrow \overline{z + w} = \bar{z} + \bar{w}$, $\overline{zw} = \bar{z}\,\bar{w}$.
If $z \in \mathbb{C} \Rightarrow z\bar{z} = a^2 + b^2 \succeq 0$, we define $|z| = \sqrt{z\bar{z}} = \sqrt{a^2 + b^2}$.

Proposition 9.4.1 *Let* $z, w \in \mathbb{C}$. *Then,*

(a) $z \neq 0 \Rightarrow |z| \succ 0$ *and* $|0| = 0$.
(b) $|\overline{z}| = |z|$.
(c) $|zw| = |z||w|$.
(d) $|Re(z)| \preceq |z|, |Im(z)| \preceq |z|$.
(e) $|z + w| \preceq |z| + |w|$, *[Triangle inequality]*.

Proof. The first three is trivial. Then,

(d) Let $z = a + ib$ $|Im(z)| = |b| = \sqrt{b^2} \preceq \sqrt{a^2 + b^2} = |z|$.
(e) $|z + w|^2 = (z + w)(\overline{z} + \overline{w}) = |z|^2 + z\overline{w} + \overline{z}w + |w|^2 \preceq (|z| + |w|)^2$
   $\overline{z}w + z\overline{w} = 2Re(z\overline{w}) \preceq |2Re(z\overline{w})| \preceq 2|z\overline{w}| \Rightarrow |z + w| \preceq |z| + |w|$.
   Take positive square roots of both sides, i.e. if $a \succeq 0$, $b \succeq 0$ and $a^2 \preceq b^2 \Rightarrow a \preceq b$. If not, $b \preceq a \Rightarrow b^2 \preceq ab$, $ba \preceq a^2 \Rightarrow b^2 \preceq a^2$. Contradiction!

□

Theorem 9.4.2 (Schwartz Inequality) *Let* $a_j, b_j \in \mathbb{C}$, $j = 1, \ldots, n$. *Then,*

$$\underbrace{\left| \sum_{j=1}^{n} a_j \overline{b_j} \right|^2}_{C} \preceq \underbrace{\left(\sum_{j=1}^{n} |a_j|^2 \right)}_{A} \underbrace{\left(\sum_{j=1}^{n} |b_j|^2 \right)}_{B}.$$

Proof. $B \succeq 0$. If $B = 0$ then $b_j = 0 \; \forall j \Rightarrow LHS = 0$; therefore, $0 \preceq 0$.
Assume $B \succ 0 \Rightarrow$

$$0 \preceq \sum_{j=1}^{n} |Ba_j - Cb_j|^2 = \sum_{j=1}^{n} (Ba_j - Cb_j)(B\overline{a_j} - \overline{Cb_j})$$

$$= \sum_{j=1}^{n} B^2 |a_j|^2 - \sum_{j=1}^{n} B\overline{C} a_j \overline{b_j} - \sum_{j=1}^{n} CB b_j \overline{a_j} + \sum_{j=1}^{n} |C|^2 |b_j|^2$$

$$= B^2 A - B|C|^2 - CB\overline{C} + |C|^2 B = B(AB - |C|^2).$$

Thus, $AB \succeq |C|^2$, since $B \succeq 0$.   □

9.5 Euclidean Space

Definition 9.5.1 *Let* $k \in \mathbb{N}$, *we define* $\mathbb{R}^k$ *as the set of all ordered k-tuples* $x = (x_1, \ldots, x_k)$ *of real numbers* $x_1, \ldots, x_k$. *We define* $(x + y) = (x_1 + y_1, \ldots, x_k + y_k)$. *If* $\alpha \in \mathbb{R}$, $\alpha x = (\alpha x_1, \ldots, \alpha x_k)$. *This way* $\mathbb{R}^k$ *becomes a vector space over* $\mathbb{R}$.
We define an inner product in $\mathbb{R}^k$ *by* $x \cdot y = \sum_{i=1}^{k} x_i y_i$. *And* $\forall x \in \mathbb{R}^k$, $x \cdot x \succeq 0$.
We define the norm of $x \in \mathbb{R}^k$ *by* $\|x\| = \sqrt{x \cdot x} = \sqrt{\sum_{n=1}^{k} x_n^2}$.

Definition 9.5.2 *An equivalence relation in X is a binary relation (where $\sim$ means equivalent) with the following properties:*

(a) $\forall x \in X$, $x \sim x$ *(reflexibility).*
(b) $x \sim y \Rightarrow y \sim x$ *(symmetry).*
(c) $x \sim y$, $y \sim z \Rightarrow x \sim z$ *(transitivity).*

Definition 9.5.3 *If $\sim$ is an equivalence relation in X, we define the equivalence class of any $x \in X$ as the following set:*

$$[x] = \{y \in X : x \sim y\}.$$

Remark 9.5.4 *If $\sim$ is an equivalence relation in X, then the collection of all equivalence classes forms a partition of X; and conversely, given any partition of X there is an equivalence relation in X such that equivalence classes are the sets in the partition.*

Remark 9.5.5 *Let C be any collection of nonempty sets. For $X, Y \in C$, define $X \sim Y$ (X and Y are numerically equivalent) if there exists a one-to one and onto function $f : X \mapsto Y$ (or $f^{-1} : Y \mapsto X$). Then, $\sim$ is an equivalence relation in C.*

9.6 Countable and Uncountable Sets

Definition 9.6.1 *Let $J_n = \{1, 2, \ldots, n\}$, $n = 1, 2, \ldots$. Let $X \neq 0$. We say*

i) X is finite if $\exists n \in \mathbb{N}$, $X \sim J_n$.
ii) X is infinite if X is not finite.
iii) X is countable if $X \sim \mathbb{N}$.
     (i.e. $\exists f : \mathbb{N} \mapsto X$, 1-1 onto, or $\exists g : X \mapsto \mathbb{N}$, 1-1 onto).
iv) X is uncountable if X is not finite and not countable.
v) X is at most countable if X is finite or countable.

Example 9.6.2 $X = \mathbb{N}$ *is countable. Let $f : \mathbb{N} \mapsto \mathbb{N}$ be the identity function.*

Example 9.6.3 $X = \mathbb{Z}$ *is countable. Define $f : \mathbb{N} \mapsto \mathbb{Z}$ as*

$$f(n) = \begin{cases} \frac{n}{2}, & \text{if } n \text{ is even;} \\ -\frac{n-1}{2}, & \text{if } n \text{ is odd.} \end{cases}$$

Example 9.6.4 $\mathbb{Q}^+$ *is countable. Let $r \in \mathbb{Q}^+$, then $r = \frac{m}{n}$ where $m, n \in \mathbb{N}$. List elements of $\mathbb{Q}^+$ in this order as in Table 9.1. If we apply the counting schema given in Figure 9.1, we get the sequence*

$$1, \frac{1}{2}, 2, \frac{1}{3}, 3, \frac{1}{4}, \frac{2}{3}, \frac{3}{2}, 4, \ldots$$

Define $f : \mathbb{N} \mapsto \mathbb{Q}^+$,

$$f(1) = 1, \ f(2) = \frac{1}{2}, \ f(3) = 2, \ f(4) = \frac{1}{3}, \ \cdots$$

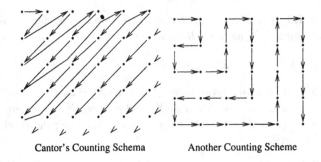

Cantor's Counting Schema   Another Counting Scheme

Fig. 9.1. Counting schema for rational's

Table 9.1. List of rational numbers

m \ n	1	2	3	4	5	$\cdots$
1	1/1	1/2	1/3	1/4	1/5	$\cdots$
2	2/1	2/2	2/3	2/4	2/5	$\cdots$
3	3/1	3/2	3/3	3/4	3/5	$\cdots$
4	4/1	4/2	4/3	4/4	4/5	$\cdots$
5	5/1	5/2	5/3	5/4	5/5	$\cdots$
$\vdots$	$\vdots$	$\vdots$	$\vdots$	$\vdots$	$\vdots$	$\ddots$

Example 9.6.5 $\mathbb{Q}$ *is countable. Since $\mathbb{Q}^+$ is countable, the elements of $\mathbb{Q}^+$ can be listed as a sequence $\{x_1, x_2, x_3, \ldots\}$. Then, $\mathbb{Q}^- = \{q : q \prec 0\}$ can be listed as $\{-x_1, -x_2, -x_3, \ldots\}$.*

$$\mathbb{Q} = 0 \quad x_1 \quad -x_1 \quad x_2 \quad -x_2 \quad x_3 \quad -x_3 \quad \cdots$$
$$\uparrow \quad \uparrow \quad \uparrow \quad \uparrow \quad \uparrow \quad \uparrow \quad \uparrow \quad \cdots$$
$$\mathbb{N} = 1 \quad 2 \quad 3 \quad 4 \quad 5 \quad 6 \quad 7 \quad \cdots$$

$f : \mathbb{N} \mapsto \mathbb{Q}$ can be defined in this way.

$$f(n) = \begin{cases} x_{\frac{n}{2}}, & \text{if } n \text{ is even} \\ -x_{\frac{n-1}{2}}, & \text{if } n \text{ is odd} \\ 0, & \text{if } n = 1. \end{cases}$$

Proposition 9.6.6 *If $\varepsilon = \{x_i, i \in I\}$ is a countable class of countable sets, then $\cup_{i \in I} x_i$ is also countable. That is, countable union of countable sets are countable.*

Proof. We have $f : \mathbb{N} \mapsto I$, 1-1, onto. Let $Y_n = X_{f(n)}$. The elements of Y_n can be listed as a sequence. $Y_n = \{X_1^n, X_2^n, \ldots\}$ $\forall n$. Use the Cantor's counting scheme for the union. Another counting schema is given in Figure 9.1. $\square$

Example 9.6.7 $X = [0, 1)$ *is not countable.*

Every $x \in [0, 1)$ has a binary expansion $x = 0.a_1 a_2 a_3 \ldots$ where $a_n = \begin{cases} 0 \\ 1 \end{cases}$

Suppose $[0, 1)$ is countable. Then, its elements can be listed as a sequence $\{X^1, X^2, X^3, \ldots\}$. Consider their binary expansions

$$X^1 = 0.a_1^1 a_2^1 a_3^1 \ldots$$

$$X^2 = 0.a_1^2 a_2^2 a_3^2 \ldots$$

$$X^3 = 0.a_1^3 a_2^3 a_3^3 \ldots$$

$$\vdots$$

Let $a_1 = \begin{cases} 0, \text{ if } a_1^1 = 1 \\ 1, \text{ if } a_1^1 = 0 \end{cases}$, $a_2 = \begin{cases} 0, \text{ if } a_2^2 = 1 \\ 1, \text{ if } a_2^2 = 0 \end{cases}$, $a_3 = \begin{cases} 0, \text{ if } a_3^3 = 1 \\ 1, \text{ if } a_3^3 = 0 \end{cases} \ldots$

Let

$$x = 0.a_1 a_2 a_3 \ldots \in [0, 1).$$

But this number is not contained in the list $\{X^1, X^2, X^3, \ldots\}$

x is different from X^1 by the first digit after 0;
x is different from X^2 by the second digit after 0;
x is different from X^3 by the third digit after 0;
$\vdots$

Therefore, $x \neq X^n$, $\forall n$; since x and X^n differ in the nth digit after zero. So, $X = [0, 1)$ is not countable.

Example 9.6.8 $X = (0, 1)$ *is not countable. Since $X = [0, 1)$ is not countable, excluding a countable number of elements (just zero) does not change uncountability. Thus, $X = (0, 1)$ is uncountable.*

Example 9.6.9 *For any open interval (a, b) we have*

$$(a, b) \sim (0, 1) \quad f : (a, b) \mapsto (0, 1).$$

Refer to Figure 9.2.

Example 9.6.10 $X = \mathbb{R}$ *is not countable. Since $\mathbb{R} \sim (-1, 1)$, by projection $\mathbb{R}$ is not countable [because $(0,1)$ is not countable]. One way of showing 1-1 correspondence between any open interval and $(0,1)$ is illustrated in Figure 9.3.*

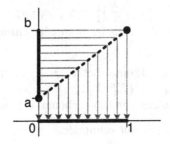

Fig. 9.2. Uncountability equivalence of (a,b) and (0,1)

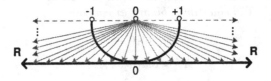

Fig. 9.3. The correspondence between (-1,1) and $\mathbb{R}$.

Example 9.6.11 $f : \mathbb{R} \mapsto (-\frac{\pi}{2}, \frac{\pi}{2})$, $f(x) = \arctan(x)$ *is a 1-1 correspondence, i.e.* $f(x)$ *is 1-1 and onto. Refer to Figure 9.4.*

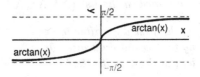

Fig. 9.4. The correspondence between $(-\frac{\pi}{2}, \frac{\pi}{2})$ and $\mathbb{R}$

Proposition 9.6.12 *If* (a, b) *is any open interval, then*

$$(0, 1) \sim (a, b) \sim \mathbb{R} \sim [0, 1).$$

Proof.

$$\exists f : (0, 1) \mapsto [0, 1) \text{ is 1-1 } (f(x) = x).$$
$$\exists g : [0, 1) \mapsto \mathbb{R} \text{ is 1-1 } (g(x) = x).$$
$$\exists h : \mathbb{R} \mapsto (0, 1) \text{ is 1-1 and onto } (f(x) = x).$$
$$[0, 1) \mapsto \mathbb{R} \mapsto (0, 1) \text{ is 1-1.}$$

By Cantor-Schruder-Bernstein Theorem $[0, 1) \sim (0, 1)$.   □

Definition 9.6.13 *Roughly speaking, the cardinality of a set (or cardinal number of a set) is the number of elements in this set.*
If $X = \emptyset$, $Card(X) = 0$,
If $X \sim J_n = \{1, 2, \ldots, n\}$, $Card(X) = n$,
If $X \sim \mathbb{N}$ (i.e. countable), $Card(X) = \aleph_0$ (aleph zero),
If $X \sim \mathbb{R}$, $Card(X) = \aleph_1$ (aleph one).

Definition 9.6.14 *Let m and n be two cardinal numbers We say $m \prec n$ if there are two sets X and $Y \ni Card(X) = m$, $Card(Y) = n$.*

Remark 9.6.15 *The list of cardinal numbers:*

$$0 \prec 1 \prec 2 \prec \cdots \prec n \prec \cdots \prec \aleph_0 \prec \aleph_1 = c.$$

Remark 9.6.16 *Question: $\exists$? a cardinal number between $\aleph_0$ and $\aleph_1$?*
The answer is still not known. Conjecture: The answer is no!
Question: Is there a cardinal number bigger than $\aleph_1$?
The answer is yes. Consider $P(\mathbb{R})$: the set of all subsets of $\mathbb{R}$ (power set of $\mathbb{R}$). $\aleph_1 = Card(\mathbb{R}) \prec Card(P(\mathbb{R}))$. We know if $Card(X) = n$, then $Card(P(X)) = 2^n$. Analogously $Card(P(\mathbb{N})) = 2^{\aleph_0} = \aleph_1$. Then, we can say that $Card(P(\mathbb{R})) = 2^{\aleph_1} = \aleph_2$.

Problems

9.1. Let A be a non-empty subset of $\mathbb{R}$ which is bounded below. Define $-A = \{-x : x \in A\}$. Show that $\inf A = -\sup(-A)$.

9.2. Let $b \succ 1$. Prove the following:
a) $\forall m, n \in \mathbb{Z}$ with $n \succ 0$, $(b^m)^{1/n} = (b^{1/n})^m$.
b) $\forall m, n \in \mathbb{Z}$ with $n \succ 0$, $(b^m)^n = b^{mn} = (b^n)^m$.
c) $\forall n \in \mathbb{Z}$ with $n \succ 0$, $1^{1/n} = 1$.
d) $\forall n, q \in \mathbb{Z}$ with $n, q \succ 0$, $b^{1/nq} = (b^{1/n})^{1/q} = (b^{1/q})^{1/n}$.
e) $\forall p, q \in \mathbb{Z}$ $b^{p+q} = b^p b^q$.

9.3. Do the following:
a) Let $m, n, p, q \in \mathbb{Z}$ with $n \prec 0, q \succ 0$ and $r = \frac{m}{n} = \frac{p}{q}$. Show that $(b^m)^{1/n} = (b^p)^{1/q}$ using the above properties.
b) Prove that $b^{r+s} = b^r b^s$ if r and s are rational.
c) Let $x \in \mathbb{R}$. Define $B(x) = \{b^t : t \in \mathbb{Q}, t \preceq x\}$. Show that if $r \in \mathbb{Q}$, $b^r = \sup B(r)$.
d) Show that $b^{x+y} = b^x b^y$ $\forall x, y \in \mathbb{R}$.

9.4. Fix $b \succ 1$ and $y \succ 0$. Show the following:
a) $\forall n \in \mathbb{N}$, $b^n - 1 \succeq n(b - 1)$.
b) $(b - 1) \succeq n(b^{1/n} - 1)$. Hint: $\forall n \in \mathbb{N}$, $b^{1/n} \succ 1$ holds. So replace $(b \succ 1)$

above by $b^{1/n} \succ 1$.

c) If $t \succ 1$ and $n \succ \frac{b-1}{t-1}$, then $b^{1/n} \prec t$.

d) If $w \ni b^w \prec y$, then $b^{w+1/n} \prec y$ for sufficiently large n.

e) If $b^w \succ y$, then $b^{w-1/n} \succ y$ for sufficiently large n.

f) Let $A = \{w \in \mathbb{R} : b^w \prec y\}$. Show that $x = \sup A$ satisfies $b^x = y$.

g) Prove that x above is unique.

9.5. Let F be an ordered field. Prove the following:

a) $x, y \in F$ and $x^2 + y^2 = 0 \Rightarrow x = 0$ and $y = 0$.

b) $x_1, x_2, \ldots, x_n \in F$ and $x_1^2 + \cdots + x_n^2 = 0 \Rightarrow x_1 = x_2 = \cdots = x_n = 0$.

9.6. Let m be a fixed integer. For $a, b \in \mathbb{Z}$, define $a \sim b$ if $a - b$ is divisible by m, i.e. there is an integer k such that $a - b = mk$.

a) Show that $\sim$ is an equivalence relation in $\mathbb{Z}$.

b) Describe the equivalence classes and state the number of distinct equivalence classes.

9.7. Do the following:

a) Let $X = \mathbb{R}$, and $x \sim y$ if $x \in [0, 1]$ and $y \in [0, 1]$. Show that $\sim$ is symmetric and transitive, but not reflexive.

b) Let $X \neq \emptyset$ and $\sim$ is a relation in X. The following seems to be a proof of the statement that if this relation is symmetric and transitive, then it is necessarily reflexive:

$$x \sim y \Rightarrow y \sim x, \quad x \sim y \text{ and } y \sim x \Rightarrow x \sim x;$$

therefore, $x \sim x$, $\forall x \in X$. In view of part a), this cannot be a valid proof. What is the flaw in the reasoning?

9.8. Prove the following:

a) If $X_1, X_2, \ldots, X_n$ are countable sets, then $X = \Pi_{i=1}^n X_i$ is also countable.

b) Every countable set is numerically equivalent to a proper subset of itself.

c) Let X and Y be non-empty sets and $f : X \mapsto Y$ be an onto function. Prove that if X is countable then Y is at most countable.

Web material

http://129.118.33.1/~pearce/courses/5364/notes_2003-03-31.pdf
http://alpha.fdu.edu/~mayans/core/real_numbers.html
http://comet.lehman.cuny.edu/keenl/realnosnotes.pdf
http://en.wikipedia.org/wiki/Complex_number
http://en.wikipedia.org/wiki/Countable
http://en.wikipedia.org/wiki/Field_(mathematics)
http://en.wikipedia.org/wiki/Numeral_system
http://en.wikipedia.org/wiki/Real_number

```
http://en.wikipedia.org/wiki/Real_numbers
http://en.wikipedia.org/wiki/Uncountable_set
http://eom.springer.de/f/f040090.htm
http://eom.springer.de/U/u095130.htm
http://kr.cs.ait.ac.th/~radok/math/mat5/algebra21.htm
http://math.berkeley.edu/~benjamin/74lecture38s05.pdf
http://mathforum.org/alejandre/numerals.html
http://mathworld.wolfram.com/CountablyInfinite.html
http://numbersorg.com/Algebra/
http://pirate.shu.edu/projects/reals/infinity/uncntble.html
http://planetmath.org/encyclopedia/MathbbR.html
http://planetmath.org/encyclopedia/Real.html
http://planetmath.org/encyclopedia/Uncountable.html
http://plato.stanford.edu/entries/set-theory/
http://www-db.stanford.edu/pub/cstr/reports/cs/tr/67/75/
     CS-TR-67-75.pdf
http://www.absoluteastronomy.com/c/countable_set
http://www.answers.com/topic/complex-number
http://www.cse.cuhk.edu.hk/~csc3640/tutonotes/tuto3.ppt
http://www.csie.nctu.edu.tw/~myuhsieh/dmath/Module-4.5-
     Countability.ppt
http://www.cut-the-knot.org/do_you_know/few_words.shtml
http://www.dpmms.cam.ac.uk/~wtg10/countability.html
http://www.eecs.umich.edu/~aey/eecs501/lectures/count.pdf
http://www.faqs.org/docs/sp/sp-121.html
http://www.faqs.org/docs/sp/sp-122.html
http://www.introducingmathematics.com/settheoryone/01.html
http://www.jcu.edu/math/vignettes/infinity.htm
http://www.math.brown.edu/~sjmiller/1/CountableAlgTran.pdf
http://www.math.niu.edu/~beachy/aaol/contents.html
http://www.math.niu.edu/~rusin/known-math/index/11-XX.html
http://www.math.niu.edu/~rusin/known-math/index/12-XX.html
http://www.math.toronto.edu/murnaghan/courses/mat240/field.pdf
http://www.math.ucdavis.edu/~emsilvia/math127/chapter1.pdf
http://www.math.ucdavis.edu/~emsilvia/math127/chapter2.pdf
http://www.math.uic.edu/~lewis/las100/uncount.html
http://www.math.uiuc.edu/~r-ash/Algebra/Chapter3.pdf
http://www.math.umn.edu/~garrett/m/intro_algebra/notes.pdf
http://www.math.unl.edu/~webnotes/classes/classAppA/classAppA.htm
http://www.math.uvic.ca/faculty/gmacgill/guide/cardinality.pdf
http://www.math.uvic.ca/faculty/gmacgill/M222F03/Countable.pdf
http://www.math.vanderbilt.edu/~schectex/courses/thereals/
http://www.math.wisc.edu/~ram/math541/
http://www.mathreference.com/set-card,cable.html
http://www.mcs.vuw.ac.nz/courses/MATH114/2006FY/Notes/11.pdf
http://www.msc.uky.edu/ken/ma109/lectures/real.htm
http://www.swarthmore.edu/NatSci/wstromq1/stat53/CountableSets.doc
http://www.topology.org/tex/conc/dgchaps.html
http://www.trillia.com/zakon-analysisI-index.html
```

Basic Topology

In this chapter, basic notions in general topology will be defined and the related theorems will be stated. This includes the following: metric spaces, open
and closed sets, interior and closure, neighborhood and closeness, compactness
and connectedness.

10.1 Metric Spaces

In $\mathbb{R}^k$, we have the notion of distance:
If $p = (x_1, x_2, \ldots, x_k)^T, q = (y_1, y_2, \ldots, y_k)^T$, $p, q \in \mathbb{R}^k$, then

$$d_2(p, q) = \sqrt{(x_1 - y_1)^2 + (x_2 - y_2)^2 + \cdots + (x_k - y_k)^2}$$

Definition 10.1.1 *Let $X \neq \emptyset$ be a set. Suppose there is a function*
$d : X \times X \Rightarrow \mathbb{R}_+ = [0, \infty)$ *with the following properties:*

 i) $d(p, q) = 0 \Leftrightarrow p = q$;
 ii) $d(p, q) = d(q, p)$, $\forall p, q$;
 iii) $d(p, q) \leq d(p, r) + d(r, q)$, $\forall p, q, r$ *[triangle inequality].*

*Then, d is called a metric (or distance function) and the pair (X, d) is called
a metric space.*

Example 10.1.2 *Let $X \neq \emptyset$ be any set. For $p, q \in X$ define*

$$d(p, q) = \begin{cases} 1, & \text{if } p \neq q \\ 0, & \text{if } p = q \end{cases}$$

is called the discrete metric.

Definition 10.1.3 *Let S be any fixed nonempty set. A function $f : S \mapsto \mathbb{R}$ is
called bounded if $f(S)$ is a bounded subset of $\mathbb{R}$.*

Example 10.1.4 *Is $f : \mathbb{R} \mapsto \mathbb{R}$, $f(s) = s^2$ bounded? (Exercise!).*
$f : \mathbb{R} \mapsto \mathbb{R}$, $f(s) = \arctan(s) = \tan^{-1}(s)$ *is bounded. See Figure 9.4.*

Definition 10.1.5 *Let $X = B(S) = $ all bounded functions $f : S \mapsto \mathbb{R}$.*
For $f, g \in B(S)$, we define the distance as $d(f, g) = \sup\{|f(s) - g(s)| : s \in S\}$.

Proposition 10.1.6 $d(f, g) \geq 0$ *is a metric, $\forall f, g \in X = B(S)$.*

Proof. by proving axioms of a metric:

(i) ($\Rightarrow$)
   if $d(f, g) = 0 \Rightarrow |f(s) - g(s)| = 0$, $\forall s \in S \Rightarrow f(s) = g(s)$, $\forall s \in S \Rightarrow f = g$.
   ($\Leftarrow$)
   if $f = g \Rightarrow d(f, g) = 0$.
(ii) trivial.
(iii) **Proposition 10.1.7** *Let $A \neq \emptyset$, $B \neq \emptyset$ be subsets of $\mathbb{R}$. Define*

$$A + B = \{a + b : a \in A, \ b \in B\}.$$

If A and B are bounded above then $A + B$ is bounded above and

$$\sup(A + B) \leq \sup A + \sup B.$$

Proof. Let $x = \sup A$, $y = \sup B$.
Given $c \in A + B$, then $\exists a \in A$, $b \in B \ni c = a + b$. Then, $c = a + b \leq x + y$.
Moreover, $\sup(A + B) \leq x + y$. □

Proposition 10.1.8 *Let C, D be nonempty subsets of $\mathbb{R}$, let D be bounded above. Suppose $\forall c \in C$, $\exists d \in D \ni c \leq d$. Then, C is also bounded above and $\sup C \leq \sup D$.*

Proof. Given $c \in C$, $\exists d \in D \ni c \leq d$. So, $\forall c \in C$, $c \leq y = \sup D$. Hence, y is an upper bound for C. Therefore, $\sup C \leq \sup D$. □

Triangular Inequality: Let $f, g, h \in B(S)$.
$C = \{|f(s) - g(s)| : s \in S\}$, then $d(f, g) = \sup C$.
$A = \{|f(s) - h(s)| : s \in S\}$, then $d(f, h) = \sup A$.
$B = \{|h(s) - g(s)| : s \in S\}$, then $d(h, g) = \sup B$.
Given $x \in C$, then $\exists s \in S \ni x = |f(s) - g(s)|$

$$x = |f(s) - g(s)| = |f(s) - h(s) + h(s) - g(s)| \leq |f(s) - h(s)| + |h(s) - g(s)|$$

$$\Rightarrow \sup C \leq \sup(A + B) \leq \sup A + \sup B. □$$

Example 10.1.9 *Let $X = \mathbb{R}^k$, $P = (x_1, \ldots, x_k)^T$ and $Q = (y_1, \ldots, y_k)^T \in \mathbb{R}^k$.*

   $d_1(p, q) = |x_1 - y_1| + \cdots + |x_k - y_k|$: l_1 *metric.*
   $d_2(p, q) = [(x_1 - y_1)^2 + \cdots + (x_k - y_k)^2]^{1/2}$: l_2 *metric.*
   $d_\infty(p, q) = max\{|x_1 - y_1|, \ldots, |x_k - y_k|\}$: l_∞ *metric.*
   See Figure 10.1.

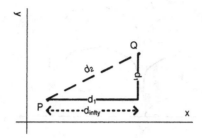

Fig. 10.1. Example 10.1.9

Definition 10.1.10 *Let (X, d) be a metric space, $p \in X$, $r > 0$.*
$B_r(p) = \{q \in X : d(p, q) < r\}$ *open ball centered at p of radius r.*
$B_r[p] = \{q \in X : d(p, q) \leq r\}$ *closed ball centered at p of radius r.*

Example 10.1.11 $X = \mathbb{R}^2$, $d = d_2$. *See Figure 10.2.*

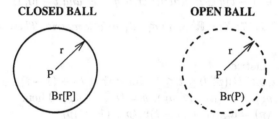

Fig. 10.2. Example 10.1.11

Example 10.1.12 *Let us have $X \neq \emptyset$, and the discrete metric.*

$$B_r(p) = \begin{cases} \{p\}, & \text{if } r < 1 \\ \{p\}, & \text{if } r = 1 \\ X, & \text{if } r > 1 \end{cases} \qquad B_r[p] = \begin{cases} \{p\}, & \text{if } r < 1 \\ X, & \text{if } r = 1 \\ X, & \text{if } r > 1 \end{cases}$$

Example 10.1.13 $X = B \subset (a, b) = \{f : (a, b) \mapsto \mathbb{R} : f \text{ is bounded}\}$

$$f, g \in X \Rightarrow d(f, g) = \sup \{|f(s) - g(s)| : s \in (a, b)\}$$

Let $f \in X$, $r > 0$, $B_r(f)$ is the set of all functions g whose graph lie within the dashed envelope in Figure 10.3.

Example 10.1.14 $X = \mathbb{R}^2$ *with d_1 metric:*

$$d_1(p, q) = |y_1 - x_1| + |y_2 - x_2|.$$

See Figure 10.4.

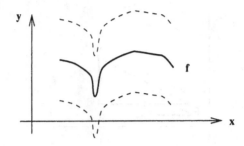

Fig. 10.3. Example 10.1.13

Example 10.1.15 $X = \mathbb{R}^2$ with d_∞ metric:

$$d_\infty(p, q) = \max\{|y_1 - x_1|, |y_2 - x_2|\}.$$

Definition 10.1.16 A subset $E \neq \emptyset$ of a vector space V is convex if $tp + (1 - t)q \in E$ whenever $p, q \in E$ and $t \in [0, 1]$.

Proposition 10.1.17 $X = \mathbb{R}^k$ with d_2, d_1 or d_∞ metric. Then, every (open) ball $B_r(p)$ is convex.

Proof. Using d_∞ metric:
Fix $B_r(p)$. Let $u, v \in B_r(p), 0 \leq t \leq 1$. Show that $tu + (1 - t)v \in B_r(p)$:
Let $p = (p_1, \ldots, p_k)$, $u = (u_1, \ldots, u_k)$, $v = (v_1, \ldots, v_k)$. Then,
$d_\infty(tu + (1 - t)v, p) = d_\infty(tu + (1 - t)v, tp + (1 - t)p)$
$\quad = \max\{|tu_i + (1 - t)v_i - tp_i - (1 - t)p_i|\}_{i=1}^{k}$
$\quad = |tu_j + (1 - t)v_j - tp_j - (1 - t)p_j| = |t(u_j - p_j) + (1 - t)(v_j - p_j)|$
$\quad \leq |t||u_j - p_j| + |1 - t||v_j - p_j| = td_\infty(u, p) + (1 - t)d_\infty(u, p) \leq tr + (1 - t)r = r.$
$\square$

Definition 10.1.18 Let (X, d) be a metric space, $E \subset X$. A point $p \in E$ is called an interior point of E if $\exists r > 0 \ni B_r(p) \subset E$. The set of all interior points of E is denoted by $int E$ or E° and is called the interior of E ($int E \subset E$).

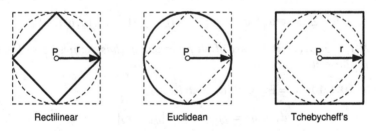

Rectilinear   Euclidean   Tchebycheff's

Fig. 10.4. Example 10.1.14

Example 10.1.19 *See Figure 10.5. $q \in intE$ but $p \notin intE$.*

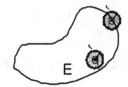

Fig. 10.5. Example 10.1.19

Example 10.1.20 *Let X be any set with at least two elements, with the discrete metric:*

$$d(p,q) = \begin{cases} 1, p \neq q \\ 0, \text{ otherwise} \end{cases}$$

Let $p \in X$, $E = \{p\}$. Then,

$$intE = E, \ r < 1 \Rightarrow B_r(p) = p \subset E \Rightarrow p \in intE.$$

Example 10.1.21 *Let $X = \mathbb{R}^2$ with d_2 metric. See Figure 10.6.*

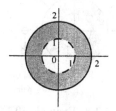

Fig. 10.6. Example 10.1.21

$$E = \{p = (x,y) \in \mathbb{R}^2 : 1 < x^2 + y^2 \leq 4\} \Rightarrow$$
$$intE = \{p = (x,y) \in \mathbb{R}^2 : 1 < x^2 + y^2 < 4\}.$$

Definition 10.1.22 *E is said to be open set if $intE = E$, i.e.*

$$\forall p \in E, \ \exists r > 0 \ni B_r(p) \subset E.$$

Example 10.1.23 *In $\mathbb{R}^2$, $E = \{p = (x,y) \in \mathbb{R}^2 : 1 < x^2 + y^2 < 4\}$ is open.*

Remark 10.1.24 *By convention, $E = \emptyset$, $E = X$ are open sets.*

Definition 10.1.25 *Let $p \in X$. A subset N of X is called a neighborhood of p if $p \in intN$.*

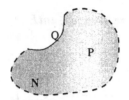

Fig. 10.7. Example 10.1.26

Example 10.1.26 *N is a neighborhood of P but it is not neighborhood of Q. See Figure 10.7.*

Definition 10.1.27 *A point $p \in X$ is called a limit point (or accumulation point) (or cluster point) of the set $E \subset X$ if every neighborhood N of p contains q of $E \ni q \neq p$. i.e. $\forall$ neighborhood N of p, $\exists q \in E \cap N$, $q \neq p$. Equivalently, $\forall r > 0$, $\exists q \in E \cap B_r(p) \ni q \neq p$.*

Example 10.1.28 $E = \{p = (x, y) \in \mathbb{R}^2 : 1 < x^2 + y^2 \leq 4\} \cup \{(3, 0)\}$. *Limit points of E are all points $p = (x, y) \ni 1 \leq x^2 + y^2 \leq 4$. See Figure 10.8.*

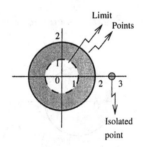

Fig. 10.8. Example 10.1.28

Definition 10.1.29 *A point $p \in E$ is called an isolated point of E if p is not a limit point of E; i.e. $\exists r > 0 \ni B_r(p) \cap E = p$.*

Example 10.1.30 $X = \mathbb{R}$, $d = d_1$:

$$E = \left\{1, \frac{1}{2}, \frac{1}{3}, \frac{1}{4}, \ldots\right\},$$

0 is the only limit point of E. $\forall p \in E$ are all isolated points.

Definition 10.1.31 *E is closed if every limit point of E belongs to E.*

Example 10.1.32 *See Figure 10.9.*

CLOSED   OPEN   Not CLOSED Not OPEN

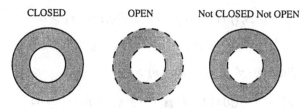

Fig. 10.9. Example 10.1.32

Definition 10.1.33 *E is perfect if it is closed and every point of E is a limit point of E; i.e. if E is closed and has no isolated points. E is bounded if $\exists M > 0 \ni \forall p, q \in E \ d[p, q] \leq M$. E is dense in X if every point of X is either a point of E or a limit point of E.*

Example 10.1.34 $X = \mathbb{R}$, $E = \mathbb{N}$ *is unbounded. Suppose it is bounded. Then, $\exists M > 0 \ni \forall x, y \in \mathbb{N}, |x - y| \leq M$. Let $n \in \mathbb{N}$ be $\ni n > M + 1 \Rightarrow |1 - n| = n - 1 \leq M \rightarrow n \leq M + 1$. Contradiction!*

Example 10.1.35 $X = \mathbb{R}$, $E = \mathbb{Q}$ *($\mathbb{Q}$ is dense in $\mathbb{R}$; i.e. given $x \in \mathbb{R}$ either $x \in \mathbb{Q}$ or x is a limit point of $\mathbb{Q}$). Let $x \in \mathbb{R}$, if $x \in \mathbb{Q}$, we are done. If $x \notin \mathbb{Q}$, we will show that x is a limit point of $\mathbb{Q}$:*
Given $r > 0$, $B_r(x) = (x - r, x + r)$. Then, $\exists y \in \mathbb{Q} \ni x - r < y < x + r \Rightarrow y \in B_r(x) \cap \mathbb{Q}$ and $y \neq x \Rightarrow x \in \mathbb{R}, y \in \mathbb{Q}$.

Let us introduce the following notation:
   E': set of all limit points of E.
   $\bar{E} = E \cup E'$, $\bar{E}$ is called the *closure* of E.

$$p \in \bar{E} \Leftrightarrow \forall r > 0, \ B_r(p) \cap E \neq \emptyset.$$

Proposition 10.1.36 *Every open ball $B_r(p)$ is an open set.*

Proof. Let $q \in B_r(p)$, we will show that $\exists s > 0 \ni B_s(q) \subset B_r(p)$:
$q \in B_r(p) \Rightarrow d(q, p) < r$, let $s = r - d(q, p) > 0$. Let $z \in B_s(q)$,

$$d(z, p) \leq d(z, q) + d(q, p) < s + d(q, p) = r \Rightarrow z \in B_r(p). \quad \square$$

Theorem 10.1.37 *p is a limit point of E if and only if every neighborhood N of p contains infinitely many points of E.*

Proof. ($\Leftarrow$): trivial.

($\Rightarrow$): Let p be the limit point of E. Let N be an arbitrary neighborhood of p. Then, $\exists r > 0 \ni B_r(p) \subset N$. Since $B_r(p)$ is a neighborhood of p

$$\exists q_1 \in B_r(p) \cap E \ni q_1 \neq p \Rightarrow d(q, p) = r_1 > 0.$$

$$\exists q_2 \in B_r(p) \cap E \ni q_2 \neq p.$$

Then, $q_2 \neq q_1$. Since $q_2 \neq p$, $r_2 = d(q_2, p) > 0$.

$$\exists q_3 \in B_{r_2}(p) \cap E \ni q_3 \neq p \neq q_2 \neq q_1; \cdots. \quad \square$$

Corollary 10.1.38 *If E is a finite set, $E' = \emptyset$.*

Theorem 10.1.39 *E is open if and only if E^c is closed.*

Proof. ($\Rightarrow$): Let E be open, Let p be a limit point of E^c. Show $p \in E^c$. Suppose not:

$$p \in E \Rightarrow \exists r > 0 \ni B_r(p) \subset E \quad [\text{because } E \text{ is open}] \quad (*)$$

Since p is a limit point of E^c, for every neighborhood N of p, $N \cap E^c \neq \emptyset$. In particular (by taking $N = B_r(p)$), $B_r(p) \cap E^c \neq \emptyset$, *Contradiction to* (*).

($\Leftarrow$): Assume E^c is closed. Show E is open; i.e. $\forall p \in E$, $\exists r > 0 \ni B_r(p) \subset E$. Let $p \in E \Rightarrow p \notin E^c \Rightarrow p$ is not a limit point of E^c. So $\exists r > 0 \ni B_r(p) \cap E^c$ does not contain any $q \neq p$ (p either). $\Rightarrow B_r(p) \cap E^c = \emptyset \Rightarrow B_r(p) \subset E$. $\quad \square$

Theorem 10.1.40 *Let $E \subset X$, then*

(a) $\bar{E}$ is closed.
(b) $E = \bar{E} \Leftrightarrow E$ is closed.
(c) $\bar{E}$ is the smallest closed set which contains E; i.e. if F is closed and $E \subset F \Rightarrow \bar{E} \subset F$.

Proof. $E \subset X$.

(a): $(\bar{E})^c$ is open.
Let $p \in (\bar{E})^c \Rightarrow p \notin \bar{E} \Rightarrow \exists r > 0 \ni B_r(p) \cap E = \emptyset \Rightarrow B_r(p) \subset (\bar{E})^c$.
Show that $B_r(p) \subset (\bar{E})^c$:
If it is not true $\exists q \in B_r(p)$ and $q \notin (\bar{E})^c \Rightarrow q \in E^c$.
Find $s > 0 \ni B_s(q) \subset B_r(p)$. Then $B_s(q) \cap E \neq \emptyset \Rightarrow B_r(q) \cap E \neq \emptyset$.
Contradiction.

(b): ($\Rightarrow$): Immediate from (a).
($\Leftarrow$): E is closed. Show $E = \bar{E}$, i.e. $\bar{E} \subset E$. Let $p \in \bar{E} = E \cup E'$, if $p \in E$, we are done.
If $p \in E' \Rightarrow p \in E$ (because E is closed).

(c): Let F be closed, $E \subset F$. Show that $\bar{E} \subset F$. Let $p \in \bar{E} = E \cup E'$, if $p \in E \Rightarrow p \in F$. If $p \in E'$ we have to show that $p \in F'$:
Given $r > 0$, show $B_r(p) \cap F$ contains a point $q \neq p$. Since $p \in E'$, $B_r(p) \cap E$ contains a point $q \neq p$. Then, $q \in B_r(p) \cap F$ (because $E \subset F$).
So, $p \in F' \Rightarrow p \in F$ (because F is closed). $\quad \square$

Let (X, d) be a metric space, then

1. The union of a finite collection of open sets is open.
2. The intersection of a <u>finite</u> collection of open sets is open (not true for infinite).
3. The intersection of any collection of closed sets is closed.
4. The union of a <u>finite</u> collection of closed sets is closed (not necessarily true for infinite).
5. E is open $\Leftrightarrow E^c$ is closed.
6. E is closed $\Leftrightarrow E = \bar{E}$.
7. $\bar{E}$ is the smallest closed set containing E.
8. $int E$ is the largest open set contained in E (i.e. if $A \subset E$ and A is open then $A \subseteq int E$).

Example 10.1.41 *Intersection of infinitely many open sets needs not to be open,* $X = \mathbb{R}$, $d(x, y) = |x - y|$: *Let* $A_n = (-\frac{1}{n}, \frac{n+1}{n})$, $n = 1, 2, \dots$. *Then,* $\bigcap_{n=1}^{\infty} A_n = [0, 1]$. *If* $0 \leq x \leq 1$ *then* $x \in (-\frac{1}{n}, \frac{n+1}{n}) = A_n$, $\forall n \Rightarrow x \in \bigcap_{n=1}^{\infty} A_n$. *Let* $x \in \bigcap_{n=1}^{\infty} A_n$, *show that* $0 \leq x \leq 1$: *If not,* $x < 0$ *or* $x > 1$. *If* $x > 1$, $\exists n \in \mathbb{N} \ni 1 < \frac{n+1}{n} < x$, $x \notin A_n$. *Case* $x < 0$ *is similar.*

Proposition 10.1.42 *Let* $\emptyset \neq E \subset \mathbb{R}$ *be bounded above. Then,* $\sup E \in \bar{E}$.

Proof. $y = \sup E$, show that $\forall r > 0$, $B_r(y) \cap E \neq \emptyset$: Since $y - r < y \Rightarrow y - r$ is not upper bound of E. $\exists x \in E \ni y \geq x > y - r \Rightarrow x \in (y - r, y + r) \cap E \Rightarrow B_r(y) \cap E \neq \emptyset$. $\square$

Let (X, d) be a metric space and $\emptyset \neq Y \subset X$, then Y is a metric space in its own right with the same distance function d. In this case, (Y, d) is a subspace of (X, d).

If $E \subset Y$, E may be open in (Y, d) but not open in (X, d).

Example 10.1.43 $X = \mathbb{R}^2$, $Y = \mathbb{R}$, $E = (a, b)$: *When considered in* $\mathbb{R}$, E *is open whereas* E *is not open in* $\mathbb{R}^2$, *as seen in Figure 10.10.*

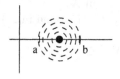

Fig. 10.10. Example 10.1.43

Definition 10.1.44 *Let* $E \subset Y \subset X$. *We say* E *is open (respectively closed) relative to* Y *if* E *is* <u>open</u> *(respectively* <u>closed</u>*) as a subset of the metric space* (Y, d).

E is open relative to $Y \Leftrightarrow \forall p \in E \; \exists r > 0 \ni B_r(p) \cap Y \subset E.$

E is closed relative to $Y \Leftrightarrow Y \setminus E = Y \cap E^c$ is open relative to $Y.$

Theorem 10.1.45 *Let $X \subset Y \subset E.$ Then,*

(a) E is open relative to $Y \Leftrightarrow \exists$ an open set F in $X \ni E = F \cap Y.$

(b) E is closed relative to $Y \Leftrightarrow \exists$ a closed set F in $X \ni E = F \cap Y.$

Proof. $X \subset Y \subset E.$

(a) $(\Rightarrow)$:

Let E be open relative to Y. Then,

$$\forall p \in E \; \exists r_p > 0 \ni B_{r_p}(p) \cap Y \subset E.$$

Let $F = \bigcup_{p \in E} B_{r_p}(p)$. F is open in X.

$$\bigcup_{p \in E} [B_{r_p}(p) \cap Y] \subset E \quad F \cap Y \subset E$$

Conversely, $q \in E$, then

$$q \in B_{r_q}(q) \subset F, \; q \in E \subset Y \Rightarrow q \in F \cap Y \Rightarrow E \subset F \cap Y$$

$(\Leftarrow)$:

$E = F \cap Y$ where F is open in X. Given $p \in E \Rightarrow p \in F$. Since F is open, $\exists r > 0 \ni B_r(p) \subset F.$

$$B_r(p) \cap Y \subset F \cap Y = E.$$

(b) $(\Rightarrow)$:

E is closed relative to $Y \Rightarrow Y \setminus E$ is open relative to Y. Then,

$\exists F \in X$ open in $X \ni Y \setminus E = F \cap Y.$

$E = Y \setminus (Y \setminus E) = Y \setminus (F \cap Y) = Y \cap (F \cap Y)^c = Y \cap F^c \cup \emptyset = Y \cap F^c.$

F^c closed in X.

$(\Leftarrow)$:

$E = F \cap Y$ where F is closed in X.

$Y \setminus E = Y \cap (F \cap Y)^c = Y \cap F^c$ (F^c open in X) $\Longrightarrow Y \setminus E$ is open relative to $Y.$

$\Rightarrow E$ is closed relative to $Y.$ $\square$

10.2 Compact Sets

Definition 10.2.1 *Let (X, d) be a metric space, $E \subset X$ be a nonempty subset of X. An open cover of E is a collection of open sets $\{G_i : i \in I\}$ in $X \ni E \subset \bigcup_i G_i.$*

Example 10.2.2 $X = \mathbb{R}^k$ *with d_2 metric:*
$E = B_1(0)$, *for* $n \in \mathbb{N}$, $G_n = B_{\frac{n}{n+1}}(0) \Rightarrow E \subset \bigcup_{n=1}^{\infty} G_n$.

Example 10.2.3 $X = \mathbb{R}$, $E = (0,1)$:
$\forall x \in (0,1), G_x = (-1, x) \Rightarrow E \subset \bigcup_{x \in (0,1)} G_x$.

Definition 10.2.4 E *is said to be compact if for every open cover $\{G_i : i \in I\}$ of E, we can find*

$$\dot{G}_{i_1}, \ldots, G_{i_n} \ni E \subset [G_{i_1} \cup G_{i_2} \cup \cdots \cup G_{i_n}].$$

Example 10.2.5 *In $X = \mathbb{R}$, $E = (0,1)$ is not compact:*
Consider $\{G_x : x \in (0,1)\}$ where $G_x = (-1, x)$. Suppose $\exists x_1, \ldots, x_n \in (0,1) \ni$ $(0,1) \subset \bigcup_{i=1}^{n} (-1, x_i)$. Let $Y = \max\{x_1, \ldots, x_n\} \Rightarrow 0 < y < 1 \Rightarrow (0,1) \subset (-1, y)$. Let $x = \frac{y+1}{2} \Rightarrow 0 < x < 1$, $x \notin (-1, y)$ Contradiction! Thus, $(0,1)$ is not compact.

Remark 10.2.6 *In the Euclidean space, open sets are not compact.*

Theorem 10.2.7 *Let $K \subset Y \subset X$. Then, K is compact relative to Y if and only if K is compact relative to X.*

Proof. ($\Rightarrow$): Suppose K is compact relative to Y. Let $\{G_i, i \in I\}$ be an open cover of K in X. Then, $K \subset \bigcup_{i \in I} G_i$, so $K = K \cap Y \subset (\bigcup_{i \in I} G_i) \cap Y = \bigcup_{i \in I} (G_i \cap Y)$: open relative to Y. Since K is open relative to Y, $\exists i_1, \ldots, i_n \ni$ $K \subset (Gi_1 \cap Y) \cup (Gi_2 \cap Y) \cup \cdots \cup (Gi_n \cap Y) \Rightarrow K \subset \bigcup_{i=1}^{n} G_i$.
($\Leftarrow$): Suppose K is compact relative to X. Let $\{E_i, i \in I\}$ be any open cover of K in Y. Then,
$\forall i \in I \; \exists$ an open set $G_i \in X \ni E_i = G_i \cap Y$. $K \subset (\bigcup_{i \in I} E_i) \subset (\bigcup_{i \in I} G_i)$.
So, $\{G_i, i \in I\}$ is an open cover in X. Then, $\exists i_1, \ldots, i_n \ni$
$K \subset G_{i_1} \cup G_{i_2} \cup \cdots \cup G_{i_n} \Rightarrow K = K \cap Y \subset (G_{i_1} \cap Y) \cup \cdots \cup (G_{i_n} \cap Y) = E_{i_1} \cup .. \cup E_{i_n}$. $\square$

Theorem 10.2.8 *Let (X, d) be a metric space and $K \subset X$ be compact. Then, K is closed.*

Proof. We will show that K^c is open.
Let $p \in K^c$ be an arbitrary fixed point. $\forall q \in K \Rightarrow d(p, q) > 0$. Let $r_q = \frac{1}{2} d(p, q) > 0$.
$\quad V_q = B_r(p), W_q = B_r(q)$. $K \subset \bigcup_{q \in K} W_q$ (because K is compact)

$$\Rightarrow \exists q_1, \ldots, q_n \in K \ni K \subset W_{q_1} \cup \cdots \cup W_{q_n} = W.$$

Let $V = V_{q_1} \cap V_{q_2} \cap \cdots \cap V_{q_n}$. If $r = Min\{r_{q_1}, \ldots, r_{q_n}\} > 0$, then $V = B_r(p)$.
Let us show that $W \cap V = \emptyset$: If not, $\exists z \in W \cap V \Rightarrow z \in W \Rightarrow z \in Wq_i$ for some $i = 1, \ldots, n$. Hence, $d(z, q_i) < r_{q_i} = \frac{1}{2} d(q, q_i)$. $z \in V \Rightarrow z \in V_{q_i}$ for the same i. Thus, $d(z, p) < r_{q_i} = \frac{1}{2} d(p, q_i)$.

$$\Rightarrow d(p, q_i) < d(p, z) + d(z, q_i) < d(p, q_i).$$

Contradiction! Therefore, $W \cap V = \emptyset$.

Thus, $V = B_r(p) \subset X^c \subset K^c \Rightarrow K^c$ is open $\Rightarrow$ K is closed. $\square$

Theorem 10.2.9 *Closed subsets of compact sets are compact.*

Corollary 10.2.10 *If F is closed and K is compact, then $F \cap K$ is compact.*

Theorem 10.2.11 *Let $\{K_i; i \in I\}$ be a collection of compact subsets of a metric space such that the intersection of every finite subcollection of K_i is nonempty. Then,*

$$\bigcap_{i \in I} K_i \neq \emptyset.$$

Proof. Assume $\bigcap_{i \in I} K_i = \emptyset$.
Fix a member of $\{K_i, i \in I\}$ and call it $\mathcal{K}$. Then,

$$\mathcal{K} \cap [\bigcap_{K_i \neq \mathcal{K}} K_i] = \emptyset \Rightarrow \mathcal{K} \subset [\bigcup_{K_i \neq \mathcal{K}} K_i^c].$$

Since $\mathcal{K}$ is compact, $\exists K_1, \ldots, K_n \ni \mathcal{K} \subset [K_1^c \cup \cdots \cup K_n^c] \Rightarrow \mathcal{K} \cap K_1 \cap \cdots \cap K_n = \emptyset$, since we intersect a finite subcollection, we have a contraposition (Contradiction). $\square$

Corollary 10.2.12 *If (K_n) is a sequence of nonempty compact sets $\ni K_1 \supset K_2 \supset \cdots$, then $\bigcap_{n=1}^{\infty} K_n \neq \emptyset$.*

Theorem 10.2.13 (Nested Intervals) *Let (I_n) be a sequence of non-empty, closed and bounded intervals in $\mathbb{R} \ni I_1 \subset I_2 \subset \cdots$, then*

$$\bigcap_{n=1}^{\infty} I_n \neq \emptyset.$$

Proof. Let $I_n = [a_n, b_n] \ni a_n \leq b_n$. Then,

$$I_1 \subset I_2 \subset \cdots \Rightarrow a_1 \leq a_2 \leq \cdots \leq a_n \leq \cdots \leq b_n \leq \cdots b_2 \leq b_1.$$

Moreover, if $k \leq n \Rightarrow I_k \subset I_n$ and $a_k \leq a_n \leq b_n \leq b_k$.
Let $E = a_1, a_2, \ldots$ is bounded above by b_1. Let $x = \sup E$, then $\forall n$, $a_n \leq x$.
Let us show that $\forall n$, $x \leq b_n$: If not, $\exists n \ni b_n \leq x \Rightarrow \exists a_k \in E \ni b_n < a_k$.

case 1: $k \leq n \Rightarrow a_k \leq a_n \leq b_n \leq a_k$, Contradiction!
case 2: $k > n \Rightarrow a_n \leq a_k \leq b_k \leq b_n \leq a_k$, Contradiction!

Thus, $\forall n$, $x \leq b_n \Rightarrow x \in I_n$, $\forall n \Rightarrow x \in \bigcap_{n=1}^{\infty} I_n \Rightarrow \bigcap_{n=1}^{\infty} I_n \neq \emptyset$. $\square$

Remark 10.2.14 *Here are some remarks:*

1. *If* $\lim_{n\to\infty}(b_n - a_n) = \lim_{n\to\infty} length(I_n) = 0$, $\Rightarrow \bigcap_{n=1}^{\infty} I_n$ *consists of one point.*
2. *If* I_n *'s are not closed, conclusion is false, e.g.* $I_n = (0, \frac{1}{n})$.
3. *If* I_n *'s are not bounded, conclusion is false, e.g.* $I_n = [n, \infty]$.

Definition 10.2.15 *Let* $a_1 \le b_1, \ldots, a_k \le b_k$ *be real numbers, then the set of all points* $p \in \mathbb{R}^k \ni p = (x_1, \ldots, x_k)$, $a_i \le x_i \le b_i$, $i = 1, \ldots, k$ *is called a k-cell. So a k-cell is*

$$[a_i, b_i] \times \cdots \times [a_k, b_k].$$

Theorem 10.2.16 *Let* $k \in \mathbb{N}$ *be fixed. Let* I_n *be a sequence of k-cells in* $\mathbb{R}^k \ni$ $I_1 \supset I_2 \supset \cdots$. *Then,* $\bigcap_{i=1}^{\infty} I_n \ne \emptyset$.

Theorem 10.2.17 *Every k-cells is compact (with* d_2 *metric).*

Proof. Let $I = [a_1, b_1] \times \cdots \times [a_k, b_k] \subset \mathbb{R}^k$ be a k-cell. If $a_1 = b_1, \ldots, a_k = b_k$, then I consists of one point. Then, I is compact. So assume for at least one j, $a_j < b_j$, $j \in \{1, \ldots k\}$. Let $\delta = [\sum_{i=1}^{k}(b_i - a_i)^2]^{\frac{1}{2}} > 0$. Suppose I is not compact. So, there is an open cover $\{G_\alpha, \alpha \in A\}$ of $I \ni \{G_\alpha\}$ does not have any finite subcollection the union of whose elements covers I.
Let $c_i = \frac{a_i + b_i}{2}$. Then, $[a_i, b_i] = [a_i, c_i] \cap [c_i, b_i]$.
This way I can be divided into 2^k k-cells $Q_j \ni \bigcup_{j=1}^{2^k} Q_c = I$.
Also, $\forall j$ we have $p, q \in Q_j$, $d(p,q) \le \frac{1}{2}\delta$.
Since I cannot be covered by a finite number of G_α's, at least one of the Q_j's, say I_1 cannot be covered by a finite number of G_α's. Subdivide I_1 into 2^k cells by halving each side. Continue this way ... We eventually get a sequence $\{I_n\}$ of k-cells such that

a) $I_1 \subset I_2 \subset \cdots$;
b) I_n cannot be covered by any finite subcollection of $\{G_\alpha, x \in A\}$, $\forall n$;
c) $p, q \in I_n \Rightarrow d(p,q) \le \frac{1}{2^n}\delta$, $\forall n$.

By a) $\bigcap_{n=1}^{\infty} I_n \ne \emptyset$. Let $p^* \in \bigcap_{n=1}^{\infty} I_n \subset I$, then $\exists \alpha_0 \in A \ni p^* \in G_{\alpha_0}$. Since G_{α_0} is open, $\exists r > 0 \ni B_r(p^*) \subset G_{\alpha_0}$. Find $n_0 \in \mathbb{N} \ni \frac{\delta}{r} < 2^{n_0}$ [i.e. $\frac{\delta}{2^{n_0}} < r$].
Show $I_{n_0} \subset G_{\alpha_0}$: $p^* G \bigcap_{i=1}^{n} I_n \subset I_{n_0}$. Let $p \in I_{n_0}$, by c) $d(p, p^*) \le \frac{1}{2^{n_0}}\delta < r$.
$\Rightarrow p \in B_r(p^*) \subset G_{\alpha_0} \Rightarrow I_{n_0} \subset G_{\alpha_0}$ and this contradicts to b). Thus, I is compact. $\square$

Theorem 10.2.18 *Consider* $\mathbb{R}^k$ *with* d_2 *metric, let* $E \subset \mathbb{R}^k$. *Then, the following are equivalent:*

(a) E is closed and bounded.
(b) E is compact.
(c) Every infinite subset of E has a limit point which is contained in E.

Remark 10.2.19 *Consider the following remarks on Theorem 10.2.18:*

1. *The equivalence of (a) and (b) is known as Heine-Barel Theorem:*
   A subset E of $\mathbb{R}^k$ is compact if and only if it is closed and bounded.
2. *(b)$\Leftrightarrow$(c) holds in every metric space.*
3. *(c)$\Rightarrow$(a) , (b)$\Rightarrow$(a) hold in every metric space.*
4. *(a)$\Rightarrow$(c) , (a)$\Rightarrow$(b) are not true in general.*

Theorem 10.2.20 (Balzano-Weierstrass) *Every bounded infinite subset of $\mathbb{R}^k$ has a limit point in $\mathbb{R}^k$.*

Proof. Let $E \subset \mathbb{R}^k$ be infinite and bounded. Since E is bounded $\exists$ a k-cell $I \ni E \subset I$. Since I is compact, E has a limit point $p \in I \subset \mathbb{R}^k$. $\square$

Theorem 10.2.21 *Let $P \neq \emptyset$ be a perfect set in $\mathbb{R}^k$. Then, P is countable.*

10.3 The Cantor Set

Definition 10.3.1 *Let*
$E_0 = [0,1]$,
$E_1 = [0,\frac{1}{3}] \cup [\frac{2}{3},1]$,
$E_2 = [0,\frac{1}{3^2}] \cup [\frac{2}{3^2},\frac{3}{3^2}] \cup [\frac{6}{3^2},\frac{7}{3^2}] \cup [\frac{8}{3^2},1]$,
$\vdots$

continue this way. Then, Cantor set C is defined as

$$C = \bigcap_{n=1}^{\infty} E_n.$$

Some properties are listed below:

1. C is compact.
2. $C \neq \emptyset$.
3. C contains no segment (α, β).
4. C is perfect.
5. C is countable.

Proof (Property 3). In the first step, $(\frac{1}{3},\frac{2}{3})$ has been removed; in the second step $(\frac{1}{3^2},\frac{2}{3^2}),(\frac{7}{3^2},\frac{8}{3^2})$ have been removed; and so on. C contains no open interval of the form $(\frac{3k+1}{3^n}, \frac{3k+2}{3^n})$, since all such intervals have been removed in the $1^{st}, \ldots, (n-1)^{st}$ steps.
Now, suppose C contains an interval (α, β) where $\alpha < \beta$. Let $a > 0$ be a constant which will be determined later. Choose $n \in \mathbb{N} \ni 3^{-n} < \frac{\beta - \alpha}{a}$. Let k be the smallest integer $\ni \alpha < \frac{3k+1}{3^n}$, i.e. $\frac{a3^n-1}{3} < k$, then $k-1 \leq \frac{a3^n-1}{3}$. Show $\frac{3kR}{3^n} < \beta$, i.e. $k < \frac{\beta 3^n - 2}{3}$ $k \leq 1 + \frac{a3^n-1}{3}$; so show $1 + \frac{3^n-1}{3} < \frac{\beta 3^n-2}{3}$, i.e.

$$1 < \frac{1}{3}[\beta 3^n - 2 - \alpha 3^n + 1] = \frac{(\beta - \alpha)3^n - 1}{3} > \frac{a3^{-n}3^n - 1}{3} > \frac{\alpha - 1}{3} > 1,$$

is what we want. So, $a > 4$. Then, $(\frac{3k+1}{3^n}, \frac{3k+2}{3^n}) \subset (\alpha, \beta) \subset C$, Contradiction! □

Proof (Property 4). Let $x \in C$ be an arbitrary point of C. Let $B_r(x) = (x - r, x + r)$ be any open ball centered at x. Find $n \in \mathbb{N} \ni \frac{1}{3^n} < r$, $x \in C = \bigcap_{m=1}^{\infty} E_m \Rightarrow x \in E_n = I_1^n \cup \cdots \cup I_{2^n}^n$, (disjoint intervals). So $x \in I_j^n$ for some $j = 1, 2, \ldots, 2^n$. Then, $x \in (x - r, x + r) \cap I_j^n$ and length$(I_j^n) = \frac{1}{j^n} < r \Rightarrow I_j^n \subset (x - r, x + r)$.

Let y be the end point of $I_j^n \ni y \neq x$. Then, $y \in C \cap (x - r, x + r) \Rightarrow x$ is a limit point of C. □

10.4 Connected Sets

Definition 10.4.1 *Let (X, d) be a metric space and $A, B \subset X$. We say A and B are separated if $\bar{A} \cap B = \emptyset$ and $A \cap \bar{B} = \emptyset$. A subset E of X is said to be disconnected if $\exists$ two nonempty separated sets $A, B \ni E = A \cup B$. $E \subset X$ is called connected if it is not a union of two nonempty separated sets, i.e. $\exists$ <u>no</u> nonempty separated subsets $A, B \ni E = A \cup B$ ($\forall$ A,B pairs).*

Example 10.4.2 $X = \mathbb{R}^2$, *with* d_2, d_1 *or* d_∞ *metric.*
Let $E = \{(x, y) : x^2 < y^2\} = \{(x, y) : |x| < |y|\}$. *See Figure 10.11.*

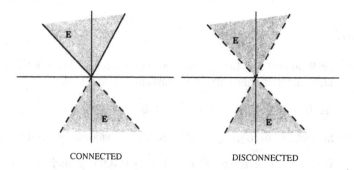

CONNECTED   DISCONNECTED

Fig. 10.11. Example 10.4.2

Theorem 10.4.3 *A subset $E \neq \emptyset$ of $\mathbb{R}$ is connected if and only if E is an interval (E is an interval if and only if $z, x \in E$ and $x < z \Rightarrow \forall y$ with $x < y < z \Rightarrow y \in E$).*

Proof. Let us mark the statement

$$z, x \in E \text{ and } x < z \Rightarrow \forall y \text{ with } x < y < z \Rightarrow y \in E \ (*).$$

($\Rightarrow$):

Let $E \neq \emptyset$ be connected. If E is not an interval $\Rightarrow$ (*) does not hold. i.e. $\exists x, z \in E \ni x < z$ and $\exists y \ni x < y < z$ and $y \notin E$. Let $A_y = (-\infty, y) \cap E, B_y = (y, +\infty) \cap E$. $A_y \neq \emptyset$ (because $x \in A_y$) and $B_y \neq \emptyset$ (because $z \in B_y$). $A_y \cup B_y = [(-\infty, y) \cup (y, \infty)] \cap E = E$. $A_y \subset (-\infty, y) \Rightarrow \bar{A}_y \subset (-\infty, y]$ and $B_y \subset (y, \infty) \Rightarrow \bar{A}_y \cap B_y \subset (-\infty, y] \cap (y, +\infty) = \emptyset \Rightarrow \bar{A}_y \cap B_y = \emptyset$.

Similarly, $A_y \cap \bar{B}_y = \emptyset \Rightarrow E$ is disconnected, Contradiction!

($\Leftarrow$):

Suppose E is disconnected. Then $\exists$ nonempty separated sets $A, B \ni A \cup B = E$. Let $x \in A, y \in B$. Assume without loss of generality $x < y$ (because $A \cap B = \emptyset$, $x \neq y$). Let $z = \sup(A \cap [x, y])$, then $z \in \overline{A \cap [x, y]} \subset \bar{A}$ (because $A \subset B \Rightarrow \bar{A} \subset \bar{B}$), $z \notin B$. Since $x \in A \cap [x, y]$, we have $x \leq z$. $z \in \overline{A \cap [x, y]} \subset \overline{[x, y]} = [x, y] \Rightarrow z \leq y \Rightarrow x \leq z \leq y$.

If $\left.\begin{array}{l} z = y \in \bar{A} \\ y \in B \end{array}\right\} \Rightarrow y \in \bar{A} \cap B = \emptyset$, Contradiction; hence, $z < y$.

So, $x \leq z < y$, and $z \in \bar{A}$.

If $z \notin A \Rightarrow x < z < y$. So $x, y \in E \ni x < y$ and $z \ni x < z < y$. $z \notin E$ because $z \notin B, z \notin A$. So (*) does not hold.

If $z \in A \Rightarrow z \notin \bar{B}$ (because sets are separated).

Claim: $(z, y) \not\subset B$. If not, $(z, y) \subset B \Rightarrow \overline{(z, y)} \subset \bar{B} \Rightarrow [z, y] \subset \bar{B} \Rightarrow z \in \bar{B}$, Contradiction.

Therefore, $\exists z_1 \in (z, y) \ni z_1 \notin B \Rightarrow x \leq z < z_1 < y \Rightarrow z_1 \in [x, y]$. If $z_1 \in A$, then $z_1 < z \Rightarrow z_1 \notin A, z_1 \notin E. \Rightarrow x, y \in E \ni x < y$ and $z_1 \ni x < z_1 < y$, Contradiction to (*)! $\square$

Problems

10.1. Let $X \neq \emptyset$ be any set. Let d, g be two metrics on X. We say the metrics d and g are equivalent if there are two constants:

$$A, B > 0 \ni Ag(p, q) \leq d(p, q) \leq Bg(p, q), \ \forall p, q \in X.$$

Show that the metrics d_1, d_2, d_∞ for $\mathbb{R}^k$ are all equivalent, i.e. find A, B.

10.2. Let (X, d) be a metric space, $p \in X, r > 0$. One is inclined to believe that $\overline{B_r(p)} = B_r[p]$; i.e. the closure of the open ball is the closed ball. Give an example to show that this is not necessarily true.

10.3. Show that a metric space (X, d) is disconnected if and only if X has a nonempty proper subset which is both open and closed.

10.4. Consider the Printed Circuit Board (PCB) given in Figure 10.12 having 36 legs separated uniformly along the sides of the wafer. Suppose that a CNC

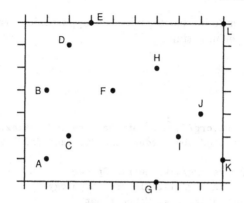

Fig. 10.12. The PCB example

machine with a robot arm makes vias (a kind of drill operation) at points
$A, B, \ldots, L$. A high volume of PCB's are processed one after another.

a) Suppose that the robot arm moves in horizontal as well as vertical direction
using a single motor. It switches its direction in an infinitesimal time unit.
The CNC programmer uses the following logic to find the sequence of vias
to be processed: Start from A, go to the closest neighbor if it has not been
processed yet. Break the ties in terms of ascending lexicographical order of
locations. Once the initial sequence (Hamiltonean tour) is obtained, examine
the nonconsecutive pair of edges of the tour if it is possible to delete these
edges and construct another tour (which is uniquely determined by the four
locations) that yields smaller tour in length. In order to check whether there
exist such an opportunity, the programmer calculates the gains associated
with all possible pairs once. Suppose that the connections between (α, β) and
(γ, δ) is broken in the current tour. Then, new connections (α, γ) and (β, δ) is
constructed in such a way that some portion of the tour is reversed and a new
tour spanning all locations is obtained. Once all the gains are calculated, all
the independent switches is made. This improvement procedure is executed
only once.

1. Find the initial tour after deciding on the appropriate metric.
2. Improve the tour.

b) What if the robot arm moves in any direction using its motor?
c) What if the robot arm moves in horizontal as well as vertical direction
using two independent but identical motors?
d) Suppose that we have N PCBs to process. All the operation times are
identical, each taking p time units. The robot arm moves at a speed of one
leg distance per unit time along each direction. Let C_1 be the cost of making
the robot arm to move along any direction using the single motor and C_2 be
the cost of adding a second motor. Using the improved solutions found, which

robot configuration is to be selected when the opportunity cost of keeping the system busy is C_o per unit time?

Web material

```
http://br.endernet.org/~loner/settheory/reals2/reals2.html
http://community.middlebury.edu/~schar/Courses/fs023.F02/paper1/
    bahls.txt
http://en.wikibooks.org/wiki/Metric_Spaces
http://en.wikipedia.org/wiki/Closure_(topology)
http://en.wikipedia.org/wiki/Compact_set
http://en.wikipedia.org/wiki/Compact_space
http://en.wikipedia.org/wiki/Discrete_space
http://en.wikipedia.org/wiki/Limit_point
http://en.wikipedia.org/wiki/Metric_space
http://eom.springer.de/c/c023470.htm
http://eom.springer.de/c/c023530.htm
http://eom.springer.de/C/c025350.htm
http://eom.springer.de/m/m063680.htm
http://homepages.cwi.nl/~bens/1metric.htm
http://homepages.nyu.edu/~eo1/Book-PDF/chapterC.pdf
http://kr.cs.ait.ac.th/~radok/math/mat6/calc13.htm
http://math.bu.edu/DYSYS/FRACGEOM/node5.html
http://mathstat.carleton.ca/~ckfong/ba4.pdf
http://mathworld.wolfram.com/ClosedSet.html
http://mathworld.wolfram.com/CompactSet.html
http://mathworld.wolfram.com/CompleteMetricSpace.html
http://mathworld.wolfram.com/MetricSpace.html
http://mathworld.wolfram.com/Topology.html
http://msl.cs.uiuc.edu/planning/node196.html
http://msl.cs.uiuc.edu/planning/node200.html
http://oregonstate.edu/~peterseb/mth614/docs/40-metric-spaces.pdf
http://pirate.shu.edu/projects/reals/topo/open.html
http://pirate.shu.edu/~wachsmut/ira/topo
http://planetmath.org/encyclopedia/
    ANonemptyPerfectSubsetOfMathbbRThatContainsNoRationalNumber.html
http://planetmath.org/encyclopedia/
    ClosedSubsetsOfACompactSetAreCompact.html
http://planetmath.org/encyclopedia/Compact.html
http://planetmath.org/encyclopedia/MetricSpace.html
http://planetmath.org/encyclopedia/NormedVectorSpace.html
http://planning.cs.uiuc.edu/node184.html
http://staff.um.edu.mt/jmus1/metrics.pdf
http://uob-community.ballarat.edu.au/~smorris/topbookchap92001.pdf
http://web01.shu.edu/projects/reals/topo/compact.html
http://web01.shu.edu/projects/reals/topo/connect.html
http://web01.shu.edu/projects/reals/topo/open.html
```

```
http://www-db.stanford.edu/~sergey/near.html
http://www-history.mcs.st-andrews.ac.uk/Extras/Kuratowski_
    Topology.html
http://www.absoluteastronomy.com/c/compact_space
http://www.absoluteastronomy.com/c/connected_space
http://www.absoluteastronomy.com/m/metric_space
http://www.all-science-fair-projects.com/science_fair_projects_
    encyclopedia/Limit_point
http://www.answers.com/topic/limit-point-1
http://www.bbc.co.uk/dna/h2g2/A1061353
http://www.cs.colorado.edu/~lizb/topology-defs.html
http://www.cs.colorado.edu/~lizb/topology.html
http://www.cs.mcgill.ca/~chundt/354review.pdf
http://www.di.ens.fr/side/slides/vermorel04metricspace.pdf
http://www.dpmms.cam.ac.uk/~tkc/Further_Analysis/Notes.pdf
http://www.fact-index.com/t/to/topology_glossary.html
http://www.hss.caltech.edu/~kcb/Notes/MetricSpaces.pdf
http://www.mast.queensu.ca/~speicher/Section8.pdf
http://www.math.buffalo.edu/~sww/0papers/COMPACT.pdf
http://www.math.ksu.edu/~nagy/real-an/
http://www.math.louisville.edu/~lee/05Spring501/chapter4.pdf
http://www.math.miami.edu/~larsa/MTH551/Notes/notes.pdf
http://www.math.niu.edu/~rusin/known-math/index/54EXX.html
http://www.math.nus.edu.sg/~matwyl/d.pdf
http://www.math.ohio-state.edu/~gerlach/math/BVtypset/node7.html
http://www.math.okstate.edu/mathdept/dynamics/lecnotes/node33.html
http://www.math.sc.edu/~sharpley/math555/Lectures/
    MetricSpaceTopol.html
http://www.math.ucdavis.edu/~emsilvia/math127/chapter3.pdf
http://www.math.unl.edu/~webnotes/classes/class34/class34.htm
http://www.mathacademy.com/pr/prime/articles/cantset/
http://www.mathreference.com/top-ms,intro.html
http://www.maths.mq.edu.au/~wchen/lnlfafolder/lfa02-ccc.pdf
http://www.maths.nott.ac.uk/personal/jff/G13MTS/
http://www.ms.unimelb.edu.au/~rubin/math127/summary2.pdf
http://www.msc.uky.edu/ken/ma570/lectures/lecture2/html/compact.htm
http://www.unomaha.edu/wwwmath/MAM/2002/Poster02/Fractals.pdf
http://www22.pair.com/csdc/car/carfre64.htm
http://wwwrsphysse.anu.edu.au/~vbr110/thesis/ch2-connected.pdf
```

11

Continuity

In this chapter, we will define the fundamental notions of limits and continuity of functions and study the properties of continuous functions. We will discuss these properties in more general context of a metric space. The concept of compactness will be introduced. Next, we will focus on connectedness and investigate the relationships between continuity and connectedness. Finally, we will introduce concepts of monotone and inverse functions and prove a set of Intermediate Value Theorems.

11.1 Introduction

Definition 11.1.1 *Let* $(X, d_X), (Y, d_Y)$ *be two metric spaces;* $E \neq \emptyset$, $E \subset X$. *Let* $f : E \mapsto Y, p \in E, q \in Y$. *We say* $\lim_{n \to p} f(x) = q$ *or* $f(x) \to q$ *as* $x \to p$ *if* $\forall \varepsilon > 0, \exists \delta > 0 \ni \forall x \in E$ *with* $d_X(x, p) < \delta$ *we have* $d_Y(f(x), q) < \varepsilon$ *(i.e.* $\forall \varepsilon > 0, \exists \delta > 0 \ni f(E \cap B_\delta^x(p)) \subset B_\varepsilon^y(q)$).

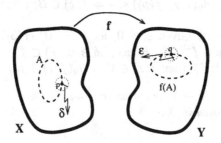

Fig. 11.1. Limit and continuity

Definition 11.1.2 *Let* $(X, d_X), (Y, d_Y)$ *be metric spaces;* $\emptyset \neq E \subset X$, *and* $f : X \mapsto Y, p \in E$. f *is said to be continuous at* p *if*

$\forall \varepsilon > 0, \exists \delta > 0 \ni \forall x \in E$ with $d_x(x, p) < \delta$ we have $d_y(f(x), f(p)) < \varepsilon$.

Remark 11.1.3 *The following characteristics are noted:*

- f has to be defined at p, but p does not need to be a limit point of E.
- If p is an isolated point of E, then f is continuous at p. That is, given $\varepsilon > 0$ (no matter what ε is), find $\delta \ni E \cap B_\delta^X(p) = \{p\}$. Then, $x \in E, d(p, x) < \delta \Rightarrow x = p$. Hence, $d_y(f(x), f(p)) = 0 < \varepsilon$.
- If p is a limit point of E, then f is continuous at $p \Leftrightarrow \lim_{x \to p} f(x) = f(p)$.

Definition 11.1.4 *If f is continuous at every point of E, we say f is continuous on E.*

Proposition 11.1.5 *Let $(X, d_X), (Y, d_Y), (Z, d_Z)$ be metric spaces and $\emptyset \neq E \subset X, f : E \mapsto Y, g : f(E) \mapsto Z$. If f is continuous at $p \in E$ and g is continuous at $f(p)$, then $g \circ f$ is continuous at p.*

Proof. Let $q = f(p)$. Let $\varepsilon > 0$ be given. Since g is continuous at q, $\exists \eta > 0 \ni \forall y \in f(E)$ with $d_Y(y, q) < \eta$ we have $d_2(g(y), g(q)) < \varepsilon$. Since f is continuous at p, $\exists \delta > 0 \ni \forall x \in E$ with $d_X(x, p) < \delta \Rightarrow$ we have $d_Y(f(x), f(p)) < \eta$. Let $x \in E$ be $\ni d_X(x, p) < \delta$. Then, $y = f(x) \in f(E)$ and $d_Y(y, q) = d_Y(f(x), f(p)) < \eta$. Hence, $d_Z(g(f(x)), g(f(p))) = d_Z(g(y), g(q)) < \varepsilon$. $\square$

Theorem 11.1.6 *Let $(X, d_X), (Y, d_Y)$ be metric spaces, and let $f : X \mapsto Y$. Then, f is continuous on X if and only if $\forall$ open set V in Y, $f^{-1}(V) = \{p \in X : f(p) \in V\}$ is open in X.*

Proof. ($\Rightarrow$):
Let V be open in Y. If $f^{-1}(V) \neq \emptyset$, let $p \in f^{-1}(V)$ be arbitrary. Show $\exists r > 0 \ni B_r^X(p) \subset f^{-1}(V)$: $p \in f^{-1}(V)$ implies $f(p) \in V$. Since V is open, $\exists s > 0 \ni B_s^Y(f(p)) \subset V$. Since f is continuous at p, for $\varepsilon = s$, $\exists r > 0 \ni \forall x \in X$ with $d_x(x, p) < r \Rightarrow d_y(f(x), f(p)) < s \Rightarrow f(x) \in B_s^Y(f(p)) \Rightarrow x \in f^{-1}(V)$.
($\Leftarrow$):
Let $p \in X$ be arbitrary. Given $\varepsilon > 0$, let $V = B_\varepsilon^Y(f(p))$ be open. Then, $f^{-1}(V)$ is open and $p \in f^{-1}(V)$. Hence, $\exists \delta \ni B_\delta(p) \subset f^{-1}(V)$. If $d_x(x, p) < \delta \Rightarrow x \in B_\delta^X(p) \subset f^{-1}(V)$, then $f(x) \in V \Rightarrow d_y(f(x), f(p)) < \varepsilon$. $\square$

Corollary 11.1.7 *$f : X \to Y$ is continuous on X if and only if $\forall$ closed set C in Y, $f^{-1}(C)$ is closed in X.*

Proof. $f^{-1}(E^C) = (f^{-1}(E))^C$. $\square$

Definition 11.1.8 *Let (X, d) be a metric space and $f_1, \ldots, f_k : X \mapsto \mathbb{R}$. Define $f : X \mapsto \mathbb{R}^k$ by $f(x) = (f_1(x), \ldots, f_k(x))^T$, then $f_1, \ldots, f_k$ are called components of f.*

Proposition 11.1.9 f *is continuous if and only if every component is continuous.*

Proof. ($\Rightarrow$): Fix j. Show that f_j is continuous: Fix $p \in X$. Show that f_j is continuous at p. Given $\varepsilon > 0 \; \exists \delta > 0 \; \ni \; \forall x$ with $d_2(x,p) < \delta$, then $|f_j(x) - f_j(p)| = d_1(f_j(x), f_j(p)) \leq d_2(f(x), f(p)) < \varepsilon$.
($\Leftarrow$): Assume that $\forall j$, f_j is continuous at $p \in X$. Show that f is continuous at p. Let $\varepsilon > 0$ be given.
f_1 is continuous at $p \Rightarrow \exists \delta_1 > 0 \ni d_2(x,p) < \delta_1 \Rightarrow |f_1(x) - f_1(p)| < \frac{\varepsilon}{\sqrt{k}}$.
f_2 is continuous at $p \Rightarrow \exists \delta_2 > 0 \ni d_2(x,p) < \delta_2 \Rightarrow |f_2(x) - f_2(p)| < \frac{\varepsilon}{\sqrt{k}}$.
$$\vdots$$
f_k is continuous at $p \Rightarrow \exists \delta_k > 0 \ni d_2(x,p) < \delta_k \Rightarrow |f_k(x) - f_k(p)| < \frac{\varepsilon}{\sqrt{k}}$.
Let $\delta = \min\{\delta_1, \ldots, \delta_k\} > 0$. Let X be $\ni d(x,p) < \delta$. Then,

$$d_2(f(x), f(p)) = \left[\sum_{j=1}^{k} |f_j(x) - f_j(p)|^2\right]^{1/2} < \left[\sum_{j=1}^{k} \left(\frac{\varepsilon}{\sqrt{k}}\right)^2\right]^{1/2} = \varepsilon. \quad \square$$

11.2 Continuity and Compactness

Theorem 11.2.1 *The continuous image of a compact space is compact, i.e. if $f : X \mapsto Y$ is continuous and (X,d) is compact, then $f(X)$ is a compact subspace of (Y, d_Y).*

Proof. Let $\{V_\alpha : \alpha \in A\}$ be any open cover of $f(X)$. Since f is continuous, $f^{-1}(V_\alpha)$ is open in X. $f(x) \subset \bigcup_{\alpha \in A} V_\alpha \Rightarrow X \subset f^{-1}(f(x)) \subset \bigcup_{\alpha \in A} f^{-1}(V_\alpha)$. Since X is compact, $\exists \alpha_1, \ldots, \alpha_n \ni X \subset [f^{-1}(V_{\alpha_1}) \cup \cdots \cup f^{-1}(V_{\alpha_n})] \Rightarrow f(x) \subset f[f^{-1}(V_{\alpha_1}) \cup \cdots \cup f^{-1}(V_{\alpha_n})] = V_{\alpha_1} \cup \cdots \cup V_{\alpha_n}$, since for $A \subset f^{-1}f(A), f^{-1}f(B) \subset B$ we have

$$f(\bigcup_\alpha A_\alpha) = \bigcup f(A_\alpha) \text{ and } f^{-1}(\bigcup_\alpha B_\alpha) = \bigcup f^{-1}(B_\alpha). \quad \square$$

Corollary 11.2.2 *A continuous real valued function on a compact metric space attains its maximum and minimum.*

Proof. $f(X)$ is a compact subset of $\mathbb{R} \Rightarrow f(X)$ is bounded. Let $m = \inf f(x)$, $M = \sup f(x)$. Then, $m, M \in \mathbb{R}$; since $f(X)$ is bounded. Also, $m, M \in \overline{f(X)}$. Furthermore, $\overline{f(x)} = f(x)$, since $f(X)$ is compact. Thus, $\exists p \in X \ni m = f(p)$ and $\exists q \in X \ni M = f(q)$. Finally, $m = f(p) \leq f(x) \leq f(q) = M, \forall x \in X$. $\square$

Theorem 11.2.3 *Let (X, d_X) be a compact metric space, (Y, d_Y) be a metric space, $f : X \mapsto Y$ be continuous, one-to-one and onto. Then, $f^{-1} : Y \mapsto X$ is continuous.*

Proof. Let $g = f^{-1} : Y \to X$. Show that $\forall$ closed set C in X, $g^{-1}(C)$ is a closed set in Y: $g^{-1}(C) = (f^{-1})^{-1}(C) = f(C)$, since X is compact. Hence, $f(C)$ is closed, thus $g^{-1}(C)$ is closed. $\square$

Remark 11.2.4 *If compactness is relaxed, the theorem is not true. For example, take $X = [0, 2\pi)$ with d_1 metric. $Y = \{(x, y) \in \mathbb{R}^2 : x^2 + y^2 = 1\}$ with d_2 metric.*

$$f : X \mapsto Y, \ f(t) = (\cos t, \sin t).$$

f is one-to-one, onto, continuous. However f^{-1} is not continuous at $P = (0, 1) = f(0)$. If we let $\varepsilon = \pi$, suppose there is a $\delta > 0 \ni \forall (x, y) \in Y$ with $d_2((x, y), (1, 0)) < \delta$, then we have

$$|f^{-1}(x, y) - f^{-1}(1, 0)| < \varepsilon.$$

However, for $(x, y) \ni \frac{3\pi}{2} < f^{-1}(x, y) < 2\pi \ (\delta = \sqrt{2})$, we have

$$|f^{-1}(x, y) - f^{-1}(1, 0)| > \frac{3\pi}{2} > \pi.$$

Thus, we do not have

$$|f^{-1}(x, y) - f^{-1}(1, 0)| < \varepsilon = \pi \ \forall (x, y) \in Y \ni d_2[(x, y), (1, 0)] < \delta.$$

11.3 Uniform Continuity

Definition 11.3.1 *Let $(X, d_X), (Y, d_Y)$ be two metric spaces, $f : X \mapsto Y$. We say f is uniformly continuous on X if*

$$\forall \varepsilon > 0, \ \exists \delta > 0 \ni \forall p, q \in X \text{ with } d_X(p, q) < \delta, \text{ we have } d_Y(f(p), f(q)) < \varepsilon.$$

Remark 11.3.2 *Uniform continuity is a property of a function on a set, whereas continuity can be defined at a single point. If f is uniformly continuous on X, it is possible for each $\varepsilon > 0$ to find <u>one</u> number $\delta > 0$ which will do for all points p of X. Clearly, every uniform continuous function is continuous.*

Example 11.3.3

$$f(x, y) = 2x + \frac{1}{y^2}, \ E = \{(x, y) \in \mathbb{R}^2 : 1 \leq y \leq 2\}.$$

Let us show that f is uniformly continuous on E. Let $\varepsilon > 0$ be given. Suppose we have found $\delta > 0$ whose value will be determined later. Let $p = (x, y), q = (u, v) \in E$ be such that $d_2(p, q) < \delta$, Show $|f(x, y) - f(u, v)| < \varepsilon$: $d_2(p, q) < \delta \Rightarrow |x - u| < \delta$ and $|y - v| < \delta \Rightarrow |f(x, y) - f(u, v)| = |2x + \frac{1}{y^2} - 2u - \frac{1}{v^2}| \leq 2|x - u| + (\frac{1}{y^2} - \frac{1}{v^2}) < 2\delta + \left| \frac{(v-y)(v+y)}{(vy)^2} \right| = 2\delta + \frac{|v-y||v+y|}{v^2 y^2}$. Since $\frac{|v-y||v+y|}{v^2 y^2} < 4\delta$, we have $|f(x, y) - f(u, v)| < 6\delta = \varepsilon$. Hence, one can safely choose $\delta = \frac{\varepsilon}{6} > 0$.

Example 11.3.4 $f(x) = \frac{1}{x}, E = (0,1) \subset \mathbb{R}$. *Let us show that f is not uniformly continuous on E but continuous on E: given $\varepsilon > 0$, let $\delta > 0$ be chosen. Let $x \in E$ and $|x - x_0| < \delta$.*

If $x_0 - \delta > 0$, then $|x - x_0| < \delta \Leftrightarrow x_0 - \delta < x < x_0 + \delta$.

$$\left|\frac{1}{x} - \frac{1}{x_0}\right| \leq \frac{|x_0 - x|}{xx_0} \leq \frac{\delta}{xx_0} < \frac{\delta}{(x_0 - \delta)x_0} \leq \varepsilon \Rightarrow \delta \leq \frac{\varepsilon x_0^2}{1 + \varepsilon x_0}.$$

Hence, f is continuous at x_0 and δ depends on ε and x_0. However, dependence on x_0 does not imply that f is not uniformly continuous, because some other calculation may yield another δ which is independent of x_0. So, we must show that the negation of uniform continuity to hold:

$$\exists \varepsilon > 0 \ni \forall \delta > 0 \ \exists x_1, x_2 \in E \ni |x_1 - x_2| < \delta \text{ but } |f(x_1) - f(x_2)| \geq \varepsilon.$$

Let $\varepsilon = 1$. Let δ be given. If $\delta \leq \frac{1}{3}$, one can find $k \ni \delta \leq \frac{1}{k+1}$ i.e. $k = \lceil \frac{1}{\delta} - 1 \rceil$. Thus, $k \geq 2$. Let $x_1 = \delta$, $x_2 = \delta + \frac{\delta}{k} \Rightarrow 0 < x_1 \leq \frac{1}{3}$, $0 < x_2 < 2\delta \leq \frac{2}{3} < 1 \Rightarrow x_1, x_2 \in E$. $|x_1 - x_2| = \frac{\delta}{k} \leq \frac{\delta}{2} < \delta, |f(x_1) - f(x_2)| = \left|\frac{1}{\delta} - \frac{1}{\delta + \frac{\delta}{k}}\right| = \frac{(\delta/k)}{\delta(\delta + \frac{\delta}{k})} = \frac{1}{\delta(k+1)} > 1$. If $\delta > \frac{1}{3} \Rightarrow$ Let $\delta' = \frac{1}{3}$. Find $x_1, x_2 \ni |x_1 - x_2| < \delta' < \delta$ and $|f(x_1) - f(x_2)| < \varepsilon$.

Theorem 11.3.5 *Let (X, d_X) be a compact metric space, (Y, d_Y) be a metric space, and $f : X \mapsto Y$ be continuous on X. Then, f is uniformly continuous.*

Remark 11.3.6 *Let $\emptyset \neq E \subset \mathbb{R}$ be non-compact. Then,*

(a) *$\exists$ a <u>continuous</u> $f : E \to \mathbb{R}$ which is <u>not bounded</u>. If E is noncompact then either E is not closed or not bounded. If E is bounded and not closed, then E has a limit point $x_0 \ni x_0 \notin E$. Let $f(x) = \frac{1}{x - x_0}, \forall x \in E$. If E is unbounded then let $f(x) = x, \forall x \in E$.*

(b) *$\exists$ a <u>continuous bounded</u> function $f : E \to \mathbb{R}$ which has <u>no maximum</u>. If E is bounded let x_0 be as in (a). Then, $f(x) = \frac{1}{1 + (x - x_0)^2}, \forall x \in E$. $\sup f(x) = 1$ but $\exists$ no $x \in E \ni f(x) = 1$.*

(c) *If E is <u>bounded</u>, $\exists$ a <u>continuous</u> function $f : E \to \mathbb{R}$ which is <u>not uniformly continuous</u>. Let x_0 be as in (a). Let $f(x) = \frac{1}{x - x_0}, \forall x \in E$ which is not uniformly continuous.*

11.4 Continuity and Connectedness

Theorem 11.4.1 *Let $(X, d_X), (Y, d_Y)$ be metric spaces, $\emptyset \neq E \in X$ be connected and let $f : X \mapsto Y$ be continuous on X. Then, $f(E)$ is connected.*

Proof. Assume that $f(E)$ is not connected, i.e.

$$\exists \text{ nonempty } A, B \subset Y \ni \overline{A} \cap B = \emptyset, \ A \cap \overline{B} = \emptyset, \ f(E) = A \cup B.$$

Let $G = E \cap f^{-1}(A)$, $H = E \cap f^{-1}(B)$, $A \neq \emptyset \Rightarrow \exists q \in A \subset f(E) \Rightarrow q = f(p)$
for some $p \in E \Rightarrow p \in f^{-1}(A) \Rightarrow p \in G \Rightarrow G \neq \emptyset$.
Assume $\overline{G} \cap H \neq \emptyset$. Let $p \in \overline{G} \cap H \Rightarrow p \in H = E \cap f^{-1}(B) \Rightarrow$

$$f(p) \in B, \ p \in G = E \cap f^{-1}(A) \subset f^{-1}(A) \ (*)$$

$A \subset \overline{A} \Rightarrow f^{-1}(A) : \text{closed} \Rightarrow \overline{f^{-1}(A)} \subset f^{-1}(\overline{A}) \Rightarrow p \in f^{-1}(\overline{A}) \Rightarrow$

$$f(p) \in \overline{A} \ (**)$$

$(*)+(**) \Rightarrow f(p) \in \overline{A} \cap B \neq \emptyset$, Contradiction. Thus, $\overline{G} \cap H = \emptyset$. Similarly,
$G \cap \overline{H} = \emptyset$.

$$E \subset f^{-1}(f(E)) = f^{-1}(A \cup B) = f^{-1}(A) \cup f^{-1}(B)$$

$$E = E \cap [f^{-1}(A) \cup f^{-1}(B)] = [E \cap f^{-1}(A)] \cup [E \cap f^{-1}(B)] = G \cup H,$$

meaning that E is not connected. Contradiction!   □

Corollary 11.4.2 (Intermediate Value Theorem) *Let $f : [a, b] \to \mathbb{R}$ be
continuous and assume $f(a) < f(b)$. Let $c \in \mathbb{R}$ be such that*

$$f(a) < c < f(b) \Rightarrow c \in f([a, b]), \ \textit{i.e. } \exists x \in (a, b) \ni f(x) = c.$$

Proof. $[a, b]$ is connected, so $f([a, b])$ is connected; thus $f([a, b])$ is an interval
$[\alpha, \beta]$. $f(a), f(b) \in [\alpha, \beta] \Rightarrow c \in f([a, b])$,

$$\exists x \in [a, b] \ni f(x) = c, f(a) < c \Rightarrow x \neq a \text{ and } f(b) > c \Rightarrow x \neq b.$$

Thus, $x \in (a, b)$.   □

Example 11.4.3 *Let $I = [0, 1], f : I \to I$ be continuous. Let us show that
$\exists x \in I \ni f(x) = x$. Let $g(x) = f(x) - x$ be continuous. Show $\exists x \in I \ni g(x) =
0$. If $\nexists$ such $x \Rightarrow \forall x \in I$ we have $g(x) > 0$ or $g(x) < 0$.*

(i) $g(x) > 0, \ \forall x \in I \Rightarrow f(x) > x, \forall x \in I$. *Then, $f(1) > 1$; a Contradiction.*
(ii) $g(x) < 0, \ \forall x \in I \Rightarrow f(x) < x, \forall x \in I$. *Then, $f(0) < 0$; a Contradiction.*

Definition 11.4.4 (Discontinuities) *Let $f : (a, b) \to X$ where (X, d) is a
metric space. Let x be $\ni a \leq x < b$ and $q \in X$. We say, $f(x+) = q$ or
$\lim_{t \to x+} f(t) = q$ if $\forall \varepsilon > 0 \ \exists \delta > 0 \ \ni \forall t$ with $x < t < x + \delta$ we have
$d(f(t), f(x)) < \varepsilon$. $f(x+) = q \Leftrightarrow \forall$ subsequence $\{t_n\}$ with $x < t_n < b, \forall n$
and $\lim_{n \to \infty} b_n = x$ we have $\lim_{n \to \infty} f(t) = q$. $f(x-) = \lim_{x \to x-} f(t)$ is
defined analogously. Let $x \in (a, b) \Rightarrow \lim_{t \to x} f(t)$ exists $\Leftrightarrow f(x+) = f(x-) =
\lim_{t \to x} f(t)$. Suppose f is discontinuous at some $x \in (a, b)$.*

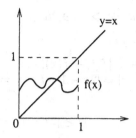

Fig. 11.2. Example 11.4.3

(i) *If $f(x+)$ or $f(x-)$ does not exist, we say the discontinuity at x is of the second kind.*

(ii) *If $f(x+)$ and $f(x-)$ both exist, we say the discontinuity at x is of the first kind or simple discontinuity.*

(iii) *If $f(x+) = f(x-)$, but f is discontinuous at x, then the discontinuity at x is said to be removable.*

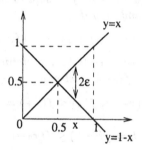

Fig. 11.3. Example 11.4.5

Example 11.4.5

$$f : \mathbb{R} \to \mathbb{R}, \quad f(x) = \begin{cases} x, & x \in \mathbb{Q} \\ 1 - x, & x \in \mathbb{R} \setminus \mathbb{Q} \end{cases}$$

f is continuous (only) at $x = \frac{1}{2}$:
Let $\varepsilon > 0$ be given. Let $\delta = \varepsilon$. Let $t \in \mathbb{R} \ni |t - x| < \delta$ where $x = \frac{1}{2}$.
$t \in \mathbb{Q} \Rightarrow |f(t) - f(x)| = |t - x| = |t - \frac{1}{2}| < \delta = \varepsilon.$
$t \in \mathbb{R} \setminus \mathbb{Q} \Rightarrow |f(t) - f(x)| = |1 - t - x| = |\frac{1}{2} - x| = |1 - x| < \delta = \varepsilon.$
Hence, f is continuous at $x = \frac{1}{2}$.
CLAIM:f is discontinuous every other point than $x = \frac{1}{2}$
(without loss of generality, we may assume that $x > \frac{1}{2}$):
Let $x \neq \frac{1}{2}$. Show $f(x+)$ does not exist. Let $\varepsilon = \frac{|2x-1|}{2}$. Assume $f(x+)$ exists,

then for this specific $\varepsilon > 0$, $\exists \delta > 0 \ni \forall t$ *with* $x < t < x + \delta$, *we have* $|f(t) - f(x)| < \varepsilon$.

CASE 1: $X \in \mathbb{Q}$.

Find $t \in \mathbb{R} \setminus \mathbb{Q} \ni x < t < x + \delta$ $|f(t) - f(x)| < \varepsilon = \frac{|2x-1|}{2}$.

But $4|f(t) - f(x)| = |1 - t - x| = |2x - 1 + t - x| = |(2x - 1) - (x - t)|$.

Since $|a - b| \geq ||a| - |b||$,

$$|f(t) - f(x)| \geq ||2x - 1| - |x - 1|| \geq |2x - 1| - |x - t| > |2x - 1| - \delta.$$

Then, we have

$$|2x - 1| - \delta < |f(t) - f(x)| < \frac{|2x - 1|}{2}.$$

$\Rightarrow \delta > \frac{|2x-1|}{2}$, *Contradiction since* $\delta > 0$ *can be taken as small as we want.*

CASE 2: $X \in \mathbb{R} \setminus \mathbb{Q}$. *Proceed in similar way, but choose* t *as rational.*

11.5 Monotonic Functions

Definition 11.5.1 *Let* $f : (a, b) \mapsto \mathbb{R}$. f *is said to be monotonically increasing (decreasing) on* (a, b) *if and only if*

$$a < x_1 < x_2 < b \Rightarrow f(x_1) \leq f(x_2) \quad (f(x_1) \geq f(x_2)).$$

Proposition 11.5.2 *Let* $f : (a, b) \mapsto \mathbb{R}$ *be monotonically increasing on* (a, b). *Then,* $\forall x \in (a, b)$, $f(x+)$ *and* $f(x-)$ *exist and*

$$\sup_{a < t < x} f(t) = f(x-) \leq f(x) \leq f(x+) = \inf_{x < t < b} f(t).$$

Furthermore, $a < x_1 < x_2 < b \Rightarrow f(x_1+) \leq f(x_2-)$.

Theorem 11.5.3 *Let* $f : (a, b) \mapsto \mathbb{R}$ *be monotonically decreasing on* (a, b), *then* $\forall x \in (a, b)$, $f(x+)$ *and* $f(x-)$ *exist and*

$$\inf_{a < t < x} f(t) = f(x-) \geq f(x) \geq f(x+) = \sup_{x < t < b} f(t).$$

Furthermore, $a < x_1 < x_2 < b \Rightarrow f(x_1+) \geq f(x_2-)$.

Proof. Let $x \in (a, b)$ be arbitrary. $\forall t$ with $0 < t < x$, we have $f(t) \geq f(x)$. So, $\{f(t) : a < t < x\}$ is bounded below by $f(x)$. Let $A = \inf\{f(t) : a < t < x\}$. We will show $A = f(x_1-)$:

Let $\varepsilon > 0$ be given. Then, $A + \varepsilon$ is no longer lower bound of $\{f(t) : a < t < x\}$. Hence, $\exists t_0 \in (a, x) \ni f(t_0) < A + \varepsilon$. Let $\delta = x - t_0$ $\forall t \ni x - \delta = t_0 < t < x \Rightarrow f(x) \leq f(t_0) < A + \varepsilon$ and $f(t) > A > A - \varepsilon$. Hence, $\forall t \in (x - \delta, x)$ we have $A - \varepsilon < f(t) < A + \varepsilon \Rightarrow |f(t) - A| < \varepsilon$. Thus, $A = f(x-)$. Therefore, $\inf_{a < t < x} f(t) = A = f(x-) \geq f(x)$. Similarly, $\sup_{x < t < b} f(x) = f(x+) \leq f(x)$. Let $a < x_1 < x_2 < b$, apply first part $b \leftarrow x_2$ and $x \leftarrow x_1$. $f(x_1+) = \sup_{x_1 < t < x_2} f(t) \geq \inf_{x_1 < t < x_2} f(t) = f(x_2+)$. $\square$

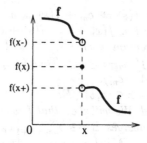

Fig. 11.4. Proof of Theorem 11.5.3

Corollary 11.5.4 *Monotonic functions have no discontinuities of the second type.*

Theorem 11.5.5 *Let $f : (a, b) \mapsto \mathbb{R}$ be monotonic. Let A be the set of discontinuous points of f, then A is at most countable.*

Proof. Assume f is decreasing, then $A = \{x \in \{a, b\} : f(x+) < f(x-)\}$. $\forall x \in A$, find $f(x) \in \mathbb{Q} \ni f(x+) < r(x) < f(x-)$ and fix $r(x)$. Define $g : A \mapsto \mathbb{Q}$ by $g(x) = r(x)$. We will show that g is one-to-one: Let $x_1 \neq x_2 \in A, x_1 < x_2 \Rightarrow r(x_1) > f(x_1+) \geq f(x_2-) > r(x_2) \Rightarrow r(x_1) \neq r(x_2)$. Thus, g is one-to-one, and A is numerically equivalent to $\mathbb{Q}$ by $g(x) = r(x)$. Therefore, A is at most countable. $\square$

Remark 11.5.6 *The points in A may not be isolated. In fact, given any countable subset E of (a, b) (E may even be dense), there is a monotonic function $f : (a, b) \mapsto \mathbb{R} \ni f$ is discontinuous at every $x \in E$ and continuous at every other point. The elements of E as a sequence $\{x_1, x_2, \ldots\}$. Let $c_n > 0 \ni \sum c_n$ is convergent. Then, every rearrangement $\sum c_{\phi(n)}$ also converges and has the same sum. Given $x \in (a, b)$ let $N_x = \{n : x_n < x\}$. This set may be empty or not. Define $f(x)$ as follows*

$$f(x) = \begin{cases} 0, & N_x = \emptyset \\ \sum_{n \in N_x} c_n, & \text{otherwise} \end{cases}$$

This function is called saltus function or pure jump function.

(a) f is monotonically increasing on (a, b):
  Let $a < x < y < b$. If $N_x = \emptyset$, $f(x) = 0$ and $f(y) \geq 0$.
  If $N_x \neq \emptyset$, $x < y \Rightarrow f(x) = \sum_{n \in N_x} c_n \leq \sum_{n \in N_y} c_n = f(y)$.
(b) f is discontinuous at every $x_m \in E$:
  Let $x_m \in E$ be fixed. $f(x_m+) = \inf_{x_m < t < b} f(t), \quad f(x_m-) = \sup_{a < s < x_m} f(s)$.
  Let $x_m < t < b$, $a < s < x_m$ be arbitrary $\Rightarrow a < s < x_m < t < b$. Then,
  $N_s \subset N_t$, $m \in N_t$, $m \notin N_s \Rightarrow m \in N_t \setminus N_s$.
  $f(t) - f(s) = \sum_{n \in N_t} c_n - \sum_{n \in N_s} c_n = \sum_{n \in N_t \setminus N_s} c_n \geq c_m \Rightarrow$

$f(t) \geq c_m + f(s)$. *Fix* $f(s) \Rightarrow c_m + f(s)$ *is a bound for all* $f(t)'s$. *So,
take the infimum over* $t's$. $f(x_m+) \geq f(s) + c_m \Leftrightarrow f(x_m+) - c_m \geq
f(s)$. *If we take supremum over* $s's$, *we will have* $f(x_m+) - c_m \geq
f(x_m-) \Rightarrow f(x_m+) - f(x_m-) \geq c_m$. *Therefore,* $f(x_m+) \neq f(x_m-)$ *(In
fact,* $f(x_m+) - f(x_m-) = c_m$*).*

(c) *f is continuous at every* $x \in (a,b) \setminus E$:

Let $x \in (a,b)\setminus E$ *be fixed. We will show that f is continuous at x. Let* $\varepsilon > 0$
be given, since $\sum c_n$ *converges,* $\exists N \ni \sum_{n=N+1}^{\infty} c_n < \infty$ $c_{N+1} + s_N = s \Rightarrow
r_{N+1} = s - s_N$. *Let* $\delta' = Min\{|x - x_1|, \ldots, |x - x_N|, x - a, b - x\}$. *Let*
$\delta = \frac{\delta'}{2}$.

Claim (i) If $x \leq x_n < x + \delta$ *then* $n \geq N + 1$. *If* $n < N + 1$, *then*
$|x - x_N| \geq \delta' = 2\delta$, *Contradiction.*

Claim (ii) If $x - \delta < x_n < x$, *then* $n \geq N + 1$. $f(x) - \varepsilon < f(x -
\delta)$, $f(x+\delta) < f(x) + \varepsilon$, $f(x) - f(x-\delta) = \sum_{n \in N_x} c_n - \sum_{n \in N_{x-\delta}} c_n =
\sum_{n \in N_x \setminus N_{x-\delta}} c_n \leq \sum_{n=N+1}^{\infty} c_n < \varepsilon$. *For the second claim,* $f(x + \delta) -
f(x) = \sum_{n \in N_{x+\delta}} c_n - \sum_{n \in N_x} c_n = \sum_{n \in N_{x+\delta} \setminus N_x} c_n \leq \sum_{n=N+1}^{\infty} c_n < \varepsilon$.
Let t *be* $\ni |t-x| < \delta$, *i.e.* $x - \delta < t < x + \delta \Rightarrow f(x-\delta) \leq f(t) \leq F(x+\delta)$.
Hence, $f(x) - \varepsilon < f(t) < f(x) + \varepsilon$, $|f(t) - f(x)| < \varepsilon$.

Problems

11.1. Let (X, d) be a metric space. A function $f : X \mapsto \mathbb{R}$ is called lower
semi-continuous (*lsc*) if $\forall b \in \mathbb{R}$ the set $\{x \in X : f(x) > b\}$ is open in X;
upper semi-continuous (*usc*) if $\forall b \in \mathbb{R}$ the set $\{x \in X : f(x) < b\}$ is open in
X. Show that
a) f is *lsc* $\Leftrightarrow \forall \varepsilon > 0, \forall x_0; \exists \delta > 0 \ni x \in B_\delta(x_0) \Rightarrow f(x) > f(x_0) - \varepsilon$.
b) f is *usc* $\Leftrightarrow \forall \varepsilon > 0, \forall x_0; \exists \delta > 0 \ni x \in B_\delta(x_0) \Rightarrow f(x) < f(x_0) + \varepsilon$.

11.2. Let (X, d_X) be a compact metric space, (Y, d_Y) be a metric space and
let $f : X \mapsto Y$ be continuous and one-to-one. Assume for some sequence $\{p_n\}$
in X and for some $q \in Y$, $\lim_{x \to \infty} f(p_n) = q$. Show that

$$\exists p \in X \ni \lim_{x \to \infty} p_n = p \text{ and } f(p) = q.$$

11.3. Give a mathematical argument to show that a heated wire in the shape
of a circle (see Figure 11.5) must always have two diametrically opposite points
with the same temperature.

Web material

http://archives.math.utk.edu/visual.calculus/1/continuous.7/

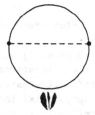

Fig. 11.5. A heated wire

```
      index.html
http://at.yorku.ca/course/atlas2/node7.html
http://at.yorku.ca/i/a/a/b/23.dir/ch2.htm
http://bvio.ngic.re.kr/Bvio/index.php/Monotonic_function
http://cepa.newschool.edu/het/essays/math/contin.htm
http://clem.mscd.edu/~talmanl/TeachCalculus/Chapter020.pdf
http://documents.kenyon.edu/math/neilsenj.pdf
http://en.wikipedia.org/wiki/Intermediate_value_theorem
http://en.wikipedia.org/wiki/List_of_general_topology_topics
http://en.wikipedia.org/wiki/Monotonic_function
http://en.wikipedia.org/wiki/Uniformly_continuous
http://eom.springer.de/T/t093150.htm
http://homepages.nyu.edu/~eo1/Book-PDF/chapterD.pdf
http://intermediate_value_theorem.iqexpand.com/
http://math.berkeley.edu/~aclayton/math104/8-10_Final_review.pdf
http://math.furman.edu/~dcs/book/c2pdf/sec25.pdf
http://math.furman.edu/~dcs/book/c3pdf/sec37.pdf
http://math.stanford.edu/~aschultz/w06/math19/
      coursenotes_and_handouts/
http://mathworld.wolfram.com/IntermediateValueTheorem.html
http://mcraefamily.com/MathHelp/CalculusTheorem1IntermediateValue.htm
http://nostalgia.wikipedia.org/wiki/Connectedness
http://ocw.mit.edu/OcwWeb/Mathematics/18-100BAnalysis-IFall2002/
      LectureNotes/index.htm
http://oregonstate.edu/instruct/mth251/cq/Stage4/Lesson/IVT.html
http://personal.stevens.edu/~nstrigul/Lecture4.pdf
http://personal.stevens.edu/~nstrigul/Lecture5.pdf
http://personal.stevens.edu/~nstrigul/Lecture8.pdf
http://personal.stevens.edu/~nstrigul/Lecture9.pdf
http://pirate.shu.edu/projects/reals/cont/proofs/ctunifct.html
http://planetmath.org/encyclopedia/IntermediateValueTheorem.html
http://planetmath.org/encyclopedia/UniformlyContinuous.html
http://poncelet.math.nthu.edu.tw/chuan/cal98/uniform.html
http://toshare.info/en/Monotonic_function.htm
http://tutorial.math.lamar.edu/AllBrowsers/2413/Continuity.asp
http://web01.shu.edu/projects/reals/cont/contin.html
http://webalt.com/Calculus-2006/HowTo/Functions/Intermediate_Value_
```

Theorem.ppt

http://whyslopes.com/Calculus-Introduction/Theorem-
One_Sided_Range.html

http://www-history.mcs.st-and.ac.uk/~john/analysis/Lectures/L20.html

http://www.absoluteastronomy.com/i/intermediate_value_theorem

http://www.answers.com/topic/continuous-function-topology

http://www.bostoncoop.net/~tpryor/wiki/index.php?title=Monotonic

http://www.calculus-help.com/funstuff/tutorials/limits/limit06.html

http://www.cut-the-knot.org/Generalization/ivt.shtml

http://www.danceage.com/biography/sdmc_Monotonic

http://www.econ.umn.edu/~mclennan/Classes/Ec5113/
ec5113-lec05-1.16.99.pdf

http://www.fastload.org/mo/Monotonic.html

http://www.geocities.com/Athens/Delphi/5136/Continuity/
continuity.html

http://www.karlscalculus.org/ivtproof.html

http://www.math.ksu.edu/~mkb9154/chapter3/ivt.html

http://www.math.ku.dk/~moller/e03/3gt/3gt.html

http://www.math.louisville.edu/~lee/RealAnalysis/realanalysis.html

http://www.math.mcgill.ca/drury/rootm.pdf

http://www.math.sc.edu/~sharpley/math554/Lectures/math554_
Lectures.html

http://www.math.ucdavis.edu/~emsilvia/math127/chapter5.pdf

http://www.math.ucsb.edu/~gizem/teaching/S117/S117.html

http://www.math.unl.edu/~webnotes/classes/class28/class28.htm

http://www.math.unl.edu/~webnotes/contents/chapters.htm

http://www.math.uu.se/~oleg/topoman.ps

http://www.mathreference.com/top-ms,ivt.html

http://www.mathreference.com/top-ms,unif.html

http://www.maths.abdn.ac.uk/~igc/tch/ma2001/notes/node38.html

http://www.maths.mq.edu.au/~wchen/lnlfafolder/lfa02-ccc.pdf

http://www.maths.nott.ac.uk/personal/jff/G12RAN/pdf/Uniform.pdf

http://www.maths.ox.ac.uk/current-students/undergraduates/handbooks-
synopses/2001/html/mods-01/node10.html

http://www.maths.ox.ac.uk/current-students/undergraduates/lecture-
material/Mods/analysis2/pdf/analysis2-notes.pdf

http://www.maths.qmul.ac.uk/~reza/MAS101/MV-WEB.pdf

http://www.maths.tcd.ie/pub/coursework/424/GpReps-II.pdf

http://www.nuprl.org/documents/real-analysis/node6.html

http://www.people.vcu.edu/~mikuleck/courses/limits/tsld028.htm

http://www.recipeland.com/facts/Monotonic

http://www.sccs.swarthmore.edu/users/02/rebecca/pdf/Math47.pdf

http://www.sosmath.com/calculus/limcon/limcon06/limcon06.html

http://www.termsdefined.net/mo/monotone-decreasing.html

http://www.thebestlinks.com/Connected_space.html

http://zeus.uwindsor.ca/math/traynor/analysis/analbook.pdf

www.isid.ac.in/~arup/courses/topology.ps

12

Differentiation

In physical terms, differentiation expresses the rate at which a quantity, y, changes with respect to the change in another quantity, x, on which it has a functional relationship. This small chapter will start with the discussion of the derivative, which is one of the two central concepts of calculus (the other is the integral). We will discuss the Mean Value Theorem and look at some applications that include the relationship of the derivative of a function with whether the function is increasing or decreasing. We will expose Taylor's theorem as a generalization of the Mean Value Theorem. In calculus, Taylor's theorem gives the approximation of a differentiable function near a point by a polynomial whose coefficients depend only on the derivatives of the function at that point. There are many OR applications of Taylor's approximation, especially in linear and non-linear optimization.

12.1 Derivatives

Definition 12.1.1 *Let $f : [a,b] \mapsto \mathbb{R}$. $\forall x \in [a,b]$, let $\phi(t) = \frac{f(t)-f(x)}{t-x}$, $a < t < b$, $t \neq x$. $f'(x) = \lim_{t \to x} \phi(t)$ provided that the limit exists. f' is called the derivative of f. If f' is defined at x, we say f is differentiable at x. If f' is defined at $\forall x \in E \subset [a,b]$, we say f is differentiable on E. Moreover, left-hand (right-hand) limits give rise to the definition of left-hand (right-hand) derivatives.*

Remark 12.1.2 *If f is defined on (a,b) and if $a < x < b$, then f' can be defined as above. However, $f'(a)$ and $f'(b)$ are not defined in general.*

Theorem 12.1.3 *Let f be defined on $[a,b]$, f is differentiable at $x \in [a,b]$ then f is continuous at x.*

Proof. As $t \to x$, $f(t) - f(x) = \frac{f(t)-f(x)}{t-x}(t-x) \to f'(x) \cdot 0 = 0$. $\quad\square$

Remark 12.1.4 *The converse is not true. One can construct continuous functions which fail to be differentiable at isolated points.*

Let us state some properties: Suppose f and g are defined on $[a, b]$ and are differentiable at $x \in [a, b]$. Then, $f + g$, $f \cdot g$ and f/g are differentiable at x, and

(a) $(f + g)'(x) = f'(x) + g'(x)$.
(b) $(f \cdot g)'(x) = f'(x)g(x) + f(x)g'(x)$.
(c) $(f/g)'(x) = \frac{f'(x)g(x) - f(x)g'(x)}{g^2(x)}$, $g(x) \neq 0$.
(d) *Chain Rule:* If $h(t) = (g \circ f)(t) = g(f(t))$, $a \leq t \leq b$, and if f is continuous at $[a, b]$, f' exists at $x \in [a, b]$, g is defined over range of f and g is differentiable at $f(x)$. Then, h is differentiable at x and

$$h'(x) = g'(f(x))f'(x).$$

Example 12.1.5 (Property (c)) *The derivative of a constant is zero. If $f(x) = x$ then $f'(x) = 1$. If $f(x) = x \cdot x = x^2$ then $f'(x) = x + x = 2x$ by property (b). In general, if $f(x) = x^n$ then $f'(x) = nx^{n-1}$, $n \in \mathbb{N}$. If $f(x) = \frac{1}{x} = x^{-1}$ then $f'(x) = \frac{0-1}{x^2} = -x^{-2}$. In this case, $x \neq 0$. if $f(x) = x^{-n}$, $n \in \mathbb{N}$ then $f'(x) = -nx^{-(n+1)}$. Thus, every polynomial is differentiable, and every rational function is differentiable except at the points where denominator is zero.*

Example 12.1.6 (Property (d)) *Let*

$$f(x) \begin{cases} x \sin \frac{1}{x}, & x \neq 0 \\ 0, & x = 0 \end{cases}$$

Then, $f'(x) = \sin \frac{1}{x} - \frac{1}{x} \cos \frac{1}{x}$, $x \neq 0$. At $x = 0$, $\frac{1}{x}$ is not defined $\frac{f(t) - f(0)}{t - 0} = \sin \frac{1}{t}$. As $t \to 0$, the limit does not exist, thus $f'(0)$ does not exist.

12.2 Mean Value Theorems

Definition 12.2.1 *Let $f : [a, b] \mapsto \mathbb{R}$. We say f has a local maximum at $p \in X$ if $\exists \delta > 0 \ni f(q) \leq f(p)$, $\forall q \in X$ with $d(p, q) < \delta$. Local minimum is defined similarly.*

Theorem 12.2.2 *Let $f : [a, b] \mapsto \mathbb{R}$. If f has a local maximum (minimum) at $x \in (a, b)$ and if $f'(x)$ exists, then $f'(x) = 0$.*

Proof. We will prove the maximum case:
Choose δ as in the definition: $a < x - \delta < x < x + \delta < b$.
If $x - \delta < t < x$, then $\frac{f(t) - f(x)}{t - x} \geq 0$. Let $t \to x \Rightarrow f'(x) \geq 0$.
If $x < t < x + \delta$, then $\frac{f(t) - f(x)}{t - x} \leq 0$. Let $t \to \infty \Rightarrow f'(x) \leq 0$.
Thus, $f'(x) = 0$. $\square$

Theorem 12.2.3 *Suppose* $f : [a,b] \mapsto \mathbb{R}$ *is differentiable and* $f'(a) < \lambda < f'(b)$ $[f'(a) > \lambda > f'(b)]$. *Then,* $\exists x \in (a,b) \ni f'(x) = \lambda$.

Proof. Let $g(t) = f(t) - \lambda t$. Then, $g'(a) < 0$ $[g'(a) > 0]$ so that $g(t_1) < g(a)$ $[g(t_1) > g(a)]$ for some $t_1 \in (a,b)$, so that $g(t_2) < g(b)$ $[g(t_2) > g(a)]$ for some $t_2 \in (a,b)$. Hence, g attains its minimum [maximum] on $[a,b]$ at some points $x \in (a,b)$. By the first mean value theorem, $g'(x) = 0$. Hence, $f'(x) = \lambda$. $\square$

Corollary 12.2.4 *If* f *is differentiable on* $[a,b]$, *then* f' *cannot have any simple discontinuities on* $[a,b]$.

Remark 12.2.5 *But* f' *may have discontinuities of the second kind.*

Theorem 12.2.6 (L'Hospital's Rule) *Suppose* f *and* g *are real and differentiable in* (a,b) *and* $g'(x) \neq 0$, $\forall x \in (a,b)$ *where* $\infty \leq a < b \leq +\infty$. *Suppose*

$$\frac{f'(x)}{g'(x)} \to A \text{ as } x \to a \ (\diamond).$$

If $f(x) \to 0$ *and* $g(x) \to 0$ *as* $x \to a$ *or if* $f(x) \to +\infty$ *and* $g(x) \to +\infty$ *as* $x \to a$, *then*

$$\frac{f(x)}{g(x)} \to A \text{ as } x \to a.$$

Proof. Let us consider the case $-\infty \leq A < +\infty$: Choose $q \in \mathbb{R} \ni A < q$, and choose $r \ni A < r < q$. By $(\diamond)$,

$$\exists c \in (a,b) \ni a < x < c \Rightarrow \frac{f'(x)}{g'(x)} < r. \ (\spadesuit)$$

If $a < x < y < c$, then by the second mean value theorem,

$$\exists y \in (x,y) \ni \frac{f(x) - f(y)}{g(x) - g(y)} = \frac{f'(t)}{g'(y)} < r. \ (\clubsuit)$$

Suppose $f(x) \to 0$ and $g(x) \to 0$ as $x \to a$. Then, $(\spadesuit)$ $\frac{f(y)}{g(y)} \leq r < q, a < y < c$. Suppose $g(x) \to +\infty$ as $x \to a$. Keeping y fixed, we can choose $c_1 \in (a,y) \ni g(x) > g(y)$ and $g(x) > 0$ if $a < x < c_1$. Multiplying $(\spadesuit)$ by $[g(x) - g(y)]/g(x)$, we have $\frac{f(x)}{g(x)} < r - r\frac{g(x)}{g(y)} + \frac{f(y)}{g(x)}$, $a < x < c_1$. If $x \to a$ $\exists c_2 \in (a,c_1) \ni \frac{f(x)}{g(x)} < q$, $a < x < c_2$. Summing with $(\clubsuit)$ $\forall q \ni A < q$ yields

$$\exists c_2 \ni \frac{f(x)}{g(x)} < q \text{ if } a < x < c_2.$$

Similarly, if $-\infty < A \leq +\infty$ and $p \ni p < A$, $\exists c_3 \ni p < \frac{f(x)}{g(x)}$, $a < x < c_3$. $\square$

12.3 Higher Order Derivatives

Definition 12.3.1 *If f has a derivative f' on an interval and if f' is itself differentiable, we denote derivative of f' as f'', and call the second derivative of f. Higher order derivatives are denoted by $f', f'', f^{(3)}, \ldots, f^{(n)}$, each of which is the derivative of the previous one.*

Theorem 12.3.2 (Taylor's Theorem) *Let $f : [a,b] \mapsto \mathbb{R}$, $n \in \mathbb{N}$, $f^{(n-1)}$ be continuous on $[a,b]$, and $f^{(n)}(t)$ exists $\forall t \in [a,b]$. Let $\alpha \neq \beta \in [a,b]$ and define*

$$p(t) = \sum_{k=0}^{n-1} \frac{f^{(k)}(\alpha)}{k!}(t - \alpha)^k.$$

Then, $\exists x \in (\alpha, \beta) \ni f(\beta) = p(\beta) + \frac{f^{(n)}(\alpha)}{n!}(\beta - \alpha)^n$.

Remark 12.3.3 *For $n = 1$, the above theorem is just the mean value theorem.*

Proof. Let $M \ni f(\beta) = p(\beta) + M(\beta - \alpha)^n$.
Let $g(t) = f(t) - p(t) - M(t-\alpha)^n$, $a \leq t \leq b$, the error function. We will show that $n! \, M = f^{(n)}(x)$ for some $x \in (a,b)$. We have $g^{(n)}(t) = f^{(n)}(t) - n! \, M$, $a < t < b$. If $\exists x \in (a,b) \ni g^{(n)}(x) = 0$, we are done.

$$p^{(k)}(\alpha) = f^{(k)}(\alpha), \ k = 0, \ldots, n-1 \Rightarrow$$

$$g(\alpha) = g'(\alpha) = g''(\alpha) = \cdots = g^{(n-1)}(\alpha) = 0.$$

Our choice of M yields $g(\beta) = 0$, thus $g'(x_1) = 0$ for some $x_1 \in (\alpha, \beta)$ by the Mean Value Theorem. This is for $g''(\cdot)$, one may continue in this manner. Thus, $g^{(n)}(x_n) = 0$, for some $x_n \in (\alpha, x_{n-1}) \subset (\alpha, \beta)$. □

Definition 12.3.4 *A function is said to be of class C^r if the first r derivatives exist and continuous. A function is said to be smooth or of class C^∞ if it is of class C^r, $\forall r \in \mathbb{N}$.*

Theorem 12.3.5 (Taylor's Theorem) *Let $f : A \mapsto \mathbb{R}$, be of class C^r for $A \subset \mathbb{R}^n$, an open set. Let $x, y \in A$ and suppose that the segment joining x and y lies in A. Then, $\exists c$ in that segment $\ni$*

$$f(y) - f(x) = \sum_{k=1}^{r-1} \frac{1}{k!} f^{(k)}(y - x, \ldots, y - x) + \frac{1}{r!}(c) f^{(r)}(y - x, \ldots, y - x)$$

where $f^{(k)}(y - x, \ldots, y - x) = \sum_{i_1, \ldots, i_k} \left[\left(\frac{\partial^k f}{\partial_{i_1} \cdots \partial_{i_k}} \right)(y_{i_1} - x_{i_1}) \cdots (y_{i_n} - x_{i_n}) \right]$. Setting $y = x + h$, we can write Taylor's formula as

$$f(x + h) = f(x) + f'(x) \cdot h + \cdots + \frac{1}{(r-1)!} f^{(r-1)}(x) \cdot (h, \ldots, h) + R_{r-1}(x, h),$$

where $R_{r-1}(x, h)$ is the remainder. Furthermore,

$$\frac{R_{r-1}(x, h)}{\|h\|^{r-1}} \to 0 \text{ as } h \to 0.$$

Problems

12.1. Suppose $f : [0, \infty) \mapsto \mathbb{R}$ is continuous, $f(0) = 0$, f is differentiable on $(0, \infty)$ and f' is nondecreasing. Prove that $g(x) = \frac{f(x)}{x}$ is nondecreasing for $x > 0$.

12.2. Let $A \subset \mathbb{R}^n$ be an open convex set and $f : A \mapsto \mathbb{R}^m$ be differentiable. If $f'(t) = 0$, $\forall t$ then show that f is constant.

12.3. Compute the second order Taylor's formula for $f(x, y) = \sin(x + 2y)$ around the origin.

12.4. Let $f \in C^2$ and $x^* \in \mathbb{R}^n$ be local minimizer.
a) Prove the first order necessary condition (x^* is a local minimizer then $\nabla f(x^*) = \theta$) using Taylor's approximation.
b) Prove the second order necessary condition (x^* is a local minimizer then $\nabla^2 f(x^*)$ is positive semi-definite) using Taylor's approximation.
c) Design an iterative procedure to find $\nabla f(x) = \theta$ in such a way that it starts from an initial point and updates as $x_k = x_{k-1} + p_k$. The problem at each iteration is to find a direction p_k that makes $\nabla f(x_{k-1})$ closer to the null vector. Use the second order Taylor's approximation to find the best p_k at any iteration.
d) Use the above results to find a local solution to

$$\min f(x_1, x_2) = x_1^4 + 2x_1^3 + 24x_1^2 + x_2^4 + 12x_2^2.$$

Start from $[1, 1]^T$.

Web material

http://archives.math.utk.edu/visual.calculus/3/index.html
http://calclab.math.tamu.edu/~belmonte/m151/L/c5/L53.pdf
http://ccrma-www.stanford.edu/~jos/mdft/Formal_Statement_Taylor_s_
    Theorem.html
http://courses.math.nus.edu.sg/ma1104/lecture_notes/Notes_1.pdf
http://d.faculty.umkc.edu/delawarer/RDvsiCalcList.htm
http://en.wikipedia.org/wiki/Derivative
http://en.wikipedia.org/wiki/L'Hopital's_rule

```
http://en.wikipedia.org/wiki/Mean_value_theorem
http://en.wikipedia.org/wiki/Taylor's_theorem
http://grus.berkeley.edu/~jrg/ay202/node191.html
http://hilltop.bradley.edu/~jhahn/Note3.pdf
http://home.uchicago.edu/~lfmedina/MathRev3.pdf
http://kr.cs.ait.ac.th/~radok/math/mat11/chap7.htm
http://kr.cs.ait.ac.th/~radok/math/mat6/calc2.htm
http://mathworld.wolfram.com/Derivative.html
http://mathworld.wolfram.com/LHospitalsRule.html
http://mathworld.wolfram.com/Mean-ValueTheorem.html
http://ocw.mit.edu/ans7870/textbooks/Strang/strangtext.htm
http://ocw.mit.edu/OcwWeb/Mathematics/18-100BAnalysis-IFall2002/
    LectureNotes/
http://people.hofstra.edu/faculty/stefan_waner/RealWorld/
    math19index.html
http://pirate.shu.edu/projects/reals/cont/derivat.html
http://saxonhomeschool.harcourtachieve.com/en-US/Products/
    sh_calculustoc.htm
http://web.mit.edu/wwmath/calculus/differentiation/
http://www-math.mit.edu/~djk/18_01/chapter26/section01.html
http://www.absoluteastronomy.com/l/lh%C3%B4pitals_rule1
http://www.analyzemath.com/calculus.html
http://www.jtaylor1142001.net/
http://www.ma.utexas.edu/cgi-pub/kawasaki/plain/derivatives/1.html
http://www.math.dartmouth.edu/~m3cod/textbooksections.htm
http://www.math.harvard.edu/computing/math/tutorial/taylor.html
http://www.math.hmc.edu/calculus/tutorials/
http://www.math.scar.utoronto.ca/calculus/Redbook/goldch7.pdf
http://www.math.tamu.edu/~fulling/coalweb/lhop.htm
http://www.math.tamu.edu/~fulling/coalweb/taylor.htm
http://www.math.tamu.edu/~tom.vogel/gallery/node12.html
http://www.math.uconn.edu/~corluy/calculus/lecturenotes/node15.html
http://www.mathdaily.com/lessons/Category:Calculus
http://www.mathreference.com/ca,tfn.html
http://www.maths.abdn.ac.uk/~igc/tch/eg1006/notes/node136.html
http://www.maths.abdn.ac.uk/~igc/tch/ma1002/appl/node54.html
http://www.maths.abdn.ac.uk/~igc/tch/ma1002/diff/node39.html
http://www.maths.abdn.ac.uk/~igc/tch/ma2001/notes/node46.html
http://www.maths.lse.ac.uk/Courses/MA203/sec4a.pdf
http://www.maths.manchester.ac.uk/~mdc/old/211/notes4.pdf
http://www.mathwords.com/index_calculus.htm
http://www.npac.syr.edu/REU/reu94/williams/ch3/chap3.html
http://www.physics.nau.edu/~hart/matlab/node52.html
http://www.sosmath.com/calculus/diff/der11/der11.html
http://www.sosmath.com/tables/derivative/derivative.html
http://www.toshare.info/en/Mean_value_theorem.htm
http://www.univie.ac.at/future.media/moe/galerie/diff1/diff1.html
http://www.wellington.org/nandor/Calculus/notes/notes.html
http://www.wikipedia.org/wiki/Mean_value_theorem
```

Power Series and Special Functions

In mathematics, power series are devices that make it possible to employ much of the analytical machinery in settings that do not have natural notions of "convergence". They are also useful, especially in combinatorics, for providing compact representations of sequences and for finding closed formulas for recursively defined sequences, known as the method of generating functions. We will discuss first the notion of series, succeeded by operations on series and tests for convergence/divergence. After power series is formally defined, we will generate exponential, logarithmic and trigonometric functions in this chapter. Fourier series, gamma and beta functions will be discussed as well.

13.1 Series

13.1.1 Notion of Series

Definition 13.1.1 *An expression*

$$\sum_{k=0}^{\infty} u_k = \sum_{0}^{\infty} u_k = u_0 + u_1 + u_2 + \cdots$$

where the numbers u_k (terms of the series) depend on the index $k = 0, 1, 2, \ldots$ is called a (number) series. The number

$$S_n = u_0 + u_1 + \cdots + u_n, \; n = 0, 1, \ldots$$

is called the nth partial sum of the above series.

We say that the series is convergent if the limit, $\lim_{n \to \infty} S_n = S$, exists. In this case, we write

$$S = u_0 + u_1 + u_2 + \cdots = \sum_{k=0}^{\infty} u_k$$

and call S the sum of the series; we also say that the series converges to S.

Proposition 13.1.2 (Cauchy's criterion) *The series*

$$\sum_{k=0}^{\infty} u_k$$

is convergent if and only if

$$\forall \epsilon > 0,\ \exists N \ni \forall n, p \in \mathbb{N}, n > N,\quad |u_{n+1} + \cdots + u_{n+p}| = |S_{n+p} - S_n| < \epsilon.$$

Remark 13.1.3 *In particular, putting $p = 1$ we see that if $\sum_{k=0}^{\infty} u_k$ is convergent its general term u_k tends to zero. This condition is necessary but not sufficient!*

Definition 13.1.4 *The series are called the remainder series of the series $\sum_{k=0}^{\infty} u_k$:*

$$u_{n+1} + u_{n+2} + \cdots = \sum_{k=1}^{\infty} u_{n+k}.$$

Since the conditions of Cauchy's criterion are the same for the series and its remainder series, they are simultaneously convergent or divergent. If they are convergent, the remainder series is

$$\lim_{n \to \infty} \sum_{k=1}^{m} u_{n+k} = \lim_{n \to \infty} (S_{n+m} - S_n) = S - S_n.$$

If the series are real and nonnegative, its partial sums form a nondecreasing sequence $S_1 \leq S_2 \leq S_3 \leq \cdots$ and if this sequence is bounded (i.e. $S_n \leq M$, $n = 1, 2, \ldots$), then the series is convergent and its sum satisfies the inequality

$$\lim_{n \to \infty} S_n = S \leq M.$$

If this sequence is unbounded the series is divergent $\lim_{n \to \infty} S_n = \infty$. In this case, we write $\sum_{k=0}^{\infty} u_k = \infty$ and say that the series with nonnegative terms is divergent to ∞ or properly divergent.

Example 13.1.5 *The nth partial sum of the series $1 + z + z^2 + \cdots$ is*

$$S_n(z) = \frac{1 - z^{n+1}}{1 - z} \quad \text{for } z \neq 1.$$

If $|z| < 1$ then $\left|z^{n+1}\right| = |z|^{n+1} \to 0$, that is $z^{n+1} \to 0$ as $n \to \infty$.
If $|z| > 1$ then $\left|z^{n+1}\right| \to \infty$.
Finally, if $|z| = 1$ then $z^{n+1} = \cos(n+1)\theta + i\sin(n+1)\theta$, where θ is the argument of z, and we see that the variable z^{n+1} has no limit as $n \to \infty$ because its real or imaginary part (or both) has no limit as $n \to \infty$. For $z = 1$, the divergence of the series is quite obvious.

We see that the series is convergent and has a sum equal to $(1 - z)^{-1}$ in the open circle $|z| < 1$ of the complex plane and is divergent all other points z.

13.1.2 Operations on Series

Proposition 13.1.6 *If $\sum_{k=0}^{\infty} u_k$ and $\sum_{k=0}^{\infty} v_k$ are convergent series and $\alpha \in \mathbb{C}$, then the series $\sum_{k=0}^{\infty} \alpha u_k$ and $\sum_{k=0}^{\infty} (u_k \pm v_k)$ are also convergent and we have*

$$\sum_{k=0}^{\infty} \alpha u_k = \alpha \sum_{k=0}^{\infty} u_k \text{ and } \sum_{k=0}^{\infty} (u_k \pm v_k) = \sum_{k=0}^{\infty} u_k \pm \sum_{k=0}^{\infty} v_k.$$

Proof. Indeed,
$\sum_0^{\infty} \alpha u_k = \lim_{n \to \infty} \sum_0^n \alpha u_k = \alpha \lim_{n \to \infty} \sum_0^n u_k = \alpha \sum_0^{\infty} u_k$, and
$\sum_0^{\infty} (u_k \pm v_k) = \lim_{n \to \infty} \sum_0^n (u_k \pm v_k) = \lim_{n \to \infty} \sum_0^n u_k \pm \lim_{n \to \infty} \sum_0^n v_k = \sum_0^{\infty} u_k \pm \sum_0^{\infty} v_k.$ $\square$

Remark 13.1.7 *It should be stressed that, generally speaking, the convergence of $\sum_0^{\infty} u_k \pm \sum_0^{\infty} v_k$ does not imply the convergence of each of the series $\sum_{k=0}^{\infty} u_k$ and $\sum_{k=0}^{\infty} v_k$, which can be confirmed by the example below:*

$$(\alpha - \alpha) + (\alpha - \alpha) + \cdots, \ \forall \alpha \in \mathbb{C}.$$

13.1.3 Tests for positive series

Theorem 13.1.8 (Comparison Tests) *Let there be given two series*

$$(i) \sum_0^{\infty} u_k \text{ and } (ii) \sum_0^{\infty} v_k$$

with nonnegative terms.

(a) *If $u_k \leq v_k$, $\forall k$, the convergence of series (ii) implies the convergence of series (i) and the divergence of series (i) implies the divergence of series (ii).*
(b) *If $\lim_{k \to \infty} \frac{u_k}{v_k} = A > 0$, then series (i) and (ii) are simultaneously convergent and divergent.*

Proof. Exercise! $\square$

Theorem 13.1.9 (D'Alembert's Test) *Let there be a positive series*

$$\sum_0^{\infty} u_k \ni u_k > 0, \ \forall k = 0, 1, \dots$$

(a) *If $\frac{u_{k+1}}{u_k} \leq q < 1$, $\forall k$, then the series $\sum_0^{\infty} u_k$ is convergent. If $\frac{u_{k+1}}{u_k} \geq 1$, then the series $\sum_0^{\infty} u_k$ is divergent.*
(b) *If $\lim_{k \to \infty} \frac{u_{k+1}}{u_k} = q$ then the series $\sum_0^{\infty} u_k$ is convergent for $q < 1$ and divergent for $q > 1$.*

Proof. We treat the cases individually.

(a) We have

$$u_n = u_0 \frac{u_1}{u_0} \frac{u_2}{u_1} \cdots \frac{u_n}{u_{n-1}}, \; \forall n = 0, 1, 2, \ldots$$

and therefore

$$\frac{u_{k+1}}{u_k} \leq q < 1 \Rightarrow u_n \leq u_0 q^n, \; q < 1, \forall n.$$

Since the series $\sum_1^\infty u_0 q^n$ is convergent, the series $\sum_0^\infty u_k$ is convergent.

$$\frac{u_{k+1}}{u_k} \geq 1 \Rightarrow u_n \geq u_0, \forall n.$$

Since the series $u_0 + u_0 + \cdots$ is divergent, so is $\sum_0^\infty u_k$.

(b) $\lim_{k \to \infty} \frac{u_{k+1}}{u_k} = q < 1 \Rightarrow \forall \epsilon > 0 \ni q + \epsilon < 1$; we have $\frac{u_{k+1}}{u_k} < q + \epsilon < 1$, $k \geq N$, where N is sufficiently large. Then, the series $\sum_N^\infty u_k$ is convergent and hence so is $\sum_0^\infty u_k$. On the other hand,

$$\lim_{k \to \infty} \frac{u_{k+1}}{u_k} = q > 1 \Rightarrow \frac{u_{k+1}}{u_k} > 1, \; \forall k \geq N$$

for sufficiently large N, and therefore $\sum_0^\infty u_k$ is divergent.   □

Theorem 13.1.10 (Cauchy's Test) *Let $\sum_0^\infty u_k$ be a series with positive terms,*

(a)

$$(u_k)^{\frac{1}{k}} < q < 1, \forall k \Rightarrow \text{ the series } \sum_0^\infty u_k \text{ is convergent.}$$

$$(u_k)^{\frac{1}{k}} \geq 1, \forall k \Rightarrow \text{ the series } \sum_0^\infty u_k \text{ is divergent.}$$

(b) If $\lim_{k \to \infty} (u_k)^{\frac{1}{k}} = q$, then the series $\sum_0^\infty u_k$ is convergent for $q < 1$ and divergent for $q > 1$.

Remark 13.1.11 *Let a series be convergent to a sum S. Then, the series obtained from this series by rearranging and renumbering its terms in an arbitrary way is also convergent and has the same sum S.*

13.2 Sequence of Functions

Definition 13.2.1 *A sequence of functions $\langle f_n \rangle$, $n = 1, 2, 3, \ldots$ converges uniformly on E to a function f if*

$$\forall \epsilon > 0, \; \exists N \in \mathbb{N} \ni n \geq N \Rightarrow |f_n(x) - f(x)| \leq \epsilon, \; \forall x \in E.$$

Similarly, we say that the series $\sum f_n(x)$ converges uniformly on E if the sequence $\langle S_n \rangle$ of partial sums converges uniformly on E.

Remark 13.2.2 *Every uniformly convergent sequence is pointwise convergent. If $\langle f_n \rangle$ converges pointwise on E, then there exist a function f such that, for every $\epsilon > 0$ and for every $x \in E$, there is an integer N, depending on ϵ and x, such that $|f_n(x) - f(x)| \le \epsilon$ holds if $n \ge N$; if $\langle f_n \rangle$ converges uniformly on E, it is possible, for each $\epsilon > 0$, to find one integer N which will do for all $x \in E$.*

Proposition 13.2.3 (Cauchy's uniform convergence) *A sequence of functions, $\langle f_n \rangle$, defined on E, converges uniformly on E if and only if*

$$\forall \epsilon > 0,\ \exists N \in \mathbb{N} \ni m \ge N, n \ge N, x \in E \Rightarrow |f_m(x) - f_n(x)| \le \epsilon.$$

Corollary 13.2.4 *Suppose $\lim_{n \to \infty} f_n(x) = f(x)$, $x \in E$. Put*

$$M_n = \sup_{x \in E} |f_n(x) - f(x)|.$$

Then, $f_n \to f$ uniformly on E if and only if $M_n \to 0$ as $n \to \infty$.

Proposition 13.2.5 (Weierstrass) *Suppose $\langle f_n \rangle$ is a sequence of functions defined on E, and $|f(x)| \le M_n$, $x \in E$, $n = 1, 2, 3, \ldots$ Then, $\sum f_n$ converges uniformly on E if $\sum M_n$ converges.*

Proposition 13.2.6

$$\lim_{t \to x} \lim_{n \to \infty} f_n(t) = \lim_{n \to \infty} \lim_{t \to x} f_n(t).$$

Remark 13.2.7 *The above assertion means the following: Suppose $f_n \to f$ uniformly on a set E in a metric space. Let x be a limit point of E, and suppose that $\lim_{t \to x} f_n(t) \to A_n$, $n = 1, 2, 3 \ldots$ Then, $\langle A_n \rangle$ converges, and $\lim_{t \to x} f(t) = \lim_{n \to \infty} A_n$.*

Corollary 13.2.8 *If $\langle f_n \rangle$ is a sequence of continuous functions on E, and if $f_n \to f$ uniformly on E, then f is continuous on E.*

Remark 13.2.9 *The converse is not true. A sequence of continuous functions may converge to a continuous function, although the convergence is not uniform.*

13.3 Power Series

Definition 13.3.1 *The functions of the form*

$$f(x) = \sum_{n=0}^{\infty} c_n x^n$$

or more generally,

$$f(x) = \sum_{n=0}^{\infty} c_n (x - a)^n$$

are called analytic functions.

Theorem 13.3.2 *Suppose the series $\sum_{n=0}^{\infty} c_n x^n$ converges for $|x| < R$, and define*

$$f(x) = \sum_{n=0}^{\infty} c_n x^n, \quad |x| < R$$

which converges uniformly on $[-R+\epsilon, R-\epsilon]$, no matter which $\epsilon > 0$ is chosen. The function f is continuous and differentiable in $(-R, R)$, and

$$f'(x) = \sum_{n=1}^{\infty} n c_n (x - a)^{n-1}, \quad |x| < R$$

Corollary 13.3.3 *f has derivatives of all orders in $(-R, R)$, which are given by*

$$f^{(k)}(x) = \sum_{n=k}^{\infty} n(n-1) \cdots (n-k+1) c_n (x-a)^{n-k}.$$

In particular,

$$f^{(k)}(0) = k! c_k, \quad k = 0, 1, 2, \dots$$

Remark 13.3.4 *The above formula is very interesting. On one hand, it shows how we can determine the coefficients of the power series representation of f. On the other hand, if the coefficients are given, the values of derivatives of f at the center of the interval $(-R, R)$ can be read off immediately.*

A function f may have derivatives of all order, but the power series need not to converge to $f(x)$ for any $x \neq 0$. In this case, f cannot be expressed as a power series about the origin.

Theorem 13.3.5 (Taylor's) *Suppose, $f(x) = \sum_{n=0}^{\infty} c_n x^n$, the series converging in $|x| < R$. If $-R < a < R$, then f can be expanded in a power series about the point $x = a$ which converges in $|x - a| < R - |a|$, and*

$$f(x) = \sum_{n=0}^{\infty} \frac{f^{(n)}(a)}{n!} (x - a)^n.$$

Remark 13.3.6 *If two power series converge to the same function in $(-R, R)$, then the two series must be identical.*

13.4 Exponential and Logarithmic Functions

We can define

$$E(z) = \sum_{n=0}^{\infty} \frac{z^n}{n!}, \quad \forall z \in \mathbb{C}.$$

It is one of the exercise questions to show that this series is convergent $\forall z \in \mathbb{C}$. If we have an absolutely convergent (if $|u_0| + |u_1| + \cdots$ is convergent) series, we can multiply the series element by element. We can safely do it for $E(z)$:

$$E(z)E(w) = \sum_{n=0}^{\infty} \frac{z^n}{n!} \sum_{m=0}^{\infty} \frac{w^m}{m!} = \sum_{n=0}^{\infty} \sum_{k=0}^{n} \frac{z^k w^{n-k}}{k!(n-k!)}$$

$$= \sum_{n=0}^{\infty} \frac{1}{n!} \sum_{k=0}^{n} \binom{n}{k} z^k w^{n-k} = \sum_{n=0}^{\infty} \frac{(z+w)^n}{n!} = E(z+w).$$

This yields

- $E(z)E(-z) = E(z-z) = E(0) = 1, \forall z \in \mathbb{C}.$
- $E(z) \neq 0, \forall z \in \mathbb{C}.$ $E(x) > 0, \forall x \in \mathbb{R}.$
  $E(x) \to +\infty$ as $x \to +\infty.$
  $0 < x < y \Rightarrow E(x) < E(y), E(-y) < E(-x).$
  Hence, $E(x)$ is strictly increasing on the real axis.
- $\lim_{h \to 0} \frac{E(z+h)-E(z)}{h} = E(z).$
- $E(z_1 + \cdots + z_n) = E(z_1) \cdots E(z_n).$ Let us take $z_1 = \cdots = z_n = 1.$ Since $E(1) = e$, we obtain $E(n) = e^n$, $n = 1, 2, 3, \ldots$ Furthermore, if $p = n|m$, where $n, m \in \mathbb{N}$, then $[E(p)]^m = E(mp) = E(n) = e^n$ so that $E(p) = e^p$, $p \in \mathbb{Q}_+.$ Since $E(-p) = e^{-p}$, $p \in \mathbb{Q}_+$, the above equality holds for all rational p.
- Since $x^y = \sup_{p \in \mathbb{Q} \ni p < y} x^p$, $\forall x, y \in \mathbb{R}$, $x > 1$, we define $e^x = \sup_{p \in \mathbb{Q} \ni p < x} e^p.$ The continuity and monotonicity properties of E show that

$$E(x) = e^x = \exp(x).$$

Thus, as a summary, we have the following proposition:

Proposition 13.4.1 *The following are true:*

(a) e^x *is continuous and differentiable for all* x,
(b) $(e^x)' = e^x$,
(c) e^x *is a strictly increasing function of* x, *and* $e^x > 0$,
(d) $e^{x+y} = e^x e^y$,
(e) $e^x \to +\infty$ *as* $x \to +\infty$, $e^x \to 0$ *as* $x \to -\infty$,
(f) $\lim_{x \to +\infty} x^n e^{-x} = 0$, $\forall n$.

Proof. We have already proved (a) to (e). Since $e^x > \frac{x^{n+1}}{(n+1)!}$, for $x > 0$, then $x^n e^{-x} < \frac{(n+1)!}{x}$ and (f) follows. □

Since E is strictly increasing and differentiable on $\mathbb{R}$, it has an inverse function L which is also strictly increasing and differentiable whose domain is $E(\mathbb{R}) = \mathbb{R}_+.$

$$E(L(y)) = y, \; y > 0 \Leftrightarrow L(E(x)) = x, \; x \in \mathbb{R}.$$

Differentiation yields

$$L'(E(x)) \cdot E(x) = 1 = L'(y) \cdot y \Leftrightarrow L'(y) = \frac{1}{y}, \; y > 0.$$

$x = 0 \Rightarrow L(1) = 0$. Thus, we have

$$L(y) = \int_1^y \frac{dx}{x} = \log y.$$

Let $u = E(x)$, $v = E(y)$;

$$L(uv) = L(E(x)E(y)) = L(E(x+y)) = x + y = L(u) + L(v).$$

We also have $\log x \to +\infty$ as $x \to +\infty$ and $\log x \to -\infty$ as $x \to 0$. Moreover,

$$x^n = E(nL(x)), \ x \in \mathbb{R}_+; n, m \in \mathbb{N}, \quad x^{\frac{1}{m}} = E\left(\frac{1}{m}L(x)\right)$$

$$x^\alpha = E(\alpha L(x)) = e^{\alpha \log x}, \ \forall \alpha \in \mathbb{Q}.$$

One can define x^α, for any real α and any $x > 0$ by using continuity and monotonicity of E and L.

$$(x^\alpha)' = E(\alpha L(x))\frac{\alpha}{x} = \alpha x^{\alpha - 1}$$

One more property of $\log x$ is

$$\lim_{x \to +\infty} x^{-\alpha} \log x = 0, \ \forall \alpha > 0.$$

13.5 Trigonometric Functions

Let us define

$$C(x) = \frac{1}{2}[E(ix) + E(-ix)], \ S(x) = \frac{1}{2i}[E(ix) - E(-ix)].$$

By the definition of $E(z)$, we know $E(\bar{z}) = \overline{E(z)}$. Then, $C(x)$, $S(x) \in \mathbb{R}$, $x \in \mathbb{R}$. Furthermore,

$$E(ix) = C(x) + iS(x).$$

Thus, $C(x)$, $S(x)$ are real and imaginary parts of $E(ix)$ if $x \in \mathbb{R}$. We have also

$$|E(ix)|^2 = E(ix)\overline{E(ix)} = E(ix)E(-ix) = E(0) = 1.$$

so that

$$|E(ix)| = 1, \ x \in \mathbb{R}.$$

Moreover,

$$C(0) = 1, \ S(0) = 0; \ \text{and} \ C'(x) = -S(x), \ S'(x) = C(x)$$

We assert that there exists positive numbers x such that $C(x) = 0$. Let x_0 be the smallest among them. We define number π by

$$\pi = 2x_0.$$

Then, $C(\frac{\pi}{2}) = 0$, and $S(\frac{\pi}{2}) = \pm 1$. Since $C(x) > 0$ in $(0, \frac{\pi}{2})$, S is increasing in $(0, \frac{\pi}{2})$; hence $S(\frac{\pi}{2}) = 1$. Therefore,

$$E\left(\frac{\pi i}{2}\right) = i,$$

and the addition formula gives

$$E(\pi i) = -1, \ E(2\pi i) = 1;$$

hence

$$E(z + 2\pi i) = E(z), \forall z \in \mathbb{C}.$$

Theorem 13.5.1 *The following are true:*

(a) The function E is periodic, with period $2\pi i$.
(b) The functions C and S are periodic, with period 2π.
(c) If $0 < t < 2\pi$, then $E(it) \neq 1$.
(d) If $z \in \mathbb{C} \ni |z| = 1$, $\exists$ unique $t \in [0, 2\pi) \ni E(it) = z$.

Remark 13.5.2 *The curve γ defined by $\gamma(t) = E(it)$, $0 \leq t \leq 2\pi$ is a simple closed curve whose range is the unit circle in the plane. Since $\gamma'(t) = iE(it)$, the length of γ is $\int_0^{2\pi} |\gamma'(t)| \, dt = 2\pi$. This is the expected result for the circumference of a circle with radius 1.*

The point $\gamma(t)$ describes a circular arc of length t_0 as t increases from 0 to t_0. Consideration of the triangle whose vertices are $z_1 = 0$, $z_2 = \gamma(t_0)$, and $z_3 = C(t_0)$ shows that $C(t)$ and $S(t)$ are indeed identical with $\cos(t)$ and $\sin(t)$ respectively, the latter are defined as ratios of sides of a right triangle.

The saying *the complex field is algebraically complete* means that every nonconstant polynomial with complex coefficients has a complex root.

Theorem 13.5.3 *Suppose $a_0, \ldots, a_n \in \mathbb{C}$, $n \in \mathbb{N}$, $a_n \neq 0$,*

$$P(z) = \sum_0^n a_k z^k.$$

Then, $P(z) = 0$ for some $z \in \mathbb{C}$.

Proof. Without loss of generality, we may assume that $a_n = 1$. Put $\mu = \inf_{z \in \mathbb{C}} |P(z)|$. If $|z| = R$ then

$$|P(z)| \geq R^n (1 - |a_{n-1}| R^{-1} - \cdots - |a_0| R^{-n}).$$

The right hand side of the above inequality tends to ∞ as $R \to \infty$. Hence, $\exists R_0 \ni |P(z)| \geq \mu$ if $|z| \geq R_0$. Since $|P|$ is continuous on the closed disc with center at the origin and radius R_0, it attains its minimum; i.e. $\exists z_0 \ni |P(z_0)| = \mu$.

We claim that $\mu = 0$. If not, put $Q(z) = \frac{P(z+z_0)}{P(z_0)}$. Then, Q is nonconstant polynomial, $Q(0) = 1$, and $|Q(z)| \geq 1$, $\forall z$. There is a smallest integer k, $1 \leq k \leq n$ such that

$$Q(z) = 1 + b_k z^k + \cdots + b_n z^n, \ b_k \neq 0.$$

By Theorem 13.5.1 (d), $\theta \in \mathbb{R} \ni e^{ik\theta} b_k = -|b_k|$. If $r > 0$ and $r^k |b_k| < 1$, we have $|1 + b_k r^k e^{ik\theta}| = 1 - r^k |b_k|$, so that

$$|Q(re^{i\theta})| \leq 1 - r^k[|b_k| - r |b_{k+1}| - \cdots - r^{n-k} |b_n|].$$

For sufficiently small r, the expression in squared braces is positive; hence $|Q(re^{i\theta})| < 1$, Contradiction. Thus, $\mu = 0 = P(z_0)$. $\square$

13.6 Fourier Series

Definition 13.6.1 *A trigonometric polynomial is a finite sum of the form*

$$f(x) = a_0 + \sum_{n=1}^{N}(a_n \cos nx + b_n \sin nx), \ x \in \mathbb{R},$$

where $a_0, a_1, \ldots, a_N, b_1, \ldots, b_N \in \mathbb{C}$. *One can rewrite*

$$f(x) = \sum_{-N}^{N} c_n e^{inx}, \ x \in \mathbb{R},$$

which is more convenient. It is clear that, every trigonometric polynomial is periodic, with period 2π.

Remark 13.6.2 *If* $n \in \mathbb{N}$, e^{inx} *is the derivative of* $\frac{e^{inx}}{in}$ *which also has period* 2π. *Hence,*

$$\frac{1}{2\pi}\int_{-\pi}^{\pi} e^{inx} \, dx = \begin{cases} 1, \ n = 0, \\ 0, \ n = \pm 1, \pm 2, \ldots \end{cases}$$

If we multiply $f(x)$ *by* e^{-imx} *where* $m \in \mathbb{Z}$, *then if we integrate, we have*

$$c_m = \frac{1}{2\pi}\int_{-\pi}^{\pi} e^{imx} \, dx$$

for $|m| \leq N$. *Otherwise,* $|m| > N$, *the integral above is zero.*

Therefore, the trigonometric polynomial is real if and only if

$$c_{-n} = \overline{c_n}, \ n = 0, \ldots, N.$$

Definition 13.6.3 *A trigonometric series is a series of the form*

$$f(x) = \sum_{-\infty}^{\infty} c_n e^{inx}, \; x \in \mathbb{R}.$$

If f is an integrable function on $[-\pi, \pi]$, the numbers c_m are called the Fourier coefficients of f, and the series formed with these coefficients is called the Fourier series of f.

13.7 Gamma Function

Definition 13.7.1 *For $0 < x < \infty$,*

$$\Gamma(x) = \int_0^\infty t^{x-1} e^{-t} \, dt.$$

is known as the gamma function.

Proposition 13.7.2 *Let $\Gamma(x)$ be defined above.*

(a) $\Gamma(x+1) = x\Gamma(x)$, $0 < x < \infty$.
(b) $\Gamma(n+1) = n!$, $n \in \mathbb{N}$. $\Gamma(1) = 1$.
(c) $\log \Gamma$ is convex on $(0, \infty)$.

Proposition 13.7.3 *If f is a positive function on $(0, \infty)$ such that*

(a) $f(x+1) = xf(x)$,
(b) $f(1) = 1$,
(c) $\log f$ is convex.

then $f(x) = \Gamma(x)$.

Proposition 13.7.4 *If $x, y \in \mathbb{R}_+$,*

$$\int_0^1 t^{x-1}(1-t)^{y-1} \, dt = \frac{\Gamma(x)\Gamma(y)}{\Gamma(x+y)}.$$

This integral is so-called beta function $\beta(x, y)$.

Remark 13.7.5 *Let $t = \sin \theta$, then*

$$2 \int_0^{\frac{\pi}{2}} (\sin \theta)^{2x-1} (\cos \theta)^{2y-1} \, d\theta = \frac{\Gamma(x)\Gamma(y)}{\Gamma(x+y)}.$$

The special case $x = y = \frac{1}{2}$ gives

$$\Gamma\left(\frac{1}{2}\right) = \sqrt{\pi}.$$

Remark 13.7.6 *Let $t = s^2$ in the definition of Γ.*

$$\Gamma(x) = 2 \int_0^\infty s^{2x-1} e^{-s^2}\, ds, \ 0 < x < \infty.$$

The special case $x = \frac{1}{2}$ gives

$$\int_{-\infty}^\infty e^{-s^2}\, ds = \sqrt{\pi}.$$

This yields

$$\Gamma(x) = \frac{2^{x-1}}{\sqrt{\pi}} \Gamma\left(\frac{x}{2}\right) \Gamma\left(\frac{x+1}{2}\right)$$

Remark 13.7.7 (Stirling's Formula) *This provides a simple approximate expression for $\Gamma(x+1)$ when x is large. The formula is*

$$\lim_{x \to \infty} \frac{\Gamma(x+1)}{\left(\frac{x}{e}\right)^x \sqrt{2\pi x}} = 1.$$

Problems

13.1. Prove Theorem 13.1.8, the comparison tests for nonnegative series.

13.2. Discuss the convergence and divergence of the following series:
a) $\sum_0^\infty \frac{x^k}{k!}$
b) $\sum_1^\infty \frac{x^k}{k^\alpha}$, where $\alpha > 0$
c) $\sum_1^\infty (e^{\frac{1}{k}} - 1)$
d) $\sum_1^\infty \ln\left(1 + \frac{1}{k}\right)$
e) $\sum_1^\infty q^{k+\sqrt{k}}$, where $q > 0$
f) $\sum_1^\infty \frac{1}{n}$

13.3. One can model every combinatorial problem (instance r) as

$$\sum_i x_i = r, \ x_i \in S_i \subseteq \mathbb{Z}_+. \text{ Let } A_{ij} = \begin{cases} 1, j \in S_i \\ 0, j \notin S_i \end{cases}$$

Then, the power series

$$g(x) = \prod_i \left(\sum_{j=0}^\infty A_{ij} x^j\right) = \sum_{k=0}^\infty a_k x^k$$

is known as the generating function, where the number of distinct solutions to $\sum_i x_i = r$ is the coefficient a_r. We know that, one can write down a generating function for every combinatorial problem in such a way that a_r is the number of solutions in a general instance r.

Use generating functions to
a) Prove the binomial theorem

$$(1+x)^n = \sum_{i=0}^{n} \binom{n}{i} x^i$$

and extend to the multinomial (you may not use the generating functions) theorem

$$(x_1 + \cdots x_k)^n = \sum_{\substack{i_1,\ldots,i_k \in \mathbb{Z}_+ \\ i_1 + \cdots + i_k = n}} \binom{n}{i_1,\ldots,i_k} x_1^{i_1} \cdots x_k^{i_k}.$$

b) Prove that

$$(1 + x + x^2 + x^3 + \ldots)^n = \sum_{i=0}^{\infty} \binom{n-1+i}{i} x^i$$

c) Find the probability of having a sum of 13 if we roll four distinct dice.
d) Solve the following difference equation: $a_n = 5a_{n-1} - 6a_{n-2}$, $\forall n = 2, 3, 4, \ldots$ with $a_0 = 2$ and $a_1 = 5$ as boundary conditions.

13.4. Consider the following air defense situation. There are $i = 1, \ldots, I$ enemy air threats each to be engaged to one of the allied $z = 1, \ldots, Z$ high value zones with a value of w_z. The probability that a threat (i) will destroy its target (z) is q_{iz}. More than one threats can engage to a single zone. On the other hand, there are $j = 1, \ldots, J$ allied air defense systems that can engage the incoming air threats. The single shot kill probability of an air defense missile fired by system j to a threat i is p_{ji}. Let the main integer decision variable be x_{ji} indicating the number of missiles fired from system j to threat i.
a) Write down the nonlinear constraint if there is a threshold value d_i, the minimum desired probability for destroying target i. Try to linearize it using one of the functions defined in this chapter.
b) Let our objective function that maximizes the expected total weighted survival of the zones be

$\max \sum_z w_z \alpha_z$ (0), where $\alpha_z = \prod_i \left[1 - q_{iz} \left(\prod_j (1 - p_{ji})^{x_{ji}} \right) \right] = \prod_i \beta_{iz}$.
Then, $\gamma_z = \log(\alpha_z) = \sum_i \log(\beta_{iz}) = \sum_i \delta_{iz}$ and we have the second objective function: $\max \sum_z w_z \gamma_z$ (0'). Isn't this equivalent to $\max \sum_z w_z \sum_i \delta_{iz}$ (0''), where $\delta_{iz} = \log \left[1 - q_{iz} \left(\prod_j (1 - p_{ji})^{x_{ji}} \right) \right]$? Since $\beta_{iz} = 1 - q_{iz} \left(\prod_j (1 - p_{ji})^{x_{ji}} \right)$ and we have

$$\max \delta_{iz} = \max \log(\beta_{iz}) \equiv \max \beta_{iz} \equiv \min(1 - \beta_{iz}) \equiv \min \log(1 - \beta_{iz}),$$

our fourth objective function (linear!) is $\min \sum_z w_z \sum_i \theta_{iz}$ $(0''')$, where $\theta_{iz} = \log(1 - \beta_{iz}) = \log(q_{iz}) + \left(\sum_j [\log(1 - p_{ji})] x_{ji} \right)$. Since we can drop the constants, $\log(q_{iz})$, in the objective function, we will have the fifth objective function as $\min \sum_z w_z \sum_i \left(\sum_j [\log(1 - p_{ji})] x_{ji} \right)$ (0^{iv}), which is not (clearly) equivalent to the initial objective function in catching the same optimum solution! Where is the flaw?

$$(0)? \equiv (0')? \equiv (0'')? \equiv (0''')? \equiv (0^{iv})?$$

Web material

http://archives.math.utk.edu/visual.calculus/6/power.1/index.html
http://archives.math.utk.edu/visual.calculus/6/series.4/index.html
http://arxiv.org/PS_cache/math-ph/pdf/0402/0402037.pdf
http://calclab.math.tamu.edu/~belmonte/m152/L/ca/LA4.pdf
http://calclab.math.tamu.edu/~belmonte/m152/L/ca/LA5.pdf
http://cr.yp.to/2005-261/bender1/IS.pdf
http://education.nebrwesleyan.edu/Research/StudentTeachers/
    secfall2001/Serinaldi/Chap%209/tsld009.htm
http://en.wikipedia.org/wiki/Power_series
http://en.wikipedia.org/wiki/Trigonometric_function#
    Series_definitions
http://en.wikipedia.org/wiki/Wikipedia:WikiProject_Mathematics/
    PlanetMath_Exchange/40-XX_Sequences,_series,_summability
http://eom.springer.de/c/c026150.htm
http://eom.springer.de/T/t094210.htm
http://faculty.eicc.edu/bwood/ma155supplemental/
    supplementalma155.html
http://home.att.net/~numericana/answer/analysis.htm
http://kr.cs.ait.ac.th/~radok/math/mat11/chap8.htm
http://kr.cs.ait.ac.th/~radok/math/mat6/calc8.htm
http://kr.cs.ait.ac.th/~radok/math/mat6/calc81.htm
http://math.fullerton.edu/mathews/c2003/
    ComplexGeometricSeriesMod.html
http://math.fullerton.edu/mathews/n2003/ComplexFunTrigMod.html
http://math.furman.edu/~dcs/book/c5pdf/sec57.pdf
http://math.furman.edu/~dcs/book/c8pdf/sec87.pdf
http://mathworld.wolfram.com/ConvergentSeries.html
http://mathworld.wolfram.com/HarmonicSeries.html
http://mathworld.wolfram.com/PowerSeries.html
http://media.pearsoncmg.com/aw/aw_thomas_calculus_11/topics/
    sequences.htm
http://motherhen.eng.buffalo.edu/MTH142/spring03/lec11.html
http://oregonstate.edu/~peterseb/mth306/docs/306w2005_prob_1.pdf
http://persweb.wabash.edu/facstaff/footer/Courses/M111-112/Handouts/

http://planetmath.org/encyclopedia/PowerSeries.html
http://planetmath.org/encyclopedia/SlowerDivergentSeries.html
http://shekel.jct.ac.il/~math/tutorials/complex/node48.html
http://sosmath.com/calculus/series/poseries/poseries.html
http://syssci.atu.edu/math/faculty/finan/2924/cal92.pdf
http://tutorial.math.lamar.edu/AllBrowsers/2414/
      ConvergenceOfSeries.asp
http://web.mat.bham.ac.uk/R.W.Kaye/seqser/intro2series
http://www.cs.unc.edu/~dorianm/academics/comp235/fourier
http://www.du.edu/~etuttle/math/logs.htm
http://www.ercangurvit.com/series/series.htm
http://www.math.cmu.edu/~bobpego/21132/seriestools.pdf
http://www.math.columbia.edu/~kimball/CalcII/w9.pdf
http://www.math.columbia.edu/~rf/precalc/narrative.pdf
http://www.math.harvard.edu/~jay/writings/p-adics1.pdf
http://www.math.hmc.edu/calculus/tutorials/convergence/
http://www.math.mcgill.ca/labute/courses/255w03/L18.pdf
http://www.math.niu.edu/~rusin/known-math/index/40-XX.html
http://www.math.princeton.edu/~nelson/104/SequencesSeries.pdf
http://www.math.ucla.edu/~elion/ta/33b.1.041/midterm2.pdf
http://www.math.unh.edu/~jjp/radius/radius.html
http://www.math.unl.edu/~webnotes/classes/class38/class38.htm
http://www.math.uwo.ca/courses/Online_calc_notes/081/unit6/Unit6.pdf
http://www.math.wpi.edu/Course_Materials/MA1023B04/seq_ser/
      node1.html
http://www.math2.org/math/expansion/tests.htm
http://www.math2.org/math/oddsends/complexity/e%5Eitheta.htm
http://www.mathreference.com/lc-ser,intro.html
http://www.maths.abdn.ac.uk/~igc/tch/ma2001/notes/node53.html
http://www.maths.mq.edu.au/~wchen/lnfycfolder/fyc19-ps.pdf
http://www.mecca.org/~halfacre/MATH/series.htm
http://www.ms.uky.edu/~carl/ma330/sin/sin1.html
http://www.pa.msu.edu/~stump/champ/10.pdf
http://www.richland.edu/staff/amoshgi/m230/Fourier.pdf
http://www.sosmath.com/calculus/improper/gamma/gamma.html
http://www.sosmath.com/calculus/powser/powser01.html
http://www.sosmath.com/calculus/series/poseries/poseries.html
http://www.stewartcalculus.com/data/CALCULUS%20Early%
      20Transcendentals/upfiles/FourierSeries5ET.pdf
http://www4.ncsu.edu/~acherto/NCSU/MA241/sections81-5.pdf
http://www42.homepage.villanova.edu/frederick.hartmann/Boundaries/
      Boundaries.pdf
www.cwru.edu/artsci/math/butler/notes/compar.pdf

Special Transformations

In functional analysis, the Laplace transform is a powerful technique for analyzing linear time-invariant systems. In actual, physical systems, the Laplace transform is often interpreted as a transformation from the time-domain point of view, in which inputs and outputs are understood as functions of time, to the frequency-domain point of view, where the same inputs and outputs are seen as functions of complex angular frequency, or radians per unit time. This transformation not only provides a fundamentally different way to understand the behavior of the system, but it also drastically reduces the complexity of the mathematical calculations required to analyze the system. The Laplace transform has many important Operations Research applications as well as applications in control engineering, physics, optics, signal processing and probability theory. The Laplace transform is used to analyze continuous–time systems whereas its discrete-time counterpart is the Z transform. The Z transform among other applications is used frequently in discrete probability theory and stochastic processes, combinatorics and optimization. In this chapter, we will present an overview of these transformations from differential/difference equation systems' viewpoint.

14.1 Differential Equations

Definition 14.1.1 *An (ordinary) differential equation is an equation that can be written as:*

$$\Phi(t, y, y', \ldots, y^{(n)}) = 0.$$

A solution of above is a continuous function $y : I \mapsto \mathbb{R}$ where I is a real interval such that $\Phi(t, y, y', \ldots, y^{(n)}) = 0$, $\forall t \in I$. A differential equation is a linear differential equation of order n if

$$y^{(n)} + \alpha_{n-1}(t)y^{(n-1)} + \cdots + \alpha_1(t)y' + \alpha_0(t)y = b(t)$$

where $\alpha_{n-1}, \cdots, \alpha_1, \alpha_0, b$ are continuous functions on I to $\mathbb{R}$. If $\forall \alpha_i = c_i$, the above has constant coefficients. If $b(t) = 0, \forall t \in I$, then the above is called *homogeneous, otherwise it is non–homogeneous. If we assume $0 \in I$, and* $y(0) = y_0$, $y'(0) = y_0'$, $\ldots$, $y^{(n-1)}(0) = y_0^{(n-1)}$ *where* $y_0, y_0', \ldots, y_0^{(n-1)}$ *are* n *specified real numbers, this is called initial value problems where* $y_0^{(*)}$*'s are the prescribed initial values.*

Example 14.1.2 (The 1st and 2nd order linear initial value problems)

$$y'(t) = a(t)y(t) + f(t), \quad y(0) = y_0;$$

and for $n = 2$, the constant coefficient problem is

$$y''(t) + \alpha_1 y'(t) + \alpha_0 y(t) = b(t); \quad y(0) = y_0, \ y'(0) = y_0'.$$

Remark 14.1.3 *Let*

$$
\begin{array}{cc}
y(t) = y_1(t) & y_1'(t) = y_2(t) \\
y'(t) = y_2(t) & y_2'(t) = y_3(t) \\
\vdots & \vdots \\
y^{(n-1)}(t) = y_n(t) & y_n'(t) = -\alpha_{n-1}y_n(t) - \cdots - \alpha_1 y_2(t) - \alpha_0 y_1(t) + b(t)
\end{array}
$$

$$
\Leftrightarrow A = \begin{bmatrix}
0 & 1 & 0 & \cdots & 0 \\
0 & 0 & 1 & \cdots & 0 \\
\vdots & \vdots & \vdots & & \vdots \\
0 & 0 & 0 & \cdots & 1 \\
-\alpha_0 & -\alpha_1 & -\alpha_2 & \cdots & -\alpha_{n-1}
\end{bmatrix}, \quad y(t) = \begin{bmatrix} y_1(t) \\ y_2(t) \\ \vdots \\ y_{n-1}(t) \\ y_n(t) \end{bmatrix},
$$

$$
y_0 = \begin{bmatrix} y_0 \\ y_0' \\ \vdots \\ y_0^{(n-2)} \\ y_0^{(n-1)} \end{bmatrix}, \quad f(t) = \begin{bmatrix} 0 \\ 0 \\ \vdots \\ 0 \\ b(t) \end{bmatrix}.
$$

We have linear differential systems problem:

$$y'(t) = Ay(t) + f(t); \quad y(0) = y_0.$$

14.2 Laplace Transforms

Definition 14.2.1 *The basic formula for the Laplace transformation y to η is*

$$\eta(s) = \int_0^\infty e^{-st} y(t)\, dt.$$

We call the function, η, the Laplace transform of y if $\exists x_0 \in \mathbb{R} \ni \eta(s)$ exists, $\forall s > x_0$. We call y as the inverse–Laplace transform of η.

$$\eta(s) = \mathcal{L}\{y(t)\}, \quad y(t) = \mathcal{L}^{-1}\{\eta(s)\}.$$

Proposition 14.2.2 *If* $y : \mathbb{R} \mapsto \mathbb{R}$ *satisfies*

(i) $y(t) = 0$ *for* $t < 0$,
(ii) $y(t)$ *is piecewise continuous*,
(iii) $y(t) = O(e^{x_0 t})$ *for some* $x_0 \in \mathbb{R}$,

then $y(t)$ *has a Laplace transform.*

Tables 14.1 and 14.2 contain Laplace transforms and its properties.

Table 14.1. A Brief Table for Laplace Transforms

	Inverse $y(t)$	Laplace Transform $\eta(s)$	Valid $s > x_0$ x_0
(1)	1	$\frac{1}{s}$	0
(2)	e^{at}	$\frac{1}{s-a}$, $a \in \mathbb{C}$	$\Re a$
(3)	t^m, $m = 1, 2, \ldots$	$\frac{m!}{s^{m+1}}$	0
(4)	$t^m e^{at}$, $m = 1, 2, \ldots$	$\frac{m!}{(s-a)^{m+1}}$, $a \in \mathbb{C}$	$\Re a$
(5)	$\sin bt$	$\frac{b}{s^2+b^2}$	0
(6)	$\cos bt$	$\frac{s}{s^2+b^2}$	0
(7)	$e^{ct} \sin dt$	$\frac{d}{(s-c)^2+d^2}$	c
(8)	$e^{ct} \cos dt$	$\frac{s-c}{(s-c)^2+d^2}$	c

Table 14.2. Properties of Laplace Transforms

	Inverse	Laplace Transform
(1)	$y(t)$	$\eta(s)$
(2)	$ay(t) + bz(t)$	$a\eta(s) + b\zeta(s)$
(3)	$y'(t)$	$s\eta(s) - y(0)$
(4)	$y^{(n)}(t)$	$s^n \eta(s) - s^{n-1}y(0)$ $- \cdots - y^{(n-1)}(0)$
(5) $y_c(t) = \begin{cases} 0, t < c \text{ where } c > 0 \\ 1, t \geq c \end{cases}$		$\frac{e^{-cs}}{s}$
(6)	$\frac{1}{a}e^{-\frac{bt}{a}} y\left(\frac{t}{a}\right)$, $a > 0$	$\eta(as + b)$
(7)	$t^m y(t)$, $m = 1, 2, \ldots$	$(-1)^m \eta^{(m)}(s)$
(8)	$t^{-1}y(t)$	$\int_s^\infty \eta(u)\, du$
(9)	$\int_0^t y(t - u)z(u)\, du$	$\eta(s)\zeta(s)$

Remark 14.2.3 *If* $a = c + id$ *is non–real,* $\mathcal{L}\{e^{at}\} = \mathcal{L}\{e^{ct} \cos dt\} + i\mathcal{L}\{e^{dt} \sin dt\}$ *then obtain Laplace transform using (2) in Table 14.1.*

Remark 14.2.4 *Proceed the following steps to solve an initial value problem:*

S1. $y(t) \mapsto \eta(s)$.

S2. Solve the resulting linear algebraic equation, call the solution $\eta(s)$ the formal Laplace transform of $y(t)$.

S3. Find the inverse–Laplace transform $y(t)$.

S4. Verify that $y(t)$ is a solution.

Example 14.2.5 *Find the solution to*

$$y'(t) = -4y(t) + f(t); \quad y(0) = 0,$$

where $f(t)$ is the unit step function

$$f(t) = \begin{cases} 0, t < 1 \\ 1, t \geq 1. \end{cases}$$

and $I = [0, \infty)$. Transforming both sides, we have

$$s\eta(s) - y(0) = -4\eta(s) + \frac{e^{-s}}{s}$$

$$s\eta(s) = -4\eta(s) + \frac{e^{-s}}{s}.$$

At the end of S2, we have $\eta(s) = \frac{e^{-s}}{s(s+4)}$.

$$\frac{1}{s(s+4)} = \frac{1}{4}\left(\frac{1}{s} - \frac{1}{s+4}\right).$$

Therefore,

$$\eta(s) = \frac{1}{4}e^{-s}\left(\frac{1}{s} - \frac{1}{s+4}\right).$$

Thus,

$$y(t) = \begin{cases} 0, & t < 1; \\ \frac{1}{4}\left(1 - e^{-4(t-1)}\right), & t \geq 1. \end{cases}$$

Example 14.2.6 *Let us solve*

$$y'(t) = ay(t) + f(t); \quad y(0) = 0$$

such that $y'(t) = f(t)$.

Let us take $y'(t) = f(t)$ then $s\eta(s) - y_0 = \phi(s)$, where $\phi(s) = \mathcal{L}\{f(t)\}$. Thus,

$$\eta(s) = y_0\frac{1}{s} + \frac{1}{s}\phi(s).$$

We use formula (9) in Table 14.2.

$$y(t) = y_0 + \int_0^t f(u)\, du.$$

If we relax $y'(t) = f(t)$, then we have

$$\eta(s) = y_0 \frac{1}{s-a} + \frac{1}{s-a}\phi(s)$$

and

$$y(t) = e^{at} y_0 + \int_0^t e^{a(t-u)} f(u) \, du;$$

where $\phi(s)$ is the Laplace transform of $f(t)$.

Remark 14.2.7 *In order to solve the matrix equation,*

$$y'(t) = Ay(t) + f(t); \quad y(0) = y_0$$

we will take the Laplace transform as

$$\eta(s)(sI - A) = y_0 + \phi(s).$$

where $\eta(s) = [\eta_1(s), \cdots, \eta_n(s)]^T$ is the vector of Laplace transforms of the components of y. If s is not an eigenvalue of A, then the coefficient matrix is nonsingular. Thus, for sufficiently large s

$$\eta(s) = (sI - A)^{-1} y_0 + (sI - A)^{-1}\phi(s)$$

where the matrix $(sI - A)^{-1}$ is called the resolvent matrix of A and

$$\mathcal{L}(e^{tA}) = (sI - A)^{-1} \text{ for } f(t) = 0.$$

Example 14.2.8 *Let us take an example problem as Matrix exponentials. The problem of finding e^{tA} for an arbitrary square matrix A of order n can be solved by finding the Jordan form. For $n \geq 3$, one should use a computer. However, we will show that how Cayley–Hamilton Theorem leads to another method for finding e^{tA} when $n = 2$. Let us take the following system of equations*

$$y_1'(t) = y_2(t) + 1 \quad y_1(0) = 3,$$
$$y_2'(t) = y_1(t) + t \quad y_2(0) = 1.$$

Then,

$$A = \begin{bmatrix} 0 & 1 \\ 1 & 0 \end{bmatrix}, \quad f(t) = \begin{bmatrix} 1 \\ t \end{bmatrix}, \quad y_0 = \begin{bmatrix} 3 \\ 1 \end{bmatrix}.$$

$$S^{-1}AS = \frac{1}{2}\begin{bmatrix} 1 & 1 \\ 1 & -1 \end{bmatrix}\begin{bmatrix} 0 & 1 \\ 1 & 0 \end{bmatrix}\begin{bmatrix} 1 & 1 \\ 1 & -1 \end{bmatrix} = \begin{bmatrix} 1 & 0 \\ 0 & -1 \end{bmatrix}.$$

$$e^{tA} = S\begin{bmatrix} e^t & 0 \\ 0 & e^{-t} \end{bmatrix}$$

$$S^{-1} = \frac{1}{2}\begin{bmatrix} e^t + e^{-t} & e^t - e^{-t} \\ e^t - e^{-t} & e^t + e^{-t} \end{bmatrix}.$$

Then, the unique solution is $y(t) = e^{tA}y_0 + p(t)$, *where*

$$e^{tA}y_0 = \begin{bmatrix} 2e^t + e^{-t} \\ 2e^t - e^{-t} \end{bmatrix} \quad and$$

$$p(t) = \frac{1}{2} \begin{bmatrix} \int_0^t [e^t(e^{-u} + ue^{-u}) + e^{-t}(e^{-u} - ue^{-u})]\, du \\ \int_0^t [e^t(e^{-u} + ue^{-u}) + e^{-t}(e^{-u} + ue^{-u})]\, du \end{bmatrix}.$$

Then, after integration we have

$$p(t) = \begin{bmatrix} e^t - e^{-t} - t \\ e^t + e^{-t} - 2 \end{bmatrix} \Rightarrow y(t) = \begin{bmatrix} 3e^t - t \\ 3e^t - 2 \end{bmatrix}.$$

One can solve the above differential equation system using Laplace trans-forms:

$$y'(t) = Ay(t) + f(t) \Leftrightarrow s \begin{bmatrix} \eta_1(s) \\ \eta_2(s) \end{bmatrix} - \begin{bmatrix} 3 \\ 1 \end{bmatrix} = \begin{bmatrix} 0 & 1 \\ 1 & 0 \end{bmatrix} \begin{bmatrix} \eta_1(s) \\ \eta_2(s) \end{bmatrix} + \begin{bmatrix} \frac{1}{s} \\ \frac{1}{s^2} \end{bmatrix}$$

$$\Leftrightarrow \begin{bmatrix} s & -1 \\ -1 & s \end{bmatrix} \begin{bmatrix} \eta_1(s) \\ \eta_2(s) \end{bmatrix} = \begin{bmatrix} 3 \\ 1 \end{bmatrix} + \begin{bmatrix} \frac{1}{s} \\ \frac{1}{s^2} \end{bmatrix} \quad (\star)$$

Then, the resolvent matrix is

$$(sI - A)^{-1} = \frac{1}{(s-1)(s+1)} \begin{bmatrix} s & 1 \\ 1 & s \end{bmatrix} \quad (\star\star)$$

If we multiply both sides of $(\star)$ *by* $(\star\star)$, *we have*

$$\eta(s) = \frac{1}{(s-1)(s+1)} \begin{bmatrix} 3s+1 \\ s+3 \end{bmatrix} + \frac{1}{s^2(s-1)(s+1)} \begin{bmatrix} s^2+1 \\ 2s \end{bmatrix}$$

$$\eta(s) = \frac{1}{s-1} \begin{bmatrix} 3 \\ 3 \end{bmatrix} + \frac{1}{s+1} \begin{bmatrix} 0 \\ 0 \end{bmatrix} + \frac{1}{s^2} \begin{bmatrix} -1 \\ 0 \end{bmatrix} + \frac{1}{s} \begin{bmatrix} 0 \\ -2 \end{bmatrix}$$

$$\Rightarrow y(t) = \begin{bmatrix} 3e^t - t \\ 3e^t - 2 \end{bmatrix}$$

In order to find e^{tA}, *we expand right hand side of* $(\star\star)$ *as*

$$\eta(s) = \frac{1}{s-1} \begin{bmatrix} \frac{1}{2} & \frac{1}{2} \\ \frac{1}{2} & \frac{1}{2} \end{bmatrix} + \frac{1}{s+1} \begin{bmatrix} \frac{1}{2} & -\frac{1}{2} \\ -\frac{1}{2} & \frac{1}{2} \end{bmatrix}$$

If we invert it, we will have the following

$$e^{tA} = \frac{1}{2} \begin{bmatrix} e^t + e^{-t} & e^t - e^{-t} \\ e^t - e^{-t} & e^t + e^{-t} \end{bmatrix}.$$

14.3 Difference Equations

Let us start with *first-order difference equations*:

$$\left.\begin{array}{r} y(k+1) = y(k) + f(k) \\ y(0) = y_0 \end{array}\right\} \quad \Delta y(k) = f(k), \; k = 1, 2, \ldots$$

The initial value problem of the above equation can be solved by the following recurrence relation:

$$y(k) = y(k+1) - f(k), \; k = -1, -2, \ldots$$

Therefore, we find

$$y(k) = \begin{cases} y_0 + \sum_{u=0}^{k-1} f(u), & k = 1, 2, 3, \ldots; \\ y_0, & k = 0; \\ y_0 - \sum_{u=k}^{-1} f(u), & k = -1, -2, \ldots \end{cases}$$

For *second-order* equations, we will consider first the *homogeneous* case:

$$y(k+2) + \alpha_1 y(k+1) + \alpha_0 y(k) = 0; \; y(0) = y_0, \; y(1) = y_1.$$

We seek constants

$$\lambda_1, \lambda_2 \ni z(k+1) = \lambda_2 z(k); \; z(0) = y_1 - \lambda_1 y_0$$

which are the roots of

$$\lambda^2 + \alpha_1 \lambda + \alpha_0 = 0.$$

If $\lambda_1 \neq \lambda_2$, then $y(k) = c_1 \lambda_1^k + c_2 \lambda_2^k$ where c_1, c_2 are the unique solutions of

$$c_1 + c_2 = y_0, \; c_1 \lambda_1 + c_2 \lambda_2 = y_1.$$

If $\lambda_1 = \lambda_2 = \lambda$, then $y(k) = c_1 \lambda^k + c_2 \lambda^k$ where c_1, c_2 are the unique solutions of

$$c_1 = y_0, \; c_1 \lambda + c_2 \lambda = y_1.$$

When the roots are non-real, $\lambda = \rho e^{i\theta}$ and $\bar{\lambda} = \rho e^{-i\theta}$, then

$$y(k) = c_1 \rho^k \cos k\theta + c_2 \rho^k \sin k\theta,$$

where c_1 and c_2 are the unique solutions of

$$c_1 = y_0; \; c_1 \cos \theta + c_2 \sin \theta = y_1.$$

If we have *systems of equations*,

$$y(k+1) = Ay(k), \; k = 0, 1, 2, \ldots; \; y(0) = y_0,$$

we, then, have as a recurrence relation

$$y(k) = A^k y_0 \text{ and } A^0 = I.$$

When A is singular, there does not exist a unique solution $y(-1)$ satisfying $Ay(-1) = y_0$. When A is non-singular,

$$y(k) = A^{-1} y(k+1).$$

Then, $y(-1) = A^{-1} y_0$, $y(-2) = A^{-2} y_0$, $\cdots$ where $A^{-k} = A^{-1} A^{-k+1} = (A^{-1})^k$, $k = 2, 3, \ldots$ Recall that, if $A = SJS^{-1}$ then $A^k = SJ^k S^{-1}$. Then,

$$y(k) = SJS^{-1} y_0, \quad k = 0, 1, \ldots$$

For the *non-homogeneous* case,

$$y(k+1) = Ay(k) + f(k).$$

If A is nonsingular,

$$y(k) = A^k y_0 + p(k),$$

where $p(k+1) = Ap(k) + f(k)$; $p(0) = 0$. This yields

$$p(k) = -\sum_{v=k}^{-1} A^{k-1-v} f(v).$$

Example 14.3.1 *For* $k = 0, 1, \ldots,$

$$y_1(k+1) = y_2(k) + 1, \quad y_1(0) = 3,$$

$$y_2(k+1) = y_1(k) + 1, \quad y_2(0) = 1.$$

$$A^k = \frac{1}{2} \begin{bmatrix} 1 + (-1)^k & 1 - (-1)^k \\ 1 - (-1)^k & 1 + (-1)^k \end{bmatrix}, \quad A^k y_0 = \begin{bmatrix} 2 + (-1)^k \\ 2 - (-1)^k \end{bmatrix},$$

$$p(k) = \frac{1}{2} \sum_{u=0}^{k-1} \begin{bmatrix} k - u + (-1)^u (2 - k + u) \\ k - u - (-1)^u (2 - k + u) \end{bmatrix}.$$

We know,

$$\frac{1}{2} \sum_{u=0}^{k-1} (k - u) = \frac{1}{2} \frac{k(k+1)}{2} \quad and$$

$$\frac{1}{2}(2 - k) \sum_{u=0}^{k-1} (-1)^u + \frac{1}{2} \sum_{u=0}^{k-1} u(-1)^u = \frac{3}{8} - \frac{3}{8}(-1)^k - \frac{k}{2}.$$

$$p(k) = \frac{1}{8} \begin{bmatrix} 2k^2 - 3(-1)^k + 3 \\ 2k^2 + 4k + 3(-1)^k - 3 \end{bmatrix} \Rightarrow$$

$$y(k) = k^2 \begin{bmatrix} \frac{1}{4} \\ \frac{1}{4} \end{bmatrix} + k \begin{bmatrix} 0 \\ \frac{1}{2} \end{bmatrix} + (-1)^k \begin{bmatrix} \frac{5}{8} \\ -\frac{5}{8} \end{bmatrix} + \begin{bmatrix} \frac{19}{8} \\ \frac{13}{8} \end{bmatrix}.$$

14.4 Z Transforms

Definition 14.4.1 *The Z Transformation y to η is*

$$\eta(z) = \sum_{u=0}^{\infty} \frac{y(u)}{z^u}, \quad \text{where } z \in \mathbb{C}.$$

We call the function η the Z transform of y if

$$\exists r \in \mathbb{R} \ni \eta(z) \text{ converges whenever } |z| > r,$$

in such cases y is the inverse Z transform of η.

$$\eta(z) = \mathcal{Z}\{y(t)\}, \quad y(t) = \mathcal{Z}^{-1}\{\eta(z)\}.$$

Proposition 14.4.2 *If y satisfies*

(i) $y(k) = 0$ for $k = -1, -2, \ldots$,
(ii) $y(k) = O(k^n)$, $n \in \mathbb{Z}_+$,

then y has a Z transform.

If $\eta(z)$ is the Z transform for some function $|z| > r$, then that function is

$$y(k) = \begin{cases} \frac{1}{2\pi i}\int_C z^{k-1}\eta(z)\,dz, & k = 0, 1, 2, \ldots \\ 0, & k = -1, -2, \ldots \end{cases},$$

where C is positively oriented cycle of radius $r' > r$ and center at $z = 0$.
   For Z transform related information, please refer to Tables 14.3 and 14.4.

Remark 14.4.3

$$\mathcal{Z}\{y(k+1)\} = y(1) + \frac{y(2)}{z} + \frac{y(3)}{z^2} + \cdots = z\eta(z) - zy(0).$$

The Laplace transform of $y'(t)$ is $s\eta(s) - y(0)$.

Remark 14.4.4 *The procedure to follow for using Z transforms to solve an initial value problem is as follows:*

S1. $y(k) \mapsto \eta(z)$.
S2. Solve the resulting linear algebraic equation $\eta(z) = \mathcal{Z}\{y(k)\}$.
S3. Find the inverse Z transform $y(k) = \mathcal{Z}^{-1}\{\eta(z)\}$.
S4. Verify that $y(k)$ is a solution.

Example 14.4.5

$$y(k+1) = ay(k) + f(k), \quad k = 0, 1, \ldots; \quad y(0) = y_0, \ a \neq 0$$

$$z\eta(z) - zy_0 = a\eta(z) + \phi(z) \Rightarrow \eta(z) = \frac{z}{z-a}y_0 + \frac{1}{z-a}\phi(z) = \eta_1(z) + \eta_2(z).$$

Table 14.3. A Brief Table for Z transforms

	Inverse $y(k)$	Valid Z transform $\eta(z)$	$\|z\| > r$ r
(1)	1	$\frac{z}{z-1}$	1
(2)	k	$\frac{z}{(z-1)^2}$	1
(3)	k^2	$\frac{z(z+1)}{(z-1)^3}$	1
(4)	k^3	$\frac{z(z^2+4z+1)}{(z-1)^4}$	1
(5) $k^{(m)}$, $m = 0,1,2,\ldots$		$\frac{m!z}{(z-1)^{m+1}}$	1
(6)	a^k	$\frac{z}{z-a}$	$\|a\|$
(7)	ka^k	$\frac{az}{(z-a)^2}$	$\|a\|$
(8)	$\frac{a^k}{k!}$	$e^{\frac{a}{z}}$	0
(9)	e^{-ak}	$\frac{z}{z-e^{-a}}$	e^{-a}
(10)	$\sin bk$	$\frac{z\sin b}{z^2-2z\cos b+e^{-2a}}$	1
(11)	$\cos bk$	$\frac{z(z-\cos b)}{z^2-2z\cos b+e^{-2a}}$	1
(12)	$e^{-ak}\sin bk$	$\frac{ze^{-a}\sin b}{z^2-2ze^{-a}\cos b+e^{-2a}}$	e^{-a}
(13)	$e^{-ak}\cos bk$	$\frac{z(z-e^{-a}\cos b)}{z^2-2ze^{-a}\cos b+e^{-2a}}$	e^{-a}

Table 14.4. Properties of Z transforms

	Inverse	Z transform
(1)	$y(k)$	$\eta(z)$
(2)	$ay_1(k) + by_2(k)$	$a\eta_1(z) + b\eta_2(z)$
(3)	$y(k+1)$	$z\eta(z) - zy(0)$
(4)	$y(k+n)$	$z^n\eta(z) - z^ny(0)$ $-z^{n-1}y(1) - \cdots - zy(n-1)$
(5)	$y(k-c)$, $c \geq 0$	$z^{-c}\eta(z)$
(6)	$a^ky(k)$	$\eta(\frac{z}{a})$
(7)	$ky(k)$	$-z\frac{d\eta(z)}{dz}$
(8)	$k^2y(k)$	$-z\frac{d}{dz}[-z\eta'(z)]$
(9)	$k^my(k)$, $m = 0,1,2,\ldots$	$\left(-z\frac{d}{dz}\right)^m \eta(z)$
(10)	$\sum_{u=0}^{k} y_1(k-u)y_2(u)$	$\eta_1(z)\eta_2(z)$
(11)	$y_1(k)y_2(k)$	$\frac{1}{2\pi i}\int_C \rho^{-1}\eta_1(\rho)\eta_2(\rho^{-1}z)\,d\rho$
(12)	$\sum_{u=0}^{k} y(u)$	$\frac{z}{z-1}\eta(z)$

We know $\dfrac{1}{z-a}\phi(z) = \dfrac{z}{z-a}\dfrac{\phi(z)}{z}$, *and* $\eta_2(z) = \dfrac{\phi(z)}{z} \Rightarrow y_2(k) = f(k-1)$.

Then, by superposition,

$$y(k) = a^k y_0 + \sum_{u=0}^{k} f(k-1-u)a^u.$$

Remark 14.4.6 *In order to solve the linear difference system*

$$y(k+1) = Ay(k) + f(k); \quad y(0) = y_0,$$

we will take the Z transform of the components of $y(k)$, then we have

$$(zI - A)\eta(z) = zy_0 + \phi(z),$$

where $\eta(z) = [\eta_1(z), \cdots, \eta_n(z)]^T$ is the vector of Z transforms of the components of y. If z is not an eigenvalue of A, then the coefficient matrix is nonsingular. Thus, for sufficiently large $|z|$, the unique solution is

$$\eta(z) = z(zI - A)^{-1}y_0 + (zI - A)^{-1}\phi(z),$$

where we have

$$\mathcal{Z}\left\{A^k\right\} = z(zI - A)^{-1}.$$

In order to find a particular solution, we solve $(zI - A)^{-1}p(z) = \phi(z)$ for $p(z)$ and find its inverse Z transform.

Example 14.4.7 *Let us take our previous example problem:*

$$y_1(k+1) = y_2(k) + 1, \quad y_1(0) = 3,$$

$$y_2(k+1) = y_1(k) + 1, \quad y_2(0) = 1.$$

$$\eta(z) = \frac{z}{(z-1)(z+1)}\begin{bmatrix} z & 1 \\ 1 & z \end{bmatrix}\begin{bmatrix} 3 \\ 1 \end{bmatrix} + \frac{1}{(z-1)(z+1)}\begin{bmatrix} z & 1 \\ 1 & z \end{bmatrix}\begin{bmatrix} \frac{z}{z-1} \\ \frac{z}{(z-1)^2} \end{bmatrix}$$

$$\mathcal{Z}\left\{A^k y_0\right\} = \frac{z}{z-1}\begin{bmatrix} 2 \\ 2 \end{bmatrix} + \frac{z}{z+1}\begin{bmatrix} 1 \\ -1 \end{bmatrix}$$

$$\Pi(z) = \frac{z}{(z-1)^3}\begin{bmatrix} \frac{1}{2} \\ \frac{1}{2} \end{bmatrix} + \frac{z}{(z-1)^2}\begin{bmatrix} \frac{1}{4} \\ \frac{3}{4} \end{bmatrix} + \frac{z}{(z-1)}\begin{bmatrix} \frac{3}{8} \\ -\frac{3}{8} \end{bmatrix} + \frac{z}{(z+1)}\begin{bmatrix} -\frac{3}{8} \\ \frac{3}{8} \end{bmatrix}.$$

$$\Rightarrow y(k) = k^2\begin{bmatrix} \frac{1}{4} \\ \frac{1}{4} \end{bmatrix} + k\begin{bmatrix} 0 \\ \frac{1}{2} \end{bmatrix} + (-1)^k\begin{bmatrix} \frac{5}{8} \\ -\frac{5}{8} \end{bmatrix} + \begin{bmatrix} \frac{19}{8} \\ \frac{13}{8} \end{bmatrix}.$$

Problems

14.1. Solve $y''(t) - y(t) = e^{2t}$; $y(0) = 2$, $y'(0) = 0$.

14.2. Solve $y(k+1) = y(k) + 2e^k$; $y(0) = 1$.

14.3. Consider a combat situation between Blue (x) and Red (y) forces in which Blue is under a directed fire from Red at a rate of 0.2 Blue-units/unit-time/Red-firer and Red is subjected to directed fire at a rate of 0.3 Red-units/unit-time/Blue-firer plus a non-combat loss (to be treated as self directed fire) at a rate of 0.1 Red-units/unit-time/Red-unit. Suppose that there are 50 Blue and 100 Red units initially. Find the surviving Red units at times $t = 0, 1, 2, 3, 4$ using the Laplace transformation.

14.4. Find the closed form solution for the Fibonacci sequence $F_{k+2} = F_{k+1} + F_k$, $F_1 = 1$, $F_2 = 1$ using the Z-transformation and calculate F_{100}.

Web material

http://ccrma.stanford.edu/~jos/filters/Laplace_Transform_
    Analysis.html
http://claymore.engineer.gvsu.edu/~jackh/books/model/chapters/
    laplace.pdf
http://cnx.org/content/m10110/latest/
http://cnx.org/content/m10549/latest/
http://dea.brunel.ac.uk/cmsp/Home_Saeed_Vaseghi/Chapter04-Z-
    Transform.pdf
http://dspcan.homestead.com/files/Ztran/zdiff1.htm
http://dspcan.homestead.com/files/Ztran/zlap.htm
http://en.wikipedia.org/wiki/Laplace_Transform
http://en.wikipedia.org/wiki/Z-transform
http://eom.springer.de/l/l057540.htm
http://eom.springer.de/Z/z130010.htm
http://fourier.eng.hmc.edu/e102/lectures/Z_Transform/
http://home.case.edu/~pjh4/MATH234/zTransform.pdf
http://homepage.newschool.edu/~foleyd/GEC06289/laplace.pdf
http://kwon3d.com/theory/filtering/ztrans.html
http://lanoswww.epfl.ch/studinfo/courses/cours_dynsys/extras/
    Smith(2002)_Introduction_to_Laplace_Transform_Analysis.pdf
http://lorien.ncl.ac.uk/ming/dynamics/laplace.pdf
http://math.fullerton.edu/mathews/c2003/ztransform/ZTransformBib/
    Links/ZTransformBib_lnk_3.html
http://math.fullerton.edu/mathews/c2003/ZTransformBib.html
http://math.ut.ee/~toomas_l/harmonic_analysis/Fourier/node35.html
http://mathworld.wolfram.com/LaplaceTransform.html
http://mathworld.wolfram.com/Z-Transform.htm
http://mywebpages.comcast.net/pgoodmann/EET357/Lectures/Lecture8.ppt
http://ocw.mit.edu/OcwWeb/Electrical-Engineering-and-Computer-
    Science/6-003Fall-2003/LectureNotes/
http://phyastweb.la.asu.edu/phy501-shumway/notes/lec20.pdf
http://planetmath.org/encyclopedia/LaplaceTransform.html
http://umech.mit.edu/weiss/PDFfiles/lectures/lec12wm.pdf
http://umech.mit.edu/weiss/PDFfiles/lectures/lec5wm.pdf

http://web.mit.edu/2.161/www/Handouts/ZLaplace.pdf
http://www.absoluteastronomy.com/z/z-transform
http://www.atp.ruhr-uni-bochum.de/rt1/syscontrol/node11.html
http://www.atp.ruhr-uni-bochum.de/rt1/syscontrol/node6.html
http://www.cbu.edu/~rprice/lectures/laplace.html
http://www.cs.huji.ac.il/~control/handouts/laplace_Boyd.pdf
http://www.dspguide.com/ch33.htm
http://www.ece.nmsu.edu/ctrlsys/help/lxprops.pdf
http://www.ece.rochester.edu/courses/ECE446/The%20z-transform.pdf
http://www.ece.utexas.edu/~bevans/courses/ee313/lectures/
    15_Z_Transform/index.html
http://www.ece.utexas.edu/~bevans/courses/ee313/lectures/
    18_Z_Laplace/index.html
http://www.ee.columbia.edu/~dpwe/e4810/lectures/L04-ztrans.pdf
http://www.efunda.com/math/laplace_transform/index.cfm
http://www.facstaff.bucknell.edu/mastascu/eControlHTML/Sampled/
    Sampled1.html
http://www.faqs.org/docs/sp/sp-142.html
http://www.geo.cornell.edu/geology/classes/brown/eas434/Notes/
    Fourier%20family.doc
http://www.intmath.com/Laplace/Laplace.php
http://www.just.edu.jo/~hazem-ot/signal1.pdf
http://www.ling.upenn.edu/courses/ling525/z.html
http://www.ma.umist.ac.uk/kd/ma2m1/laplace.pdf
http://www.maths.abdn.ac.uk/~igc/tch/engbook/node59.html
http://www.maths.manchester.ac.uk/~kd/ma2m1/laplace.pdf
http://www.plmsc.psu.edu/~www/matsc597/fourier/laplace/laplace.html
http://www.realtime.net/~drwolf/papers/dissertation/node117.html
http://www.roymech.co.uk/Related/Control/Laplace_Transforms.html
http://www.sosmath.com/diffeq/laplace/basic/basic.html
http://www.swarthmore.edu/NatSci/echeeve1/Ref/Laplace/Table.html
http://www.u-aizu.ac.jp/~qf-zhao/TEACHING/DSP/lec04.pdf
http://www.u-aizu.ac.jp/~qf-zhao/TEACHING/DSP/lec05.pdf
www.brunel.ac.uk/depts/ee/Research_Programme/COM/Home_Saeed_Vaseghi/
    Chapter04-Z-Transform.pdf
www.ee.ucr.edu/~yhua/ee141/lecture4.pdf

Solutions

Problems of Chapter 1

1.1

(a) Since, f is continuous at x:

$$\forall \epsilon_1 > 0 \; \exists \delta_1 > 0 \ni \forall y \ni |x - y| < \delta_1 \Rightarrow |f(x) - f(y)| < \epsilon_1.$$

g is continuous at x:

$$\forall \epsilon_2 > 0 \; \exists \delta_2 > 0 \ni \forall y \ni |x - y| < \delta_2 \Rightarrow |g(x) - g(y)| < \epsilon_2.$$

Fix ϵ_1 and ϵ_2 at $\frac{\epsilon}{2}$.

$$\exists \delta_1 > 0 \ni \forall y \ni |x - y| < \delta_1 \Rightarrow |f(x) - f(y)| < \frac{\epsilon}{2}$$

$$\exists \delta_1 > 0 \ni \forall y \ni |x - y| < \delta_2 \Rightarrow |g(x) - g(y)| < \frac{\epsilon}{2}$$

Let $\delta = \min\{\delta_1, \delta_2\} > 0$.

$$\forall y \ni |x - y| < \delta \Rightarrow |f(x) - f(y)| < \frac{\epsilon}{2}, \; |g(x) - g(y)| < \frac{\epsilon}{2}$$

$$|(f + g)(x) - (f + g)(y)| = |f(x) + g(x) - f(y) - g(y)| \leq$$

$$|f(x) - f(y)| + |g(x) - g(y)| < \frac{\epsilon}{2} + \frac{\epsilon}{2} = \epsilon$$

Thus, $\forall \epsilon > 0 \; \exists \delta > 0 \ni \forall y \ni |x - y| < \delta \Rightarrow |(f + g)(x) - (f + g)(y)| < \epsilon$.
Therefore, $f + g$ is continuous at x.

(b) f is continuous at x:

$$\forall \epsilon_1 > 0 \; \exists \delta > 0 \ni \forall y \ni |x - y| < \delta \Rightarrow |f(x) - f(y)| < \epsilon.$$

Fix $\epsilon = \bar{\epsilon}$. Then,

$$\exists \delta > 0 \; (\text{say } \bar{\delta}) \ni \forall y \; (\text{can fix at } \bar{y}) \ni |x - y| < \delta \Rightarrow |f(x) - f(y)| < \bar{\epsilon}.$$

We have $|x - \bar{y}| < \bar{\delta} \Rightarrow |f(x) - f(\bar{y})| < \bar{\epsilon}$.

$$\forall y \ni |x - y| < \acute{\delta}, \; |f(x) - f(y)| \leq c|x - y|.$$

Choose $\bar{y} \ni |x - \bar{y}| < \acute{\delta}, \; |f(x) - f(\bar{y})| \leq c|x - \bar{y}| \leq c\acute{\delta}$.

If $\left\{ \begin{matrix} 0 < \delta < \frac{\bar{\epsilon}}{c} \\ \delta < \bar{\delta} \end{matrix} \right\}$, we will reach the desired condition. One can choose $0 < \delta <$
$\min\{\bar{\delta}, \frac{\bar{\epsilon}}{c}\}$.

$$\forall \bar{y} \ni |x - \bar{y}| < \delta < \acute{\delta}, \; |f(x) - f(y)| \leq c|x - \bar{y}| \leq c\delta < \bar{\epsilon}.$$

1.2 *Observation*: Every time we break a piece, the total number of pieces is increased by one. When there is no pieces to break, each piece is a small square. At the beginning when we had the whole chocolate with n squares after $b=0$ breaks, we had $p=1$ piece. After one break ($b=1$), we got $p=2$ pieces. Therefore, p is always greater by one than b, i.e. $p = b + 1$. In the end,

$$p = b + 1 = n.$$

The above argument constitutes a direct proof. Let us use induction to prove that the above observation $b = n - 1$ is correct.

1. $n = 2 \Rightarrow b = 1$, i.e. if there are only two squares, we clearly need one break.
2. Assume that for $2 \le k \le n - 1$ squares it takes only $k - 1$ breaks. In order to break the chocolate bar with n squares, we first split into two with k_1 and k_2 squares ($k_1 + k_2 = n$). By the induction hypothesis, it will take $k_1 - 1$ breaks to split the first bar and $k_2 - 1$ to split the second. Thus, the total is

$$b = 1 + (k_1 - 1) + (k_2 - 1) = k_1 + k_2 - 1 = n - 1.$$

1.3

(a) $\binom{n}{r} = \binom{n}{n-r}$:

Full Forward Method:

$$\binom{n}{r} = \frac{n!}{(n-r)!\,r!} = \frac{n!}{r!\,(n-r)!} = \binom{n}{n-r}$$

Combinatorial Method:
$\binom{n}{r}$ denotes the number of different ways of selecting r objects out of n objects in an urn. If we look at the same phenomenon from the viewpoint of the objects left in the urn, the number of different ways of selecting $n - r$ objects out of n is $\binom{n}{n-r}$. These two must be equal since we derive them from two viewpoints of the same phenomenon.

(b) $\binom{n}{r} = \binom{n-1}{r} + \binom{n-1}{r-1}$:

Full Backward Method:

$$\binom{n-1}{r} + \binom{n-1}{r-1} = \frac{(n-1)!}{(n-1-r)!\,r(r-1)!} + \frac{(n-1)!}{(n-r)(n-r-1)!\,(r-1)!} =$$

$$= \frac{(n-1)!\,[n-r+r]}{(n-r)!\,r!} = \binom{n}{r}$$

Combinatorial Method:
$\binom{n}{r}$ denotes the number of different ways of selecting r balls out of n objects in

an urn. Let us fix a ball, call it *super ball*. Two mutually exclusive alternatives exist; we either select the super ball or it stays in the urn. Given that the super ball is selected, the number of different ways of choosing $r-1$ balls out of $n-1$ is $\binom{n-1}{r-1}$. In the case that the super ball is not selected, $\binom{n-1}{r}$ denotes the number of ways of choosing r balls out of $n-1$. By the rule of sum, the right hand side is equal to the left hand side.

(c) $\binom{n}{0} + \binom{n}{1} + \cdots + \binom{n}{n} = 2^n$:

We will use the corollary to the following theorem.

Theorem S.1.1 (Binomial Theorem)

$$(1+x)^n = \binom{n}{0}x^0 + \binom{n}{1}x^1 + \cdots + \binom{n}{n}x^n$$

Corollary S.1.2 *Let $x = 1$ in the Binomial Theorem. Then,*

$$(1+1)^n = 2^n = \binom{n}{0} + \binom{n}{1} + \cdots + \binom{n}{n}$$

Combinatorial Method:
2^n is the number of subsets of a set of size n. $\binom{n}{0} = 1$ is for the empty set, $\binom{n}{n} = 1$ is for the set itself, and $\binom{n}{r}$, $r = 2, \ldots, n-1$ is the number of proper subsets of size r.

(d) $\binom{n}{m}\binom{m}{r} = \binom{n}{r}\binom{n-r}{m-r}$:

Forward – Backward Method:

$$\binom{n}{m}\binom{m}{r} = \frac{n!}{(n-m)!\,m!}\frac{m!}{(m-r)!\,r!} = \frac{n!}{(n-m)!\,(m-r)!\,r!}$$

$$\binom{n}{r}\binom{n-r}{m-r} = \frac{n!}{r!\,(n-r)!}\frac{(n-r)!}{(n-m)!\,(m-r)!} = \frac{n!}{r!\,(n-m)!\,(m-r)!}$$

Combinatorial Method:
$\binom{n}{m}$ denotes the number of different ways of selecting m Industrial Engineering students out of n M.E.T.U. students and $\binom{m}{r}$ denotes the number of different ways of selecting r Industrial Engineering students taking the Mathematics for O.R. course out of m I.E. students. On the other hand, $\binom{n}{r}$ denotes the number of ways of selecting r Industrial Engineering students taking Mathematics for O.R. from among n M.E.T.U. students and $\binom{n-r}{m-r}$ denotes the number of different ways of selecting $m-r$ Industrial Engineering students

not taking Mathematics for O.R. out of $n - r$ M.E.T.U. students not taking Mathematics for O.R. These two are equivalent.

(e) $\binom{n}{0} + \binom{n+1}{1} + \cdots + \binom{n+r}{r} = \binom{n+r+1}{r}$:

Trivial:
Apply item (b) r-times to the right hand side.

Combinatorial Method:
The right hand side, $\binom{n+r+1}{r}$, denotes the number of different ways of selecting r balls out of $m = n+2$ balls with repetition, known as the multi-set problem. Let | be the column separator if we reserve a column for each of m objects, let $\sqrt{}$ be used as the tally mark if the object in the associated column is selected. Then, we have a string of size $r + (m - 1)$ in which there are r tally marks and $m - 1$ column separators. For instance, if we have three objects $\{x, y, z\}$, and we sample four times, "$\sqrt{}|\sqrt{}\sqrt{}|\sqrt{}$" means x and z are selected once and y is selected twice. Then, the problem is equivalent to selecting the places of r tally marks in the string of size $r + (m - 1)$, which is $\binom{r+m-1}{r}$.

Let us fix the super ball again. The left hand side is the list of the number of times that the super ball is selected in the above multi-set problem instance. That is, $\binom{n}{0}$ refers to the case in which the super ball is not selected, $\binom{n+1}{1}$ refers to the case in which the super ball is selected once, and $\binom{n+r}{r}$ refers to the case in which the super ball is always selected.

These two are equivalent.

Problems of Chapter 2

2.1 (a)

$$[A\|I_9] = \begin{array}{c} \\ a \\ b \\ c \\ d \\ e \\ f \\ g \\ h \\ i \end{array} \left[\begin{array}{ccccccccccccc|ccccccccc} 1\,2\,3\,4\,5\,6\,7\,8\,9\,10\,11\,12\,13 \\ 1\,1\,0\,0\,0\,0\,0\,0\,0\ 0\ \ 0\ \ 0\ \ 0 & 1 \\ 1\,0\,0\,0\,0\,0\,0\,1\,0\ 0\ \ 0\ \ 0\ \ 0 & & 1 \\ 0\,1\,1\,0\,0\,0\,0\,0\,0\ 0\ \ 0\ \ 0\ \ 0 & & & 1 \\ 0\,0\,1\,1\,0\,0\,0\,1\,0\ 0\ \ 0\ \ 0\ \ 0 & & & & 1 \\ 0\,0\,0\,1\,1\,0\,0\,0\,1\ 0\ \ 0\ \ 0\ \ 0 & & & & & 1 \\ 0\,0\,0\,0\,1\,1\,0\,0\,0\ 0\ \ 1\ \ 1 & & & & & & 1 \\ 0\,0\,0\,0\,0\,1\,1\,0\,0\ 0\ \ 1\ \ 0\ \ 0 & & & & & & & 1 \\ 0\,0\,0\,0\,0\,0\,0\,1\,0\ 1\ \ 1\ \ 0\ \ 1 & & & & & & & & 1 \\ 0\,0\,0\,0\,0\,0\,1\,1\,0\ 0\ \ 0\ \ 1\ \ 0 & & & & & & & & & 1 \end{array}\right]$$

$$a+b \to b;\ a+b \to a;\ b+c \to c;\ c+d \to d;$$

$$d+e \to e;\ e+f \to f;\ f+g \to g;\ g+i \to i;\ h+i \to i$$

$$[A\|I_9] \longrightarrow \left[\begin{array}{c|c} I_8 \| N \| C \\ \hline 0 \ \| D \end{array}\right]$$

$$= \begin{array}{c} a \\ b \\ c \\ d \\ e \\ f \\ g \\ h \\ i \end{array} \left[\begin{array}{cccccccc|ccccc|ccccccccc} 1\,0\,0\,0\,0\,0\,0\,0 & 1\,0\,0\,0\,0 & 0\,1\,0\,0\,0\,0\,0\,0 \\ 0\,1\,0\,0\,0\,0\,0\,0 & 1\,0\,0\,0\,0 & 1\,1\,0\,0\,0\,0\,0\,0 \\ 0\,0\,1\,0\,0\,0\,0\,0 & 1\,0\,0\,0\,0 & 1\,1\,1\,0\,0\,0\,0\,0 \\ 0\,0\,0\,1\,0\,0\,0\,0 & 0\,0\,0\,0\,0 & 1\,1\,1\,1\,0\,0\,0\,0 \\ 0\,0\,0\,0\,1\,0\,0\,0 & 0\,1\,0\,0\,0 & 1\,1\,1\,1\,1\,0\,0\,0 \\ 0\,0\,0\,0\,0\,1\,0\,0 & 0\,1\,0\,1\,1 & 1\,1\,1\,1\,1\,1\,0\,0\,0 \\ 0\,0\,0\,0\,0\,0\,1\,0 & 0\,1\,1\,1\,1 & 1\,1\,1\,1\,1\,1\,1\,0\,0 \\ 0\,0\,0\,0\,0\,0\,0\,1 & 0\,1\,1\,0\,1 & 0\,0\,0\,0\,0\,0\,0\,1\,0 \\ 0\,0\,0\,0\,0\,0\,0\,0 & 0\,0\,0\,0\,0 & 1\,1\,1\,1\,1\,1\,1\,1\,1 \end{array}\right]$$

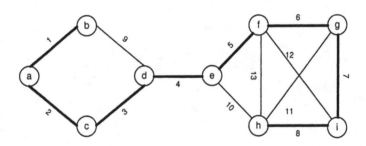

Fig. S.1. The tree T in Problem 2.1

Each basis corresponds to a spanning tree T in $G = (V, E)$, where $T \subset E$ connects every vertex and $\|T\| = \|V\| - 1$. Here, we have $T =$

$\{1, 2, 3, 4, 5, 6, 7, 8\}$. See Figure S.1.

(b) Each row represents a fundamental cocycle (cut) in the graph. In the tree, we term one node as root (node i), and we can associate an edge of the tree with every node like $1 \to b$, $2 \to a$, $3 \to c$, $4 \to d$, $\cdots$, $8 \to h$ as if we hanged the tree to the wall by its root. Then, if the associated edge (say edge 6) in the tree for the node (say f) in the identity part of z_i is removed, we partition the nodes into two sets as $V_1 = \{a, b, c, d, e, f\}$ and $V_2 = \{g, h, i\}$. The nonzero entries in z_f correspond to edges $10, 12, 13$, defining the set of edges connecting nodes in different parts of this partition or the cut. The set of such edges are termed as fundamental cocycle. See Figure S.2.

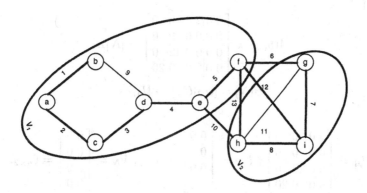

Fig. S.2. The cocycle defined by cutting edge $6 \to f$ in Problem2.1

(c) Each column represents a fundamental cycle. If we add the edge identified by I_5 part into T, we will create a cycle defined by the nonzero elements of y^j. See Figure S.3.

(d) The first 8 columns of A form a basis for column space $\mathcal{R}(A)$. The columns of matrix Y is a basis for the null space $\mathcal{N}(A)$. The rows of C constitute a basis for the row space $\mathcal{R}(A^T)$. Finally, the row(s) of matrix D is (are) the basis vectors for the left-null space $\mathcal{N}(A^T)$.

Remark S.2.1 *If our graph $G = (V, E)$ is bipartite, i.e. $V = V_1 \bigcup V_2 \ni V_1 \bigcap V_2 = \emptyset$, $V_1 \neq \emptyset \neq V_2$ and $\forall e = (v_1, v_2) \in E$, $v_1 \in V_1$, $v_2 \in V_2$, and we solve $\max c^T x$ s.t. $Ax = b$, $x \geq 0$ using standard simplex algorithm over $GF(2)$, we will have exactly what we know as the transportation simplex method. Furthermore, for general graphs $G = (V, E)$, if we solve $\max c^T x$ s.t.*

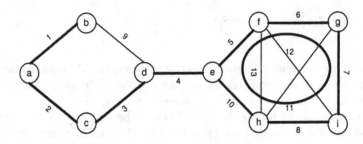

Fig. S.3. The fundamental cycle defined by edge 10 in Problem2.1

$Ax = b$, $x \geq 0$ *using a standard simplex algorithm over* $GF(2)$, *we will get the network simplex method.*

2.2 (a)

$$A(5,2) = \begin{bmatrix} 0 & 0 & 2 & 0 & 0 & 0 \\ 0 & 0 & 0 & 6 & 0 & 0 \\ 0 & 0 & 0 & 0 & 12 & 0 \\ 0 & 0 & 0 & 0 & 0 & 20 \end{bmatrix} = [N|B]$$

$$[B|N] = [U_B|U_N] \rightarrow [I_4|V_N]$$

where

$$U_B = \begin{bmatrix} 2 & 0 & 0 & 0 \\ 0 & 6 & 0 & 0 \\ 0 & 0 & 12 & 0 \\ 0 & 0 & 0 & 20 \end{bmatrix}, \quad U_N = \begin{bmatrix} 0 & 0 \\ 0 & 0 \\ 0 & 0 \\ 0 & 0 \end{bmatrix} = 0_{4\times 2}, \quad V_N = \begin{bmatrix} 0 & 0 \\ 0 & 0 \\ 0 & 0 \\ 0 & 0 \end{bmatrix} = 0_{4\times 2}.$$

Then,

$$\mathcal{R}(A) = Span\{2e_1, 6e_2, 12e_3, 20e_4\} = Span\{e_1, e_2, e_3, e_4\} = \mathbb{R}^4.$$

The rank of $A(n,k)$ is $r = 4$.

$$\mathcal{R}(A^T) = Span\{2e_3, 6e_4, 12e_5, 20e_6\} = Span\{e_3, e_4, e_5, e_6\} = \mathbb{R}^4.$$

$$\mathcal{N}(A) = Span\left\{ \begin{pmatrix} 1 \\ 0 \\ 0 \\ 0 \\ 0 \\ 0 \end{pmatrix}, \begin{pmatrix} 0 \\ 1 \\ 0 \\ 0 \\ 0 \\ 0 \end{pmatrix} \right\} = Span\{e_1, e_2\} = \mathbb{R}^2.$$

$$\mathcal{N}(A^T) = \{\theta\}, \quad dim\mathcal{N}(A^T) = 0.$$

Thus, $\mathbb{R}^6 = \mathcal{R}(A^T) \oplus \mathcal{N}(A) = \mathbb{R}^4 \oplus \mathbb{R}^2$ and $\mathbb{R}^4 = \mathcal{R}(A) \oplus \mathcal{N}(A^T) = \mathbb{R}^4 \oplus \emptyset = \mathbb{R}^4$.

(b) *Differentiator:*

$$A(n,k) = \begin{bmatrix} 0 \cdots 0 & \prod_{i=1}^{k} i & 0 & 0 & 0 & 0 \\ \vdots \ddots \vdots & 0 & \ddots & 0 & 0 & 0 \\ 0 \cdots 0 & 0 & 0 & \prod_{i=j}^{k+j-1} i & 0 & 0 \\ \hline \vdots \ddots \vdots & 0 & 0 & 0 & \ddots & 0 \\ 0 \cdots 0 & 0 & 0 & 0 & 0 & \prod_{i=n-k+1}^{n} i \end{bmatrix} = [N(n,k)|B(n,k)]$$

$$[B(n,k)|N(n,k)] \to [I_{n-k+1}|0]$$

Then,

$$\mathcal{R}(A) = Span \left\{ \left(\prod_{i=1}^{k} i \right) e_1, \cdots, \left(\prod_{i=n-k+1}^{n} i \right) e_{n-k+1} \right\}$$

$$= Span \{e_1, \cdots, e_{n-k+1}\} = \mathbb{R}^{n-k+1}.$$

$$\mathcal{R}(A^T) = Span \left\{ \left(\prod_{i=1}^{k} i \right) e_{k+1}, \cdots, \left(\prod_{i=n-k+1}^{n} i \right) e_n \right\}$$

$$= Span \{e_{k+1}, \cdots, e_n\} = \mathbb{R}^{n-k+1}.$$

$$\mathcal{N}(A) = Span \left\{ \begin{pmatrix} 1 \\ 0 \\ \vdots \\ 0 \\ \hline 0 \\ \vdots \\ 0 \end{pmatrix}, \cdots, \begin{pmatrix} 0 \\ \vdots \\ 0 \\ 1 \\ \hline 0 \\ \vdots \\ 0 \end{pmatrix} \right\} = Span \{e_1, \cdots, e_k\} = \mathbb{R}^k.$$

$$\mathcal{N}(A^T) = \{\theta\}, \quad dim\mathcal{N}(A^T) = 0.$$

(c) *Integrator:*

$$B(n,k) = \begin{bmatrix} 0 & \cdots & 0 & \cdots & 0 \\ \vdots & \ddots & \vdots & \ddots & \vdots \\ 0 & \cdots & 0 & \cdots & 0 \\ \hline \frac{1}{\prod_{i=1}^{k} i} & 0 & 0 & 0 & 0 \\ 0 & \ddots & 0 & 0 & 0 \\ 0 & 0 & \frac{1}{\prod_{i=j}^{k+j-1} i} & 0 & 0 \\ 0 & 0 & 0 & \ddots & 0 \\ 0 & 0 & 0 & 0 & \frac{1}{\prod_{i=n-k+1}^{n} i} \end{bmatrix}.$$

After permuting some rows, we have

$$PB(n,k) = \begin{bmatrix} \frac{1}{\prod_{i=1}^{k} i} & 0 & 0 & 0 & 0 \\ 0 & \ddots & 0 & 0 & 0 \\ 0 & 0 & \frac{1}{\prod_{i=j}^{k+j-1} i} & 0 & 0 \\ 0 & 0 & 0 & \ddots & 0 \\ 0 & 0 & 0 & 0 & \frac{1}{\prod_{i=n-k+1}^{n} i} \\ \hline 0 & \cdots & 0 & \cdots & 0 \\ \vdots & \ddots & \vdots & \ddots & \vdots \\ 0 & \cdots & 0 & \cdots & 0 \end{bmatrix}.$$

$$[PB(n,k)] \rightarrow \begin{bmatrix} U_B \\ 0 \end{bmatrix} \rightarrow \begin{bmatrix} I_{n-k+1} \\ 0 \end{bmatrix},$$

where

$$U_B = \begin{bmatrix} \frac{1}{\prod_{i=1}^{k} i} & 0 & 0 & 0 & 0 \\ 0 & \ddots & 0 & 0 & 0 \\ 0 & 0 & \frac{1}{\prod_{i=j}^{k+j-1} i} & 0 & 0 \\ 0 & 0 & 0 & \ddots & 0 \\ 0 & 0 & 0 & 0 & \frac{1}{\prod_{i=n-k+1}^{n} i} \end{bmatrix}.$$

Thus,

$$\mathcal{R}(B) = \mathbb{R}^{n-k+1} = \mathcal{R}(B^T).$$

Furthermore,

$$\mathcal{N}(B^T) = \mathbb{R}^k \text{ and } \mathcal{N}(B) = \{\theta\}.$$

2.3

1. Let $n = 4$ and characterize bases for the four fundamental subspaces related to $A = [y_1|y_2|\cdots|y_n]$.

$$[A\|I_4] = \begin{bmatrix} -1 & 0 & 0 & 1 & \| & 1 & 0 & 0 & 0 \\ 1 & -1 & 0 & 0 & \| & 0 & 1 & 0 & 0 \\ 0 & 1 & -1 & 0 & \| & 0 & 0 & 1 & 0 \\ 0 & 0 & 1 & -1 & \| & 0 & 0 & 0 & 1 \end{bmatrix} \rightarrow \begin{bmatrix} 1 & 0 & 0 & -1 & \| & -1 & 0 & 0 & 0 \\ 0 & 1 & 0 & -1 & \| & -1 & -1 & 0 & 0 \\ 0 & 1 & -1 & 0 & \| & 0 & 0 & 1 & 0 \\ 0 & 0 & 1 & -1 & \| & 0 & 0 & 0 & 1 \end{bmatrix} \rightarrow$$

$$\begin{bmatrix} 1 & 0 & 0 & -1 & \| & -1 & 0 & 0 & 0 \\ 0 & 1 & 0 & -1 & \| & -1 & -1 & 0 & 0 \\ 0 & 0 & -1 & 1 & \| & 1 & 1 & 1 & 0 \\ 0 & 0 & 1 & -1 & \| & 0 & 0 & 0 & 1 \end{bmatrix} \rightarrow \begin{bmatrix} 1 & 0 & 0 & -1 & \| & -1 & 0 & 0 & 0 \\ 0 & 1 & 0 & -1 & \| & -1 & -1 & 0 & 0 \\ 0 & 0 & 1 & -1 & \| & -1 & -1 & -1 & 0 \\ 0 & 0 & 0 & 0 & \| & 1 & 1 & 1 & 1 \end{bmatrix} = \begin{bmatrix} I_3 | V_N & \| & S_I \\ \hline O & \| & S_{II} \end{bmatrix},$$

where $V_N = \begin{bmatrix} -1 \\ -1 \\ -1 \end{bmatrix}$, $S_I = \begin{bmatrix} -1 & 0 & 0 & 0 \\ -1 & -1 & 0 & 0 \\ -1 & -1 & -1 & 0 \end{bmatrix}$, $S_{II} = \begin{bmatrix} 1 & 1 & 1 & 1 \end{bmatrix}$.

Thus, $\mathcal{R}(A) = Span\{y_1, y_2, y_3\}$. $\mathcal{N}(A) = Span\{t\}$, where

$$t = \begin{bmatrix} -V_N \\ \hline I_{4-3} \end{bmatrix} = \begin{bmatrix} 1 \\ 1 \\ 1 \\ \hline 1 \end{bmatrix}.$$

Moreover,

$$\mathcal{R}(A^T) = Span\left\{ \begin{bmatrix} -1 \\ 0 \\ 0 \\ 1 \end{bmatrix}, \begin{bmatrix} 1 \\ -1 \\ 0 \\ 0 \end{bmatrix}, \begin{bmatrix} 0 \\ 1 \\ -1 \\ 0 \end{bmatrix} \right\} = Span\{-y_4, -y_1, -y_2\}.$$

And finally, $\mathcal{N}(A^T) = Span\{S_{II}\} = Span\left\{ \begin{bmatrix} 1 & 1 & 1 & 1 \end{bmatrix}^T \right\}$.

The case for $n = 3$ is illustrated in Figure S.4. y_1 is on the plane defined by $Span\{e_1, e_2\}$, y_2 is on the plane defined by $Span\{e_2, e_3\}$ and y_3 is on the $Span\{e_1, e_3\}$. Let us take $\{y_1, y_2\}$ in the basis for $\mathcal{R}(A)$, which defines the red plane on the right hand side of the figure. The normal to the plane is defined by the basis vector of $\mathcal{N}(A) = Span\{[1, 1, 1]^T\}$. We have $\mathcal{N}(A) = (\mathcal{R}(A))^\perp$ since $\mathcal{N}(A) \equiv \mathcal{N}(A^T)$ (therefore, $\mathcal{R}(A^T) \equiv \mathcal{R}(A)$ by the Fundamental Theorem of Linear Algebra–Part 2) in this particular exercise.

2. Let us discuss the general case. Let $e = (1, \cdots, 1)^T$

$$[A\|I_n] \rightarrow \begin{bmatrix} I_{n-1} | V_N & \| & S_I \\ \hline O & \| & S_{II} \end{bmatrix},$$

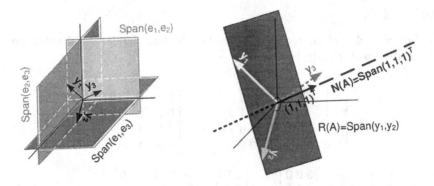

Fig. S.4. The range and null spaces of $A = [y_1|y_2|y_3]$

where $V_N = \begin{bmatrix} -1 \\ \vdots \\ -1 \end{bmatrix} = -e$, $S_I = \begin{bmatrix} -1 & 0 & 0 & | & 0 \\ \vdots & \ddots & 0 & | & \vdots \\ -1 & \cdots & -1 & | & 0 \end{bmatrix}$, $S_{II} = [1, \cdots, 1] = e^T$.

Thus, $\mathcal{R}(A) = Span\{y_1, \cdots, y_{n-1}\}$. $\mathcal{N}(A) = Span\{t\}$, where

$$t = \begin{bmatrix} -V_N \\ \hline I_1 \end{bmatrix} = \begin{bmatrix} 1 \\ \vdots \\ \hline 1 \end{bmatrix} = e.$$

Moreover,

$$\mathcal{R}(A^T) = Span\{-y_n, -y_1, \cdots, -y_{n-2}\}.$$

And finally, $\mathcal{N}(A^T) = Span\{S_{II}\} = Span\{[1, \cdots, 1]^T\} = Span\{e\}$. We have $\mathcal{N}(A) = (\mathcal{R}(A))^\perp$ since $\mathcal{N}(A) \equiv \mathcal{N}(A^T)$ (therefore, $\mathcal{R}(A^T) \equiv \mathcal{R}(A)$ by the Fundamental Theorem of Linear Algebra–part 2) in this particular exercise.

Problems of Chapter 3

3.1

$$A = \begin{bmatrix} 1 & 2 & 0 & -1 \\ 1 & -1 & 3 & 2 \\ 1 & -1 & 3 & 2 \\ -1 & 1 & -3 & 1 \end{bmatrix} = \begin{bmatrix} a^1 & a^2 & a^3 & a^4 \end{bmatrix}.$$

$$v_1 = a^1 = \begin{bmatrix} 1 \\ 1 \\ 1 \\ -1 \end{bmatrix} \Rightarrow v_1^T v_1 = 4, \ v_1^T a^2 = -1, \ v_1^T a^3 = 9, \ v_1^T a^4 = 2.$$

$$v_2 = a^2 - \frac{-1}{4} v_1 = \begin{bmatrix} \frac{9}{4} \\ -\frac{3}{4} \\ -\frac{3}{4} \\ \frac{3}{4} \end{bmatrix} \Rightarrow v_2^T v_2 = \frac{27}{4}, \ v_2^T a^3 = -9, \ v_2^T a^4 = -\frac{9}{2}.$$

$$v_3 = a^3 - \frac{-\frac{27}{4}}{\frac{27}{4}} v_2 - \frac{-9}{4} v_1 = \begin{bmatrix} 0 \\ 0 \\ 0 \\ 0 \end{bmatrix}.$$

This result is acceptable since $a^3 = 2a^1 - a^2$; hence it is dependent on a^1 and a^2.

$$v_4 = a^4 - \frac{-\frac{9}{2}}{\frac{27}{4}} v_2 - \frac{-2}{4} v_1 = \begin{bmatrix} 0 \\ 1 \\ 1 \\ 2 \end{bmatrix}.$$

Thus,

$$q_1 = \frac{v_1}{\|v_1\|} = \begin{bmatrix} \frac{1}{2} \\ \frac{1}{2} \\ \frac{1}{2} \\ -\frac{1}{2} \end{bmatrix}, \ q_2 = \frac{v_2}{\|v_2\|} = \begin{bmatrix} \frac{\sqrt{3}}{2} \\ -\frac{\sqrt{3}}{6} \\ -\frac{\sqrt{3}}{6} \\ \frac{\sqrt{3}}{6} \end{bmatrix}, \ q_4 = \frac{v_4}{\|v_4\|} = \begin{bmatrix} 0 \\ -\frac{\sqrt{6}}{6} \\ -\frac{\sqrt{6}}{6} \\ -\frac{\sqrt{6}}{3} \end{bmatrix}$$

$$a^1 = 2q_1 v_1$$
$$a^2 = -\frac{1}{2}q_1 + \frac{3\sqrt{3}}{2}q_2 = -\frac{1}{4}v_1 + v_2$$
$$a^3 = 2(2q_1) - (-\frac{1}{2}q_1 + \frac{3\sqrt{3}}{2})q_2 = \frac{9}{2}q_1 + \frac{-3\sqrt{3}}{2}q_2 = 2v_1 - v_2 \quad \Leftrightarrow$$
$$a^4 = q_1 - \sqrt{3}q_2 - \sqrt{6}q_4 = \frac{1}{2}v_1 + \frac{-2}{3}v_2 + v_4$$

$$Q = \begin{bmatrix} \frac{1}{2} & \frac{\sqrt{3}}{2} & 0 \\ \frac{1}{2} & -\frac{\sqrt{3}}{6} & \frac{\sqrt{6}}{6} \\ \frac{1}{2} & -\frac{\sqrt{3}}{6} & \frac{\sqrt{6}}{6} \\ -\frac{1}{2} & \frac{\sqrt{3}}{6} & \frac{\sqrt{6}}{3} \end{bmatrix}, \ R = \begin{bmatrix} 2 & -\frac{1}{2} & \frac{9}{2} & 1 \\ 0 & \frac{3\sqrt{3}}{2} & -\frac{3\sqrt{3}}{2} & -\sqrt{3} \\ 0 & 0 & 0 & \sqrt{6} \end{bmatrix}.$$

3.2

$$y = \beta_0 + \beta_1 x + \epsilon \Rightarrow E[y] = \beta_0 + \beta_1 x.$$

Data:

$$\left.\begin{array}{c} y_1 = \beta_0 + \beta_1 x_1 \\ y_2 = \beta_0 + \beta_1 x_2 \\ \vdots \\ y_m = \beta_0 + \beta_1 x_m \end{array}\right\} \Leftrightarrow \begin{bmatrix} 1 & x_1 \\ 1 & x_2 \\ \vdots & \vdots \\ 1 & x_m \end{bmatrix} \begin{bmatrix} \beta_0 \\ \beta_1 \end{bmatrix} - \begin{bmatrix} y_1 \\ y_2 \\ \vdots \\ y_m \end{bmatrix} \Leftrightarrow \Lambda\beta = y.$$

The problem is to minimize $SSE = \|y - A\beta\|^2 = \sum_{i=1}^{m}(y_i - \beta_0 - \beta_1 x_i)^2$. The solution is to choose $\bar{\beta} = \begin{bmatrix} \bar{\beta}_0 \\ \bar{\beta}_1 \end{bmatrix}$ such that $A\bar{\beta}$ is as close as possible to y.

$$A = \begin{bmatrix} 1 & x_1 \\ 1 & x_2 \\ \vdots & \vdots \\ 1 & x_m \end{bmatrix} \Rightarrow A^T A = \begin{bmatrix} m & \sum x_i \\ \sum x_i & \sum x_i^2 \end{bmatrix}, \ \det(A^T A) = m \sum x_i^2 - \left(\sum x_i\right)^2.$$

$$(A^T A)^{-1} = \frac{1}{m \sum x_i^2 - (\sum x_i)^2} \begin{bmatrix} \sum x_i^2 & -\sum x_i \\ -\sum x_i & m \end{bmatrix}.$$

$$\bar{\beta} = (A^T A)^{-1} A^T y$$

$$\bar{\beta} = \frac{1}{m \sum x_i^2 - (\sum x_i)^2} \begin{bmatrix} \sum x_i^2 & -\sum x_i \\ -\sum x_i & m \end{bmatrix} \begin{bmatrix} 1 & 1 & \cdots & 1 \\ x_1 & x_2 & \cdots & x_m \end{bmatrix} \begin{bmatrix} y_1 \\ y_2 \\ \vdots \\ y_m \end{bmatrix}$$

$$\bar{\beta} = (A^T A)^{-1} A^T y = \frac{1}{m \sum x_i^2 - (\sum x_i)^2} \begin{bmatrix} \sum x_i^2 & -\sum x_i \\ -\sum x_i & m \end{bmatrix} \begin{bmatrix} \sum y_i \\ \sum x_i y_i \end{bmatrix}$$

$$\bar{\beta} = \begin{bmatrix} \bar{\beta}_0 \\ \bar{\beta}_1 \end{bmatrix} = (A^T A)^{-1} A^T y = \frac{1}{m \sum x_i^2 - (\sum x_i)^2} \begin{bmatrix} \sum x_i^2 \sum y_i - \sum x_i \sum x_i y_i \\ -\sum x_i \sum y_i + m \sum x_i y_i \end{bmatrix}.$$

We know from statistics that

$$\bar{\beta}_1 = \frac{SS_{xy}}{SS_{xx}}, \ \bar{\beta}_0 = \bar{y} - \bar{\beta}_1 \bar{x},$$

where

$$\bar{x} = \frac{\sum x_i}{m}, \ \bar{y} = \frac{\sum y_i}{m}, \ SS_{xy} = \sum (x_i - \bar{x})(y_i - \bar{y}), \ SS_{xx} = \sum (x_i - \bar{x})(x_i - \bar{x}).$$

Since

$$SS_{xx} = \sum (x_i - \bar{x})^2 = \sum x_i^2 - 2\bar{x} \sum x_i + m\bar{x}^2$$

$$SS_{xx} = \sum x_i^2 - 2m\bar{x}^2 + m\bar{x}^2 = \sum x_i^2 - m\bar{x}^2,$$

$$\bar{\beta}_1 = \frac{SS_{xy}}{SS_{xx}} = \frac{-m\,SS_{xy}}{-m\,SS_{xx}} = \frac{-\sum x_i \sum y_i + m\sum x_i y_i}{m\sum x_i^2 - (\sum x_i)^2},$$

which is dictated by the matrix equation above.

$$\bar{\beta}_0 = \frac{\bar{y}\,SS_{xx} - \bar{x}\,SS_{xy}}{SS_{xx}} = \frac{\bar{y}\sum x_i^2 - m\bar{y}\bar{x}^2 - \bar{x}\sum x_i y_i + m\bar{y}\bar{x}^2}{SS_{xx}},$$

$$\bar{\beta}_0 = \frac{\bar{y}\sum x_i^2 - \bar{x}\sum x_i y_i}{SS_{xx}} = \frac{\sum y_i \sum x_i^2 - \sum x_i \sum x_i y_i}{m\,SS_{xx}},$$

$$\bar{\beta}_0 = \frac{\sum x_i^2 \sum y_i - \sum x_i \sum x_i y_i}{m\sum x_i^2 - (\sum x_i)^2},$$

which is dictated by the matrix equation above.

We may use calculus to solve min SSE:

$$SSE = \|y - A\beta\|^2 = \sum_{i=1}^{m}(y_i - [\beta_0 + \beta_1 x_i])^2$$

$$SSE = \sum y_i^2 - 2\sum y_i \beta_0 - 2\beta_1 \sum y_i x_i + m\beta_0^2 + 2\beta_0\beta_1 \sum x_i + \beta_1^2 \sum x_i^2.$$

$$\frac{\partial SSE}{\partial \beta_0} = -2\sum y_i + 2m\beta_0 + 2\beta_1 \sum x_i \doteq 0$$

$$\Leftrightarrow \beta_0 = \frac{\sum y_i - \beta_1 \sum x_i}{m} = \bar{y} - \beta_1 \bar{x}.$$

$$\frac{\partial SSE}{\partial \beta_1} = -2\sum x_i y_i + 2\beta_0 \sum x_i + 2\beta_1 \sum x_i^2 \doteq 0$$

$$\Leftrightarrow \sum x_i y_i - (\bar{y} - \beta_1 \bar{x})\sum x_i - \beta_1 \sum x_i^2 \doteq 0$$

$$\beta_1 = \frac{\sum x_i y_i - \bar{y}\sum x_i}{\sum x_i^2 - \bar{x}\sum x_i} = \frac{\sum x_i y_i - m\bar{x}\bar{y}}{\sum x_i^2 - m\bar{x}^2} = \frac{SS_{xy}}{SS_{xx}}.$$

As it can be observed above, the matrix system and the calculus minimization yield the same solution!

Let the example data be (1,1), (2,4), (3,4), (4,4), (5,7). Then,

$$A\beta = y \Leftrightarrow \begin{bmatrix} 1 & 1 \\ 1 & 2 \\ 1 & 3 \\ 1 & 4 \\ 1 & 5 \end{bmatrix} \begin{bmatrix} \beta_0 \\ \beta_1 \end{bmatrix} = \begin{bmatrix} 1 \\ 4 \\ 4 \\ 4 \\ 7 \end{bmatrix}.$$

$$A^T A = \begin{bmatrix} 5 & 15 \\ 15 & 55 \end{bmatrix} = 5\begin{bmatrix} 1 & 3 \\ 3 & 11 \end{bmatrix}, \ \det(A^T A) = 10.$$

$$\bar{\beta} = \begin{bmatrix} \bar{\beta}_0 \\ \bar{\beta}_1 \end{bmatrix} = (A^T A)^{-1} A^T y = \frac{1}{10} \begin{bmatrix} 11 & -3 \\ -3 & 1 \end{bmatrix} \begin{bmatrix} 1 & 1 & 1 & 1 & 1 \\ 1 & 2 & 3 & 4 & 5 \end{bmatrix} \begin{bmatrix} 1 \\ 4 \\ 4 \\ 4 \\ 7 \end{bmatrix}$$

$$\bar{\beta} = \begin{bmatrix} \bar{\beta}_0 \\ \bar{\beta}_1 \end{bmatrix} = (A^T A)^{-1} A^T y = \frac{1}{10} \begin{bmatrix} 11 & -3 \\ -3 & 1 \end{bmatrix} \begin{bmatrix} 20 \\ 72 \end{bmatrix} = \begin{bmatrix} 0.4 \\ 1.2 \end{bmatrix} = \begin{bmatrix} \bar{\beta}_0 \\ \bar{\beta}_1 \end{bmatrix}.$$

$$\bar{x} = 3, \ \bar{y} = 4,$$

$$SS_{xy} = (1-3)(1-4)+(2-3)(4-4)+(3-3)(4-4)+(4-3)(4-4)+(5-3)(7-4) = 12,$$

$$SS_{xx} = (1-3)^2 + (2-3)^2 + (3-3)^2 + (4-3)^2 + (5-3)^2 = 10.$$

$$\bar{\beta}_1 = \frac{12}{10}, \ \bar{\beta}_0 = 4 - 1.2(3) = 0.4.$$

3.3

(a) Let us interchange the first two equations to get $A_1' = LU$:

$$A_1' = \begin{bmatrix} 2 & 1 & 3 \\ 1 & 3 & 2 \\ 3 & 2 & 1 \end{bmatrix} = \begin{bmatrix} \frac{2}{3} & -\frac{1}{7} & 1 \\ \frac{1}{3} & 1 & 0 \\ 1 & 0 & 0 \end{bmatrix} \begin{bmatrix} 3 & 2 & 1 \\ 0 & \frac{7}{3} & \frac{5}{3} \\ 0 & 0 & \frac{18}{7} \end{bmatrix}.$$

Here, the form of L is a bit different, but serves for the purpose. We solve $LUx = b_1' = [19, 8, 3]^T$ in two stages: $Lc = b'$, then $Ux = c$.

$$Lc = b_1' \Leftrightarrow \begin{bmatrix} \frac{2}{3} & -\frac{1}{7} & 1 \\ \frac{1}{3} & 1 & 0 \\ 1 & 0 & 0 \end{bmatrix} \begin{bmatrix} c_1 \\ c_2 \\ c_3 \end{bmatrix} = \begin{bmatrix} 19 \\ 8 \\ 3 \end{bmatrix} \Leftrightarrow \begin{cases} & \Rightarrow c_3 = 18. \\ & \Rightarrow c_2 = 7 \\ c_1 = 3 \end{cases}$$

$$Ux = c \Leftrightarrow \begin{bmatrix} 3 & 2 & 1 \\ 0 & \frac{7}{3} & \frac{5}{3} \\ 0 & 0 & \frac{18}{7} \end{bmatrix} \begin{bmatrix} x_1 \\ x_2 \\ x_3 \end{bmatrix} = \begin{bmatrix} 3 \\ 7 \\ 18 \end{bmatrix} \Leftrightarrow \begin{cases} & \Rightarrow x_1 = 0. \\ & \Rightarrow x_2 = -2 \\ x_3 = 7 \end{cases}$$

Final check:

$$A_1 x = \begin{bmatrix} 1 & 3 & 2 \\ 2 & 1 & 3 \\ 3 & 2 & 1 \end{bmatrix} \begin{bmatrix} 0 \\ -2 \\ 7 \end{bmatrix} = \begin{bmatrix} 8 \\ 19 \\ 3 \end{bmatrix} \checkmark$$

(b) Let us take the first three columns of A_2 as the basis:

$$B = \begin{bmatrix} 2 & 1 & 3 \\ 1 & 3 & 2 \\ 3 & 2 & 1 \end{bmatrix}, \quad N = \begin{bmatrix} 1 & 0 \\ 0 & 1 \\ 1 & 0 \end{bmatrix}, \quad x_B = \begin{bmatrix} x_1 \\ x_2 \\ x_3 \end{bmatrix}, \quad x_N = \begin{bmatrix} x_4 \\ x_5 \end{bmatrix}.$$

Let $x_N = \theta$. Then, $Bx_B = b_2$ is solved by LU decomposition as above:

$$Lc = b_2 \Leftrightarrow \begin{bmatrix} \frac{2}{3} & -\frac{1}{7} & 1 \\ \frac{1}{3} & 1 & 0 \\ 1 & 0 & 0 \end{bmatrix} \begin{bmatrix} c_1 \\ c_2 \\ c_3 \end{bmatrix} = \begin{bmatrix} 8 \\ 19 \\ 3 \end{bmatrix} \Leftrightarrow \begin{cases} & \Rightarrow c_3 = \frac{60}{7}. \\ & \Rightarrow c_2 = 18 \\ c_1 = 3 & \end{cases}$$

$$Ux = c \Leftrightarrow \begin{bmatrix} 3 & 2 & 1 \\ 0 & \frac{7}{3} & \frac{5}{3} \\ 0 & 0 & \frac{18}{7} \end{bmatrix} \begin{bmatrix} x_1 \\ x_2 \\ x_3 \end{bmatrix} = \begin{bmatrix} 3 \\ 18 \\ \frac{60}{7} \end{bmatrix} \Leftrightarrow \begin{cases} & \Rightarrow x_1 = -\frac{11}{3}. \\ & \Rightarrow x_2 = \frac{16}{3} \\ x_3 = \frac{10}{3} & \end{cases}$$

$x_B = [\frac{-11}{3}, \frac{16}{3}, \frac{10}{3}]^T$. If $x_N \neq \theta$, then $x_B = [\frac{-11}{3}, \frac{16}{3}, \frac{10}{3}]^T - B^{-1}Nx_N$. Let $x_N = [1,1]^T$. Then,

$$x_B = \begin{bmatrix} x_1 \\ x_2 \\ x_3 \end{bmatrix} = \begin{bmatrix} -\frac{11}{3} \\ \frac{16}{3} \\ \frac{10}{3} \end{bmatrix} - \begin{bmatrix} \frac{4}{9} & -\frac{5}{18} \\ -\frac{2}{9} & \frac{7}{18} \\ \frac{1}{9} & \frac{1}{18} \end{bmatrix} \begin{bmatrix} 1 \\ 1 \end{bmatrix} = \frac{1}{6} \begin{bmatrix} -23 \\ 31 \\ 19 \end{bmatrix}$$

Final check:

$$A_2 x = \begin{bmatrix} 2 & 1 & 3 & 1 & 0 \\ 1 & 3 & 2 & 0 & 1 \\ 3 & 2 & 1 & 1 & 0 \end{bmatrix} \begin{bmatrix} -\frac{11}{3} \\ \frac{16}{3} \\ \frac{10}{3} \\ 0 \\ 0 \end{bmatrix} = \begin{bmatrix} 8 \\ 19 \\ 3 \end{bmatrix} \checkmark$$

$$A_2 x = \begin{bmatrix} 2 & 1 & 3 & 1 & 0 \\ 1 & 3 & 2 & 0 & 1 \\ 3 & 2 & 1 & 1 & 0 \end{bmatrix} \frac{1}{6} \begin{bmatrix} -23 \\ 31 \\ 19 \\ 6 \\ 6 \end{bmatrix} = \begin{bmatrix} 8 \\ 19 \\ 3 \end{bmatrix} \checkmark$$

(c)

$$A_3 = \begin{bmatrix} 1 & 2 \\ 4 & 5 \\ 7 & 8 \\ 10 & 11 \end{bmatrix}, \quad A_3^T = \begin{bmatrix} 1 & 4 & 7 & 10 \\ 2 & 5 & 8 & 11 \end{bmatrix}, \quad A_3^T A_3 = \begin{bmatrix} 166 & 188 \\ 188 & 214 \end{bmatrix}.$$

$A_3^T A_3$ is clearly invertible, and $(A_3^T A_3)^{-1} = \begin{bmatrix} \frac{107}{90} & -\frac{47}{45} \\ -\frac{47}{45} & \frac{83}{90} \end{bmatrix}$.

$$(A_3^T A_3)^{-1} A^T = \begin{bmatrix} \frac{107}{90} & -\frac{47}{45} \\ -\frac{47}{45} & \frac{83}{90} \end{bmatrix} \begin{bmatrix} 1 & 4 & 7 & 10 \\ 2 & 5 & 8 & 11 \end{bmatrix} = \begin{bmatrix} -\frac{9}{10} & -\frac{7}{15} & -\frac{1}{30} & \frac{2}{5} \\ \frac{4}{5} & \frac{13}{30} & \frac{1}{15} & -\frac{3}{10} \end{bmatrix}$$

$$\bar{x} = (A_3^T A_3)^{-1} A^T b_3 = \begin{bmatrix} x_1 \\ x_2 \end{bmatrix} = \begin{bmatrix} -\frac{9}{10} & -\frac{7}{15} & -\frac{1}{30} & \frac{2}{5} \\ \frac{4}{5} & \frac{13}{30} & \frac{1}{15} & -\frac{3}{10} \end{bmatrix} \begin{bmatrix} 2 \\ 5 \\ 6 \\ 8 \end{bmatrix} = \begin{bmatrix} -\frac{34}{30} \\ \frac{53}{30} \end{bmatrix} \quad (\heartsuit)$$

The $A_3 = QR$ decomposition is given below:

$$Q = \begin{bmatrix} -0.07762 & -0.83305 & -0.39205 & -0.38249 \\ -0.31046 & -0.45124 & 0.23763 & 0.80220 \\ -0.54331 & -0.06942 & 0.70087 & -0.45693 \\ -0.77615 & 0.31239 & -0.54646 & 0.03722 \end{bmatrix}$$

$$R = \begin{bmatrix} -12.8840 & -14.5920 \\ 0.0000 & -1.0413 \\ 0.0000 & 0.0000 \\ 0.0000 & 0.0000 \end{bmatrix}$$

The equivalent system $R\bar{x} = Q^T b_3$ is solved below:

$$Q^T b_3 = \begin{bmatrix} -0.07762 & -0.31046 & -0.54331 & -0.77615 \\ -0.83305 & -0.45124 & -0.06942 & 0.31239 \\ -0.39205 & 0.23763 & 0.70087 & -0.54646 \\ -0.38249 & 0.80220 & -0.45693 & 0.03722 \end{bmatrix} \begin{bmatrix} 2 \\ 5 \\ 6 \\ 8 \end{bmatrix} = \begin{bmatrix} -11.1770 \\ -1.8397 \\ 0.2376 \\ 0.8022 \end{bmatrix}$$

$$R\bar{x} = \begin{bmatrix} -12.8840 & -14.5920 \\ 0.0000 & -1.0413 \\ 0.0000 & 0.0000 \\ 0.0000 & 0.0000 \end{bmatrix} \begin{bmatrix} x_1 \\ x_2 \end{bmatrix} = \begin{bmatrix} -11.1770 \\ -1.8397 \\ 0.2376 \\ 0.8022 \end{bmatrix}$$

$$\Leftrightarrow \begin{cases} \Rightarrow x_1 = \frac{-11.177 - 1.7667(-14.592)}{-12.884} = -1.1333 \\ x_2 = \frac{-1.8397}{-1.0413} = 1.7667 \end{cases} \quad (\Diamond)$$

The two solutions, $(\heartsuit)$ and $(\Diamond)$, are equivalent.

$$A_3 x = \begin{bmatrix} 1 & 2 \\ 4 & 5 \\ 7 & 8 \\ 10 & 11 \end{bmatrix} \begin{bmatrix} -1.1333 \\ 1.7667 \end{bmatrix} = \begin{bmatrix} 2.4201 \\ 4.3503 \\ 6.2805 \\ 8.2107 \end{bmatrix} \neq \begin{bmatrix} 2 \\ 5 \\ 6 \\ 8 \end{bmatrix} = b_3$$

$$\|A_3 x - b\| = \left\| \begin{bmatrix} 2.4201 \\ 4.3503 \\ 6.2805 \\ 8.2107 \end{bmatrix} - \begin{bmatrix} 2 \\ 5 \\ 6 \\ 8 \end{bmatrix} \right\| = \left\| \begin{bmatrix} 0.4201 \\ -0.6497 \\ 0.2805 \\ 0.8495 \end{bmatrix} \right\| = 0.8695 \text{ is the minimum}$$

error.

(d)

$$A_4 = \begin{bmatrix} -1 & 0 & 0 & 1 \\ 1 & -1 & 0 & 0 \\ 0 & 1 & -1 & 0 \\ 0 & 0 & -1 & 1 \end{bmatrix}, \quad A_4^T = \begin{bmatrix} -1 & 1 & 0 & 0 \\ 0 & -1 & 1 & 0 \\ 0 & 0 & -1 & -1 \\ 1 & 0 & 0 & 1 \end{bmatrix}, \quad A_4^T A_4 = \begin{bmatrix} 2 & -1 & 0 & -1 \\ -1 & 2 & -1 & 0 \\ 0 & -1 & 2 & -1 \\ -1 & 0 & -1 & 2 \end{bmatrix}.$$

Clearly, $A_4^T A_4$ is not invertible. Then, we resort to the singular value decomposition $A_4 = Q_1 \Sigma Q_2^T$, where

$$Q_1 = \begin{bmatrix} -\frac{1}{2} & -\frac{\sqrt{2}}{2} & 0 & \frac{1}{2} \\ \frac{1}{2} & 0 & -\frac{\sqrt{2}}{2} & \frac{1}{2} \\ -\frac{1}{2} & \frac{\sqrt{2}}{2} & 0 & \frac{1}{2} \\ \frac{1}{2} & 0 & \frac{\sqrt{2}}{2} & \frac{1}{2} \end{bmatrix}, \quad \Sigma = \begin{bmatrix} 4 & 0 & 0 & 0 \\ 0 & 2 & 0 & 0 \\ 0 & 0 & 2 & 0 \\ 0 & 0 & 0 & 0 \end{bmatrix}, \quad Q_2^T = \begin{bmatrix} -\frac{1}{2} & -\frac{\sqrt{2}}{2} & 0 & \frac{1}{2} \\ \frac{1}{2} & 0 & -\frac{\sqrt{2}}{2} & \frac{1}{2} \\ -\frac{1}{2} & \frac{\sqrt{2}}{2} & 0 & \frac{1}{2} \\ \frac{1}{2} & 0 & \frac{\sqrt{2}}{2} & \frac{1}{2} \end{bmatrix}.$$

Then, $\bar{x} = Q_2 \Sigma^\dagger Q_1^T b_4$ finds the solution:

$$\bar{x} = \begin{bmatrix} -\frac{1}{2} & \frac{1}{2} & -\frac{1}{2} & \frac{1}{2} \\ -\frac{\sqrt{2}}{2} & 0 & \frac{\sqrt{2}}{2} & 0 \\ 0 & -\frac{\sqrt{2}}{2} & 0 & \frac{\sqrt{2}}{2} \\ \frac{1}{2} & \frac{1}{2} & \frac{1}{2} & \frac{1}{2} \end{bmatrix} \begin{bmatrix} \frac{1}{4} & 0 & 0 & 0 \\ 0 & \frac{1}{2} & 0 & 0 \\ 0 & 0 & \frac{1}{2} & 0 \\ 0 & 0 & 0 & 0 \end{bmatrix} \times$$

$$\times \begin{bmatrix} -\frac{1}{2} & \frac{1}{2} & -\frac{1}{2} & \frac{1}{2} \\ -\frac{\sqrt{2}}{2} & 0 & \frac{\sqrt{2}}{2} & 0 \\ 0 & -\frac{\sqrt{2}}{2} & 0 & \frac{\sqrt{2}}{2} \\ \frac{1}{2} & \frac{1}{2} & \frac{1}{2} & \frac{1}{2} \end{bmatrix} \begin{bmatrix} 2 \\ 4 \\ 3 \\ 3 \end{bmatrix} \Rightarrow \begin{bmatrix} x_1 \\ x_2 \\ x_3 \\ x_4 \end{bmatrix} = \begin{bmatrix} 0.22855 \\ -0.42678 \\ -0.25000 \\ 0.12500 \end{bmatrix}$$

Problems of Chapter 4

4.1 In order to prove that

$$\det A = a_{i1}A_{i1} + a_{i2}A_{i2} + \cdots + a_{in}A_{in},$$

(property 11) where A_{ij}'s are cofactors ($A_{ij} = (-1)^{i+j} \det M_{ij}$, where the minor M_{ij} is formed from A by deleting row i and column j); without loss of generality, we may assume that $i = 1$.

Let us apply some row operations,

$$A = \begin{bmatrix} a_{11} & a_{22} & a_{13} & \cdots & a_{1n} \\ a_{21} & a_{22} & a_{23} & \cdots & a_{2n} \\ a_{31} & a_{22} & a_{33} & \cdots & a_{3n} \\ \vdots & \vdots & \vdots & \ddots & \vdots \\ a_{n1} & a_{n2} & a_{n3} & \cdots & a_{nn} \end{bmatrix} \rightarrow \begin{bmatrix} a_{11} & a_{22} & a_{13} & \cdots & a_{1n} \\ 0 & \alpha_{22} & \alpha_{23} & \cdots & \alpha_{2n} \\ 0 & \alpha_{32} & \alpha_{33} & \cdots & \alpha_{3n} \\ \vdots & \vdots & \vdots & \ddots & \vdots \\ 0 & \alpha_{n2} & \alpha_{n3} & \cdots & \alpha_{nn} \end{bmatrix},$$

where $\alpha_{ij} = \frac{-a_{1j}a_{i1}+a_{ij}a_{11}}{a_{11}}$, $i, j = 2, \ldots, n$. In particular, $\alpha_{22} = \frac{-a_{12}a_{21}+a_{22}a_{11}}{a_{11}}$.
Furthermore,

$$A \rightarrow \begin{bmatrix} a_{11} & a_{22} & a_{13} & \cdots & a_{1n} \\ 0 & \alpha_{22} & \alpha_{23} & \cdots & \alpha_{2n} \\ 0 & \alpha_{22} & \alpha_{33} & \cdots & \alpha_{3n} \\ \vdots & \vdots & \vdots & \ddots & \vdots \\ 0 & \alpha_{n2} & \alpha_{n3} & \cdots & \alpha_{nn} \end{bmatrix} \rightarrow \begin{bmatrix} a_{11} & a_{22} & a_{13} & \cdots & a_{1n} \\ 0 & \alpha_{22} & \alpha_{23} & \cdots & \alpha_{2n} \\ 0 & 0 & \beta_{33} & \cdots & \beta_{3n} \\ \vdots & \vdots & \vdots & \ddots & \vdots \\ 0 & 0 & \beta_{n3} & \cdots & \beta_{nn} \end{bmatrix},$$

where $\beta_{ij} = \frac{-\alpha_{2j}\alpha_{i2}+\alpha_{ij}\alpha_{22}}{\alpha_{22}}$, $i, j = 2, \ldots, n$. In particular,

$$\beta_{33} = \frac{-\alpha_{23}\alpha_{32} + \alpha_{33}\alpha_{22}}{\alpha_{22}} =$$

$$\frac{(a_{12}a_{31} - a_{32}a_{11})(a_{23}a_{11} - a_{13}a_{21}) + (a_{33}a_{11} - a_{13}a_{31})(a_{22}a_{11} - a_{12}a_{21})}{a_{11}(a_{22}a_{11} - a_{12}a_{21})}.$$

If we open up the parentheses in the numerator, the terms without a_{11} cancel each other, and if we factor a_{11} out and cancel with the same term in the denominator, we will have

$$\beta_{33} = \frac{a_{12}a_{23}a_{31} + a_{13}a_{32}a_{21} - a_{11}a_{23}a_{32} - a_{13}a_{31}a_{22} - a_{12}a_{21}a_{33} + a_{11}a_{22}a_{33}}{-a_{12}a_{21} + a_{22}a_{11}}.$$

If we further continue the row operations to reach the upper triangular form, we will have

$$A \rightarrow \cdots \rightarrow \begin{bmatrix} a_{11} & a_{22} & a_{13} & \cdots & a_{1n} \\ 0 & \alpha_{22} & \alpha_{23} & \cdots & \alpha_{2n} \\ 0 & 0 & \beta_{33} & \cdots & \beta_{3n} \\ \vdots & \vdots & \vdots & \ddots & \vdots \\ 0 & 0 & 0 & \cdots & \zeta_{nn} \end{bmatrix}.$$

Let $\zeta_{nn} = \frac{Z}{ZZ}$. Thus,

$$\det A = a_{11} \cdot \alpha_{22} \cdot \beta_{33} \cdots \zeta_{nn} = a_{11} \cdot \left[\frac{-a_{12}a_{21} + a_{22}a_{11}}{a_{11}} \right].$$

$$\cdot \left[\frac{a_{12}a_{23}a_{31} + a_{13}a_{32}a_{21} - a_{11}a_{23}a_{32} - a_{13}a_{31}a_{22} - a_{12}a_{21}a_{33} + a_{11}a_{22}a_{33}}{-a_{12}a_{21} + a_{22}a_{11}} \right]$$

$$\cdots \left[\frac{Z}{ZZ} \right].$$

Since the denominator of one term cancels the numerator of the previous term,

$$\det A = Z = \sum_{p \in P} a_{1p_1} a_{2p_2} \cdots a_{np_n} \det[e_{p_1}, e_{p_2}, \ldots, e_{p_n}], \, (\star)$$

where P has all $n!$ permutations $(p_1, \ldots, p_n)$ of the numbers $\{1, 2, \ldots, n\}$, e_{p_i} is the p_i^{th} canonical unit vector and $\det P_p = \det[e_{p_1}, e_{p_2}, \ldots, e_{p_n}] = \pm 1$ such that the sign depends on whether the number of exchanges in the permutation matrix P_p is even or odd.

Consider the terms in the above formula for $\det A$ involving a_{11}. They occur when the choice of the first column is $p_1 = 1$ yielding some permutation $\acute{p} = (p_2, \ldots, p_n)$ of the remaining numbers $\{2, 3, \ldots, n\}$. We collect all these terms as A_{11} where the cofactor for a_{11} is

$$A_{11} = \sum_{\acute{p} \in \acute{P}} a_{2p_2} \cdots a_{np_n} \det P_{\acute{p}}.$$

Hence, $\det A$ should depend linearly on the row $(a_{11}, a_{12}, \ldots, a_{1n})$:

$$\det A = a_{11}A_{11} + a_{12}A_{12} + \cdots + a_{1n}A_{1n}.$$

Let us prove Property 11 using the induction approach. The base condition was already be shown to be true by the example in the main text. We may use $(\star)$ as the induction hypothesis for $n = k$.

Claim: $\sum_{\acute{p} \in \acute{P}} a_{2p_2} \cdots a_{np_n} \det P_{\acute{p}} = (-1)^{1+1} \det M_{11}$. We will use induction for proving the claim.

Base($n = 3$): $A_{11} = a_{22}a_{33} - a_{23}a_{32} = (-1)^2 \begin{vmatrix} a_{22} & a_{23} \\ a_{32} & a_{33} \end{vmatrix}.$

Induction($n = k + 1$): $\sum_{\acute{p} \in \acute{P}} a_{2,p_2} \cdots a_{k+1,p_{k+1}} \det P_{\acute{p}} = (-1)^{1+1} \det M_{11}.$

Using the induction hypothesis for $n = k$ in $(\star)$ we have:

$$\det M_{11} = a_{22}\grave{A}_{22} + \cdots + a_{2n}\grave{A}_{2n},$$

in which we may use the induction hypothesis of the claim for the cofactor $\grave{A}_{2j}$. The rest is almost trivial.

4.2 Let

$$A = \begin{bmatrix} 1 & 1 & -1 & -1 & -1 \\ 2 & 1 & 1 & 2 & 1 \\ 0 & 1 & 1 & 0 & -1 \\ 1 & -1 & 1 & 3 & 1 \\ 2 & -2 & 2 & 2 & 4 \end{bmatrix} \Rightarrow d(s) = (s-2)^5, \ k = 1, \ \lambda_1 = 2, \ n_1 = 5.$$

$$A_1 = A - 2I = \begin{bmatrix} -1 & 1 & -1 & -1 & -1 \\ 2 & -1 & 1 & 2 & 1 \\ 0 & 1 & -1 & 0 & -1 \\ 1 & -1 & 1 & 1 & 1 \\ 2 & -2 & 2 & 2 & 2 \end{bmatrix}$$

$$\Rightarrow \dim\mathcal{N}(A_1) = 5 - rank(A_1) = 5 - 3 = 2.$$

$$A_1^2 = 0 \Rightarrow \dim\mathcal{N}(A_1^2) = 5 \Rightarrow m_1 = 2, \ m(s) = (s-2)^2.$$

Choose $v_2 \in \mathcal{N}(A_1^2) \ni A_1 v_2 \neq \theta$.

$$v_2 = e_1^5 = (1,0,0,0,0)^T \Rightarrow v_1 = A_1 v_2 = (-1,2,0,1,2)^T.$$

Choose $v_4 \neq \alpha v_2 \ni \alpha \neq 0, \ v_4 \in \mathcal{N}(A_1^2) \ni A_1 v_2 \neq \theta$.

$$v_4 = e_2^5 = (0,1,0,0,0)^T \Rightarrow v_3 = A_1 v_4 = (1,-1,1,-1,2)^T.$$

Choose $v_5 \in \mathcal{N}(A_1)$ independent from v_1 and v_3.

$$v_5 = (1,0,0,-1,0)^T.$$

Thus,

$$S = \begin{bmatrix} -1 & 1 & 1 & 0 & 1 \\ 2 & 0 & -1 & 1 & 0 \\ 0 & 0 & 1 & 0 & 0 \\ 1 & 0 & -1 & 0 & -1 \\ 2 & 0 & -2 & 0 & 0 \end{bmatrix} \Rightarrow S^{-1}AS = \begin{bmatrix} 2 & 1 & & & \\ & 2 & & & \\ & & 2 & 1 & \\ & & & 2 & \\ & & & & 2 \end{bmatrix}.$$

4.3

$$A = \begin{bmatrix} \frac{1}{10} & \frac{1}{10} & 0 \\ 0 & \frac{1}{10} & \frac{1}{10} \\ 0 & 0 & \frac{1}{10} \end{bmatrix} \Rightarrow d(s) = \left(s - \frac{1}{10}\right)^3, \ k = 1, \ \lambda = \frac{1}{10}, \ n = 3.$$

$$A_1 = A - \frac{1}{10}I = \begin{bmatrix} 0 & \frac{1}{10} & 0 \\ 0 & 0 & \frac{1}{10} \\ 0 & 0 & 0 \end{bmatrix}$$

$$\Rightarrow \dim\mathcal{N}(A_1) = 3 - rank(A_1) = 3 - 2 = 1.$$

$$A_2 = A_1^2 = (A - \frac{1}{10}I)^2 = \begin{bmatrix} 0 & 0 & \frac{1}{10} \\ 0 & 0 & 0 \\ 0 & 0 & 0 \end{bmatrix}$$

$$\Rightarrow dim\mathcal{N}(A_1) = 3 - rank(A_1) = 3 - 1 = 2.$$

$$A_1^3 = 0 \Rightarrow dim\mathcal{N}(A_1^3) = 3 \Rightarrow m = 3, m(s) = \left(s - \frac{1}{10} \right)^3.$$

Choose $v_3 \in \mathcal{N}(A_1^3) \ni v_2 = A_1 v_3 \neq \theta \neq A_1^2 v_3 = v_1.$

$$v_3 = e_3^3 = (0,0,1)^T \Rightarrow v_2 = A_1 v_3 = \left(0, \frac{1}{10}, 0 \right)^T \Rightarrow v_1 = A_1 v_2 = \left(\frac{1}{100}, 0, 0 \right)^T.$$

Thus,

$$S = \begin{bmatrix} \frac{1}{100} & 0 & 0 \\ 0 & \frac{1}{10} & 0 \\ 0 & 0 & 1 \end{bmatrix} \Rightarrow S^{-1}AS = \begin{bmatrix} \frac{1}{10} & 1 & \\ & \frac{1}{10} & 1 \\ & & \frac{1}{10} \end{bmatrix}.$$

$$A^{10} = \frac{1}{10^{10}} \begin{bmatrix} 1 & 10 & 45 \\ 0 & 1 & 10 \\ 0 & 0 & 1 \end{bmatrix} = \begin{bmatrix} \frac{1}{100} & 0 & 0 \\ 0 & \frac{1}{10} & 0 \\ 0 & 0 & 1 \end{bmatrix} \begin{bmatrix} \frac{1}{10} & 1 & \\ & \frac{1}{10} & 1 \\ & & \frac{1}{10} \end{bmatrix}^{10} \begin{bmatrix} 100 & & \\ & 10 & \\ & & 1 \end{bmatrix} = S\Lambda^{10}S^{-1}.$$

Note that the calculation of Λ^{10} is as hard as that of A^{10} since Λ is not diagonal. However, because (easy to prove by induction)

$$\begin{bmatrix} \lambda & 1 & \\ & \lambda & 1 \\ & & \lambda \end{bmatrix}^n = \begin{bmatrix} \lambda^n & \binom{n}{1}\lambda^{n-1} & \binom{n}{2}\lambda^{n-2} \\ & \lambda^n & \binom{n}{1}\lambda^{n-1} \\ & & \lambda^n \end{bmatrix},$$

we have

$$\Lambda^{10} = \begin{bmatrix} \left(\frac{1}{10}\right)^{10} & 10\left(\frac{1}{10}\right)^9 & 45\left(\frac{1}{10}\right)^8 \\ & \left(\frac{1}{10}\right)^{10} & 10\left(\frac{1}{10}\right)^9 \\ & & \left(\frac{1}{10}\right)^{10} \end{bmatrix} = \frac{1}{10^{10}} \begin{bmatrix} 1 & 100 & 4500 \\ 0 & 1 & 100 \\ 0 & 0 & 1 \end{bmatrix}.$$

Hence, it is still useful to have Jordan decomposition.

4.4 (a)

$$\frac{dX_1}{dt} = -0.03Y_1 - 0.02Y_2 \qquad \frac{dX_2}{dt} = -0.04Y_1 - 0.01Y_2$$

$$\frac{dY_1}{dt} = -0.05X_1 - 0.02X_2 \qquad \frac{dY_2}{dt} = -0.03X_1 - 0.00X_2$$

Let $W^T = [X_1, X_2, Y_1, Y_2]$. Then, the above equation is rewritten as

$$\frac{dW}{dt} = AW,$$

where

$$A = \begin{bmatrix} 0 & 0 & -\frac{3}{100} & -\frac{1}{50} \\ 0 & 0 & -\frac{1}{25} & -\frac{1}{100} \\ -\frac{1}{20} & -\frac{1}{50} & 0 & 0 \\ -\frac{3}{100} & 0 & 0 & 0 \end{bmatrix}$$

and the initial condition is $W_0 = [100, 60, 40, 30]^T$.

(b) $A = S\Lambda S^{-1}$, where

$$S = \begin{bmatrix} 0.46791 & -0.46791 & -0.20890 & -0.20890 \\ 0.54010 & -0.54010 & 0.69374 & 0.69374 \\ 0.64713 & 0.64713 & 0.33092 & -0.33092 \\ 0.26563 & 0.26563 & -0.60464 & 0.60464 \end{bmatrix}$$

$$\Rightarrow S^{-1} = \begin{bmatrix} 0.79296 & 0.23878 & 0.63090 & 0.34529 \\ -0.79296 & -0.23878 & 0.63090 & 0.34529 \\ -0.61736 & 0.53484 & 0.27717 & -0.67525 \\ -0.61736 & 0.53484 & -0.27717 & 0.67525 \end{bmatrix}$$

and $\Lambda = \begin{bmatrix} -0.052845 & 0.000000 & 0.000000 & 0.000000 \\ 0.000000 & 0.052845 & 0.000000 & 0.000000 \\ 0.000000 & 0.000000 & -0.010365 & 0.000000 \\ 0.000000 & 0.000000 & 0.000000 & 0.010365 \end{bmatrix}$

The solution is $W = Se^{\Lambda t}S^{-1}W_0$:

$$\begin{bmatrix} X_1(t) \\ X_2(t) \\ Y_1(t) \\ Y_2(t) \end{bmatrix} = \begin{bmatrix} 0.46791 & -0.46791 & -0.20890 & -0.20890 \\ 0.54010 & -0.54010 & 0.69374 & 0.69374 \\ 0.64713 & 0.64713 & 0.33092 & -0.33092 \\ 0.26563 & 0.26563 & -0.60464 & 0.60464 \end{bmatrix}$$

$$\begin{bmatrix} e^{-0.052845\,t} & & & \\ & e^{0.052845\,t} & & \\ & & e^{-0.010365\,t} & \\ & & & e^{0.010365\,t} \end{bmatrix}$$

$$\begin{bmatrix} 0.79296 & 0.23878 & 0.63090 & 0.34529 \\ -0.79296 & -0.23878 & 0.63090 & 0.34529 \\ -0.61736 & 0.53484 & 0.27717 & -0.67525 \\ -0.61736 & 0.53484 & -0.27717 & 0.67525 \end{bmatrix} \begin{bmatrix} 100 \\ 60 \\ 40 \\ 30 \end{bmatrix}$$

Since $S^{-1}W_0 = \begin{bmatrix} 129.220 \\ -58.028 \\ -38.816 \\ -20.475 \end{bmatrix}$, we have

$$
\begin{bmatrix} X_1(t) \\ X_2(t) \\ Y_1(t) \\ Y_2(t) \end{bmatrix} = \begin{bmatrix} .46791 & -.46791 & -.20890 & -.20890 \\ .54010 & -.54010 & .69374 & .69374 \\ .64713 & .64713 & .33092 & -.33092 \\ .26563 & .26563 & -.60464 & .60464 \end{bmatrix} \begin{bmatrix} (129.22)e^{-0.052845\,t} \\ (-58.028)e^{0.052845\,t} \\ (-38.816)e^{-0.010365\,t} \\ (-20.475)e^{0.010365\,t} \end{bmatrix}
$$

(c)

$$
\begin{bmatrix} X_1(0) \\ X_2(0) \\ Y_1(0) \\ Y_2(0) \end{bmatrix} = \begin{bmatrix} 100.0000 \\ 60.0000 \\ 40.0000 \\ 30.0000 \end{bmatrix}, \quad \begin{bmatrix} X_1(1) \\ X_2(1) \\ Y_1(1) \\ Y_2(1) \end{bmatrix} = \begin{bmatrix} 98.3222 \\ 58.2381 \\ 33.8610 \\ 27.0258 \end{bmatrix}, \quad \begin{bmatrix} X_1(2) \\ X_2(2) \\ Y_1(2) \\ Y_2(2) \end{bmatrix} = \begin{bmatrix} 96.8859 \\ 56.7490 \\ 27.8324 \\ 24.0983 \end{bmatrix},
$$

$$
\begin{bmatrix} X_1(3) \\ X_2(3) \\ Y_1(3) \\ Y_2(3) \end{bmatrix} = \begin{bmatrix} 95.6871 \\ 55.5282 \\ 21.8967 \\ 21.2102 \end{bmatrix}, \quad \begin{bmatrix} X_1(4) \\ X_2(4) \\ Y_1(4) \\ Y_2(4) \end{bmatrix} = \begin{bmatrix} 94.7227 \\ 54.5719 \\ 16.0369 \\ 18.3547 \end{bmatrix}, \quad \begin{bmatrix} X_1(5) \\ X_2(5) \\ Y_1(5) \\ Y_2(5) \end{bmatrix} = \begin{bmatrix} 93.9900 \\ 53.8772 \\ 10.2360 \\ 15.5246 \end{bmatrix}.
$$

Problems of Chapter 5

5.1

Proof. Let $Q^{-1}AQ = \Lambda$ and $Q^{-1} = Q^T$,

$$x = Qy \Rightarrow R(x) = \frac{y^T \Lambda y}{y^T y} = \frac{\lambda_1 y_1^2 + \cdots + \lambda_n y_n^2}{y_1^2 + \cdots + y_n^2}$$

$y_1 = 1, y_2 = \cdots = y_n = 0 \Rightarrow \lambda_1 \leq R(x)$ since

$$\lambda_1(y_1^2 + \cdots + y_n^2) \leq \lambda_1 y_1^2 + \cdots + \lambda_n y_n^2 \Leftarrow \lambda_1 = \min\{\lambda_i\}_{i=1}^n.$$

Similarly, $\lambda_n(A) = \max_{\|x\|=1} x^T A x.$ $\quad\square$

5.2

i. $x^T A x \geq 0, \ \forall x \neq \theta;$

$$x^T A x = [x_1 \ x_2 \ x_3] \frac{1}{100} \begin{bmatrix} 2 & 1 & 0 \\ 1 & 2 & 1 \\ 0 & 1 & 1 \end{bmatrix} \begin{bmatrix} x_1 \\ x_2 \\ x_3 \end{bmatrix}$$

$$= \frac{1}{100} \left[2x_1^2 + x_1 x_2 + x_1 x_2 + 2x_2^2 + x_2 x_3 + x_2 x_3 + x_3^2 \right]$$

$$= \frac{1}{100} \left[(x_1 + x_2)^2 + (x_2 + x_3)^2 + x_1^2 \right] > 0, \ \forall x \neq \theta!$$

ii. All the eigen values of A satisfy $\lambda_i \geq 0;$

$$\det(sI - A) = \frac{1}{100} \begin{vmatrix} 100s - 2 & -1 & 0 \\ -1 & 100s - 2 & -1 \\ 0 & -1 & 100s - 1 \end{vmatrix} = 0 \Leftrightarrow$$

$$s^3 - 0.05s^2 + 0.0006s - 0.000001 = (s - 0.002)(s - 0.01552)(s - 0.03248) = 0$$

$$\Rightarrow \lambda_1 = 0.002 > 0, \ \lambda_2 = 0.01552 > 0, \ \lambda_3 = 0.03248 > 0!$$

iii. All the submatrices A_k have nonnegative determinants;
Since each entry of A is nonnegative, all 1×1 minors are OK.

$$\begin{vmatrix} 2 & 1 \\ 1 & 2 \end{vmatrix} = 3 > 0, \quad \begin{vmatrix} 2 & 0 \\ 1 & 1 \end{vmatrix} = 2 > 0, \quad \begin{vmatrix} 1 & 0 \\ 2 & 1 \end{vmatrix} = 1 > 0$$

$$\begin{vmatrix} 2 & 1 \\ 0 & 1 \end{vmatrix} = 2 > 0, \quad \begin{vmatrix} 2 & 0 \\ 0 & 1 \end{vmatrix} = 2 > 0, \quad \begin{vmatrix} 1 & 0 \\ 1 & 1 \end{vmatrix} = 1 > 0$$

$$\begin{vmatrix} 1 & 2 \\ 0 & 1 \end{vmatrix} = 1 > 0, \quad \begin{vmatrix} 1 & 1 \\ 0 & 1 \end{vmatrix} = 1 > 0, \quad \begin{vmatrix} 2 & 1 \\ 1 & 1 \end{vmatrix} = 2 > 0$$

All 2×2 minors are OK.

$$\begin{vmatrix} 2 & 1 & 0 \\ 1 & 2 & 1 \\ 0 & 1 & 1 \end{vmatrix} = 1 = 10^6 \det(A) > 0!$$

The 3×3 minor, itself, is OK as well.

iv. All the pivots (without row exchanges) satisfy $d_i \geq 0$;

$$\begin{bmatrix} 2 & 1 & 0 \\ 1 & 2 & 1 \\ 0 & 1 & 1 \end{bmatrix} \hookrightarrow \begin{bmatrix} 2 & 1 & 0 \\ 0 & \frac{3}{2} & 1 \\ 0 & 1 & 1 \end{bmatrix} \hookrightarrow \begin{bmatrix} 2 & 1 & 0 \\ 0 & \frac{3}{2} & 1 \\ 0 & 0 & \frac{1}{3} \end{bmatrix}$$

$$\Rightarrow d_1 = \frac{2}{100} > 0, \ d_2 = \frac{3}{200} > 0, \ d_3 = \frac{1}{300} > 0!$$

v. $\exists$ a possibly singular matrix $W \ni A = W^T W$;

$$A = \frac{1}{100} \begin{bmatrix} 2 & 1 & 0 \\ 1 & 2 & 1 \\ 0 & 1 & 1 \end{bmatrix} = \left\{ \frac{1}{10} \begin{bmatrix} 1 & 1 & 0 \\ 0 & 1 & 1 \\ 0 & 0 & 1 \end{bmatrix} \right\} \left\{ \frac{1}{10} \begin{bmatrix} 1 & 0 & 0 \\ 1 & 1 & 0 \\ 0 & 1 & 1 \end{bmatrix} \right\} = W^T W$$

and $W = \frac{1}{10} \begin{bmatrix} 1 & 0 & 0 \\ 1 & 1 & 0 \\ 0 & 1 & 1 \end{bmatrix}$ is nonsingular!

5.3

$$\nabla f(x) = \begin{bmatrix} \frac{\partial f}{\partial x_1} \\ \frac{\partial f}{\partial x_2} \end{bmatrix} = \begin{bmatrix} x_1^2 + x_1 + 2x_2 \\ 2x_1 + x_2 - 1 \end{bmatrix} \doteq \begin{bmatrix} 0 \\ 0 \end{bmatrix}$$

$$\Rightarrow \begin{cases} (x_1 - 1)(x_1 - 2) = 0 \\ x_2 = 1 - 2x_1 \end{cases}$$

Therefore,

$$x_A = \begin{bmatrix} 1 \\ -1 \end{bmatrix}, \ x_B = \begin{bmatrix} 2 \\ -3 \end{bmatrix}$$

are stationary points inside the region defined by $-4 \leq x_2 \leq 0 \leq x_1 \leq 3$. Moreover, we have the following boundaries

$$x_I = \begin{bmatrix} 0 \\ x_2 \end{bmatrix}, \ x_{II} = \begin{bmatrix} 3 \\ x_2 \end{bmatrix} \text{ and } x_{III} = \begin{bmatrix} x_1 \\ -4 \end{bmatrix}, \ x_{IV} = \begin{bmatrix} x_1 \\ 0 \end{bmatrix}$$

defined by

$$x_C = \begin{bmatrix} 0 \\ 0 \end{bmatrix}, \ x_D = \begin{bmatrix} 0 \\ -4 \end{bmatrix}, \ x_E = \begin{bmatrix} 3 \\ 0 \end{bmatrix}, \ x_F = \begin{bmatrix} 3 \\ -4 \end{bmatrix}.$$

Let the Hessian matrix be

$$\nabla^2 f(x) = \begin{bmatrix} \frac{\partial^2 f}{\partial x_1 \partial x_1} & \frac{\partial^2 f}{\partial x_1 \partial x_2} \\ \frac{\partial^2 f}{\partial x_2 \partial x_1} & \frac{\partial^2 f}{\partial x_2 \partial x_2} \end{bmatrix} = \begin{bmatrix} 2x_1 + 1 & 2 \\ 2 & 1 \end{bmatrix}.$$

Then, we have

$$\nabla^2 f(x_A) = \begin{bmatrix} 3 & 2 \\ 2 & 1 \end{bmatrix} \text{ and } \nabla^2 f(x_B) = \begin{bmatrix} 5 & 2 \\ 2 & 1 \end{bmatrix}.$$

Let us check the positive definiteness of $\nabla^2 f(x_A)$ using the definition:

$$v^T \nabla^2 f(x_A) v = [v_1, v_2] \begin{bmatrix} 3 & 2 \\ 2 & 1 \end{bmatrix} \begin{bmatrix} v_1 \\ v_2 \end{bmatrix} = 3v_1^2 + 4v_1 v_2 + v_2^2.$$

If $v_1 = -0.5$ and $v_2 = 1.0$, we will have $v^T \nabla^2 f(x_A) v < 0$. On the other hand, if $v_1 = 1.5$ and $v_2 = 1.0$, we will have $v^T \nabla^2 f(x_A) v > 0$. Thus, $\nabla^2 f(x_A)$ is indefinite. Let us check $\nabla^2 f(x_B)$:

$$v^T \nabla^2 f(x_B) v = [v_1, v_2] \begin{bmatrix} 5 & 2 \\ 2 & 1 \end{bmatrix} \begin{bmatrix} v_1 \\ v_2 \end{bmatrix} = 5v_1^2 + 4v_1 v_2 + v_2^2 = v_1^2 + (2v_1 + v_2)^2 > 0.$$

Thus, $\nabla^2 f(x_B)$ is positive definite and $x_B = \begin{bmatrix} 2 \\ -3 \end{bmatrix}$ is a local minimizer with $f(x_B) = 19.166667$.

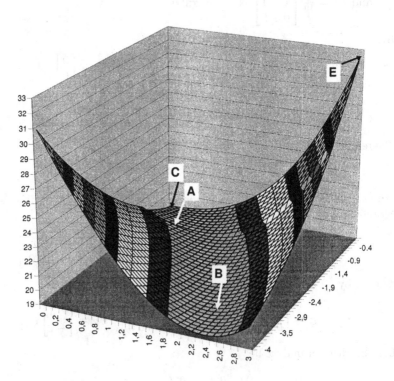

Fig. S.5. Plot of $f(x_1, x_2) = \frac{1}{3}x_1^3 + \frac{1}{2}x_1^2 + 2x_1 x_2 + \frac{1}{2}x_2^2 - x_2 + 19$

Let us check the boundary defined by x_I:

$$f(0, x_2) = \frac{1}{2}x_2^2 - x_2 + 19 \Rightarrow \frac{df(0, x_2)}{dx_2} = x_2 - 1 \doteq 0 \Rightarrow x_2 = 1.$$

Since $\frac{d^2 f(0, x_2)}{dx_2^2} = 1 > 0$, $x_2 = 1 > 0$ is the local minimizer outside the feasible region. As the first derivative is negative for $-4 \leq x_2 \leq 0$, we will check $x_2 = 0$ for minimizer and $x_2 = -4$ for maximizer (see Figure S.5).

Let us check the boundary defined by x_{II}:

$$f(3, x_2) = \frac{1}{2}x_2^2 + 5x_2 + \frac{65}{2} \Rightarrow \frac{df(3, x_2)}{dx_2} = x_2 + 5 \doteq 0 \Rightarrow x_2 = -5.$$

Since $\frac{d^2 f(0, x_2)}{dx_2^2} = 1 > 0$, $x_2 = -5 < -4$ is the local minimizer outside the feasible region. As the first derivative is positive for $-4 \leq x_2 \leq 0$, we will check $x_2 = -4$ for minimizer and $x_2 = 0$ for maximizer (see Figure S.5).

Let us check the boundary defined by x_{III}:

$$f(x_1, 0) = \frac{1}{3}x_1^3 + \frac{1}{2}x_1^2 + 19 \Rightarrow \frac{df(x_1, 0)}{dx_1} = x_1^2 + x_1 \doteq 0 \Rightarrow x_1 = 0, -1.$$

Since $\frac{d^2 f(x_1, 0)}{dx_1^2} = 2x_1 + 1$, $x_1 = 0$ is the local minimizer ($\frac{d^2 f(0, 0)}{dx_1^2} = 1 > 0$) on the boundary, and $x_1 = -1$ is the local maximizer ($\frac{d^2 f(-1, 0)}{dx_1^2} = -1 < 0$) outside the feasible region. As the first derivative is positive for $0 \leq x_2 \leq 3$, we will check $x_2 = 3$ for maximizer (see Figure S.5).

Let us check the boundary defined by x_{IV}:

$$f(x_1, -4) = \frac{1}{3}x_1^3 + \frac{1}{2}x_1^2 - 8x_1 + 31 \Rightarrow \frac{df(x_1, -4)}{dx_1} = x_1^2 + x_1 - 8 \doteq 0$$

$$\Rightarrow x_1 = \frac{-1 \pm \sqrt{1 + 32}}{2}.$$

Since $\frac{d^2 f(x_1, -4)}{dx_1^2} = 2x_1 + 1$ again, the positive root $x_1 = \frac{-1 + \sqrt{33}}{2} = 2.3723$ is the local minimizer ($\frac{d^2 f(2.3723, 0)}{dx_1^2} > 0$), and the negative root is the local maximizer but it is outside the feasible region. As the first derivative is positive for $0 \leq x_2 \leq 3$, we will check $x_2 = 3$ for maximizer again (see Figure S.5).

To sum up, we have to consider $(2, -3)$, $(0, 0)$ and $(2.3723, -4)$ for the minimizer; $(3, 0)$ and $(0, -4)$ for the maximizer:

$$f(2, -3) = 19.16667, \quad f(0, 0) = 19, \quad f(2.3723, -4) = 19.28529$$

$$\Rightarrow (0, 0) \text{ is the minimizer!}$$

$$f(3, 0) = 32.5, \quad f(0, -4) = 31 \Rightarrow (3, 0) \text{ is the maximizer!}$$

Problems of Chapter 6

6.1 The norm of a matrix A is defined as $\|A\| = \sqrt{\text{largest eigen value of } A^T A}$. If Q is orthogonal then $Q^T = Q^{-1} \Leftrightarrow Q^T Q = I$ and the unique eigen value of $Q^T Q$ is 1. Hence

$$\|Q\| = \|Q^T\| = 1.$$

Furthermore,

$$c = \|Q\| \, \|Q^{-1}\| = \|Q\|^2 = 1.$$

Hence for orthogonal matrices,

$$c = \|Q\| = 1.$$

Let $\mathcal{Q} = \alpha Q$. Then $\mathcal{Q}^T = \mathcal{Q}^{-1} = \frac{1}{\alpha} Q^T = \frac{1}{\alpha} Q^{-1} \Leftrightarrow \mathcal{Q}^T \mathcal{Q} = \alpha Q \frac{1}{\alpha} Q^T = I$. Thus,

$$\|\mathcal{Q}\| = \|\mathcal{Q}^T\| = 1,$$

and

$$c = \|\mathcal{Q}\| \, \|\mathcal{Q}^{-1}\| = \alpha \|Q\| \frac{1}{\alpha} \|Q\| = \|Q\|^2 = 1.$$

For orthogonal matrices, $\|Q\| = c(Q) = 1$. Orthogonal matrices and their multipliers (αQ) are *only* perfect condition matrices. It is left as an exercise to prove the *only* part.

6.2 $A = Q_0 R_0$, where

$$Q_0 = \begin{bmatrix} -0.4083 & -0.3762 & -0.5443 & 0.5452 & -0.3020 & 0.0843 \\ 0.9129 & -0.1882 & -0.2434 & 0.2438 & -0.1351 & 0.0377 \\ 0 & 0.9111 & -0.2696 & 0.2701 & -0.1496 & 0.0418 \\ 0 & 0 & -0.7562 & -0.5672 & 0.3142 & -0.0877 \\ 0 & 0 & 0 & -0.4986 & -0.8349 & 0.2331 \\ 0 & 0 & 0 & 0 & 0.2689 & 0.9632 \end{bmatrix}$$

$$R_0 = \begin{bmatrix} -1.2247 & 83.7098 & -73.0929 & 0 & 0 & 0 \\ 0 & -87.8778 & 87.3454 & 3.8183 & 0 & 0 \\ 0 & 0 & -5.5417 & -3.0895 & -0.1695 & 0 \\ 0 & 0 & 0 & -0.4497 & -0.1898 & 0.0050 \\ 0 & 0 & 0 & 0 & -0.0372 & 0.0095 \\ 0 & 0 & 0 & 0 & 0 & 0.0016 \end{bmatrix}$$

$$A_1 = R_0 Q_0 = \begin{bmatrix} -76.9159 & 80.2207 & 0 & 0 & 0 & 0 \\ 80.2207 & 94.3687 & -5.0493 & 0 & 0 & 0 \\ 0 & -5.0493 & 3.8305 & 0.3400 & 0 & 0 \\ 0 & 0 & 0.3400 & 0.3497 & 0.0185 & 0 \\ 0 & 0 & 0 & 0.0185 & 0.0336 & 0.0004 \\ 0 & 0 & 0 & 0 & 0.0004 & 0.0016 \end{bmatrix}$$

$A_1 = Q_1 R_1$, where

$$Q_1 = \begin{bmatrix} -0.6921 & -0.5964 & -0.3911 & 0.1109 & -0.0079 & -0.0001 \\ 0.7218 & -0.5718 & -0.3750 & 0.1063 & -0.0076 & -0.0001 \\ 0 & 0.5633 & -0.7948 & 0.2253 & -0.0161 & -0.0002 \\ 0 & 0 & -0.2734 & -0.9595 & 0.0685 & 0.0009 \\ 0 & 0 & 0 & -0.0712 & -0.9974 & -0.2331 \\ 0 & 0 & 0 & 0 & -0.0135 & 0.9999 \end{bmatrix}$$

$$R_1 = \begin{bmatrix} -111.1369 & 123.6364 & -3.6447 & 0 & 0 & 0 \\ 0 & -8.9636 & 5.0452 & 0.1915 & 0 & 0 \\ 0 & 0 & -1.2438 & -3.0895 & -0.0051 & 0 \\ 0 & 0 & 0 & -0.4497 & -0.0202 & 0 \\ 0 & 0 & 0 & 0 & -0.0322 & -0.0005 \\ 0 & 0 & 0 & 0 & 0 & 0.0016 \end{bmatrix}$$

$$A_2 = R_1 Q_1 = \begin{bmatrix} 166.1589 & -6.4701 & 0 & 0 & 0 & 0 \\ -6.4701 & 7.9677 & -0.7006 & 0 & 0 & 0 \\ 0 & -0.7006 & 1.0885 & 0.0711 & 0 & 0 \\ 0 & 0 & 0.0711 & 0.2511 & 0.0023 & 0 \\ 0 & 0 & 0 & 0.0023 & 0.0322 & 0 \\ 0 & 0 & 0 & 0 & 0 & 0.0016 \end{bmatrix}$$

$$\vdots$$

$$A_6 = R_5 Q_5 = \begin{bmatrix} 166.4231 & 0 & 0 & 0 & 0 & 0 \\ 0 & 7.7768 & -0.0002 & 0 & 0 & 0 \\ 0 & -0.0002 & 1.0218 & 0.0002 & 0 & 0 \\ 0 & 0 & 0.0002 & 0.2447 & 0 & 0 \\ 0 & 0 & 0 & 0 & 0.0321 & 0 \\ 0 & 0 & 0 & 0 & 0 & 0.0016 \end{bmatrix}$$

$A_6 = Q_6 R_6$, where

$$Q_6 = \begin{bmatrix} -1.0000 & 0 & 0 & 0 & 0 & 0 \\ 0 & -1.0000 & 0 & 0 & 0 & 0 \\ 0 & 0 & -1.0000 & 0.0002 & 0 & 0 \\ 0 & 0 & -0.0002 & -1.0000 & 0 & 0 \\ 0 & 0 & 0 & 0 & -1.0000 & 0 \\ 0 & 0 & 0 & 0 & 0 & 1.0000 \end{bmatrix}$$

$$R_6 = \begin{bmatrix} -166.4231 & 0 & 0 & 0 & 0 & 0 \\ 0 & -7.7768 & 0.0002 & 0 & 0 & 0 \\ 0 & 0 & -1.0218 & -0.0003 & 0 & 0 \\ 0 & 0 & 0 & -0.2447 & 0 & 0 \\ 0 & 0 & 0 & 0 & -0.0321 & 0 \\ 0 & 0 & 0 & 0 & 0 & 0.0016 \end{bmatrix}$$

$$A_7 = R_6 Q_6 = \begin{bmatrix} 166.4231 & 0 & 0 & 0 & 0 & 0 \\ 0 & 7.7768 & 0 & 0 & 0 & 0 \\ 0 & -0.0002 & 1.0218 & 0.0001 & 0 & 0 \\ 0 & 0 & 0.0001 & 0.2447 & 0 & 0 \\ 0 & 0 & 0 & 0 & 0.0321 & 0 \\ 0 & 0 & 0 & 0 & 0 & 0.0016 \end{bmatrix}$$

$A_7 = Q_7 R_7$, where

$$Q_7 = \begin{bmatrix} -1.0000 & 0 & 0 & 0 & 0 & 0 \\ 0 & -1.0000 & 0 & 0 & 0 & 0 \\ 0 & 0 & -1.0000 & 0.0001 & 0 & 0 \\ 0 & 0 & -0.0001 & -1.0000 & 0 & 0 \\ 0 & 0 & 0 & 0 & -1.0000 & 0 \\ 0 & 0 & 0 & 0 & 0 & 1.0000 \end{bmatrix}$$

$$R_7 = \begin{bmatrix} -166.4231 & 0 & 0 & 0 & 0 & 0 \\ 0 & -7.7768 & 0.0002 & 0 & 0 & 0 \\ 0 & 0 & -1.0218 & -0.0001 & 0 & 0 \\ 0 & 0 & 0 & -0.2447 & 0 & 0 \\ 0 & 0 & 0 & 0 & -0.0321 & 0 \\ 0 & 0 & 0 & 0 & 0 & 0.0016 \end{bmatrix}$$

$$A_8 = R_7 Q_7 = \begin{bmatrix} 166.4231 & 0 & 0 & 0 & 0 & 0 \\ 0 & 7.7768 & 0 & 0 & 0 & 0 \\ 0 & 0 & 1.0218 & 0 & 0 & 0 \\ 0 & 0 & 0 & 0.2447 & 0 & 0 \\ 0 & 0 & 0 & 0 & 0.0321 & 0 \\ 0 & 0 & 0 & 0 & 0 & 0.0016 \end{bmatrix}$$

The diagonal entries are the eigen values of A.

6.3 (a) Take $A(2)$.

1.

$$A(2) = \begin{bmatrix} 1 & \frac{1}{2} \\ \frac{1}{2} & \frac{1}{3} \end{bmatrix}$$

$$[A(2)|I_2] = \begin{bmatrix} 1 & \frac{1}{2} & 1 & 0 \\ \frac{1}{2} & \frac{1}{3} & 0 & 1 \end{bmatrix} \hookrightarrow \begin{bmatrix} 1 & \frac{1}{2} & 1 & 0 \\ 0 & \frac{1}{12} & -\frac{1}{2} & 1 \end{bmatrix}$$

$$\hookrightarrow \begin{bmatrix} 1 & 0 & 4 & -6 \\ 0 & 1 & -6 & 12 \end{bmatrix} = [I_2|A(2)^{-1}].$$

$$x_I = A(2)^{-1} b_I = \begin{bmatrix} 4 & -6 \\ -6 & 12 \end{bmatrix} \begin{bmatrix} 1.0 \\ 0.5 \end{bmatrix} = \begin{bmatrix} 1 \\ 0 \end{bmatrix},$$

$$x_{II} = A(2)^{-1}b_{II} = \begin{bmatrix} 4 & -6 \\ -6 & 12 \end{bmatrix}\begin{bmatrix} 1.5 \\ 1.0 \end{bmatrix} = \begin{bmatrix} 0 \\ 3 \end{bmatrix}.$$

$$\Delta_b = b_I - b_{II} = \begin{bmatrix} -0.5 \\ -0.5 \end{bmatrix} \Rightarrow \|\Delta_b\| = \sqrt{\frac{1}{2}},\ \|b_I\| = \sqrt{\frac{5}{4}} \Rightarrow \frac{\|\Delta_b\|}{\|b_I\|} = \sqrt{\frac{4}{10}}.$$

$$\Delta_x = x_I - x_{II} = \begin{bmatrix} 1 \\ -3 \end{bmatrix} \Rightarrow \|\Delta_x\| = \sqrt{10},\ \|x_I\| = \sqrt{1} \Rightarrow \frac{\|\Delta_x\|}{\|x_I\|} = \sqrt{\frac{10}{1}}.$$

Then, the relative error for this case is $\dfrac{\sqrt{\frac{10}{1}}}{\sqrt{\frac{4}{10}}} = 5.0$.

2. The maximum error is the condition number.

$$\det(sI - A(2)) = \begin{vmatrix} s-1 & -\frac{1}{2} \\ -\frac{1}{2} & s-\frac{1}{3} \end{vmatrix} = (s-1)\left(s-\frac{1}{3}\right) - \frac{1}{4} \doteq 0$$

$$\Rightarrow \lambda_1 = \frac{4 - \sqrt{13}}{6},\ \lambda_2 = \frac{4 + \sqrt{13}}{6}.$$

Therefore, $c[A(2)] = \dfrac{\lambda_2}{\lambda_1} = \dfrac{4+\sqrt{13}}{4-\sqrt{13}} = \dfrac{7.605551}{0.394449} = 19.2815$ is the upper bound.

3.

$$A(2) + \Delta_{A(2)} = \begin{bmatrix} 1 & \frac{1}{2} \\ \frac{1}{2} & \frac{1}{3} \end{bmatrix} + \begin{bmatrix} 0 & -\frac{1}{2} \\ -\frac{1}{2} & \frac{2}{3} \end{bmatrix} = I_2$$

$$I_2 x_{III} = b_I \Rightarrow x_{III} = b_I = \begin{bmatrix} 1.0 \\ 0.5 \end{bmatrix}.$$

$$\Delta_x = x_{III} - x_I = \begin{bmatrix} 2.0 \\ 0.5 \end{bmatrix} \Rightarrow \frac{\|\Delta_x\|}{\|x_I + \Delta_x\|} = \frac{\sqrt{4.25}}{\sqrt{1.25}} = 1.84391$$

$$\|A(2)\| = \lambda_2 = \frac{4+\sqrt{13}}{6} = 1.26759$$

$\|\Delta_{A(2)}\|$ is the largest eigenvalue of $\begin{bmatrix} 0 & -\frac{1}{2} \\ -\frac{1}{2} & \frac{2}{3} \end{bmatrix}$, which is 0.9343. Then,

$$\frac{\|\Delta_{A(2)}\|}{\|A(2)\|} = \frac{0.9343}{1.2676} = 0.7371 \Rightarrow \frac{\frac{\|\Delta_x\|}{\|x_I + \Delta_x\|}}{\frac{\|\Delta_{A(2)}\|}{\|A(2)\|}} = \frac{1.84391}{0.7371} = 2.5017$$

4. The maximum error is $\|A(2)\|\,\|A(2)^{-1}\|$, where $\|A(2)^{-1}\|$ is the largest eigenvalue of $A(2)^{-1}$ as calculated below:

$$\det(sI - A(2)^{-1}) = \begin{vmatrix} s-4 & 6 \\ 6 & s-12 \end{vmatrix} = (s-4)(s-12) - 36 \doteq 0$$

$$\Rightarrow \mu_1 = \frac{16 - \sqrt{208}}{2}, \ \mu_2 = \frac{16 + \sqrt{208}}{2}$$

$$\Rightarrow \|A(2)^{-1}\| = \mu_2 = \frac{16 + \sqrt{208}}{6} = 15.2111$$

Then, $\|A(2)\| \, \|A(2)^{-1}\| = 1.2676(15.2111) = 19.2815 = c[A(2)]$.
We know, $\mu_1 = \frac{1}{\lambda_2}$ and $\mu_2 = \frac{1}{\lambda_1}$. Consequently,

$$c[A(2)^{-1}] = \frac{\mu_2}{\mu_1} = 19.2815 = \frac{\frac{1}{\lambda_1}}{\frac{1}{\lambda_2}} = \frac{\lambda_2}{\lambda_1} = c[A(2)]$$

(b) Take $A(3)$.

$$A(3) = \begin{bmatrix} 1 & \frac{1}{2} & \frac{1}{3} \\ \frac{1}{2} & \frac{1}{3} & \frac{1}{4} \\ \frac{1}{3} & \frac{1}{4} & \frac{1}{5} \end{bmatrix} \Rightarrow A(3)^T A(3) = \begin{bmatrix} \frac{49}{36} & \frac{3}{4} & \frac{21}{40} \\ \frac{3}{4} & \frac{61}{144} & \frac{3}{10} \\ \frac{21}{40} & \frac{3}{10} & \frac{769}{3600} \end{bmatrix}$$

$$\det(sI - A(3)^T A(3)) = \begin{vmatrix} s - \frac{49}{36} & -\frac{3}{4} & -\frac{21}{40} \\ -\frac{3}{4} & s - \frac{61}{144} & -\frac{3}{10} \\ -\frac{21}{40} & -\frac{3}{10} & s - \frac{769}{3600} \end{vmatrix} \doteq 0$$

$$\Rightarrow (s - 3/415409)(s - 255/17041)(s - 1192/601) \doteq 0 \Rightarrow$$

$$\nu_1 = \frac{3}{415409}, \ \nu_2 = \frac{255}{17041}, \ \nu_3 = \frac{1192}{601}$$

$$\Rightarrow c[A(3)^T A(3)] = \frac{\nu_3}{\nu_1} = \frac{\frac{1192}{601}}{\frac{3}{415409}} = 274635.3$$

$$\det(sI - A(3)) = \begin{vmatrix} 1 & \frac{1}{2} & \frac{1}{3} \\ \frac{1}{2} & \frac{1}{3} & \frac{1}{4} \\ \frac{1}{3} & \frac{1}{4} & \frac{1}{5} \end{vmatrix} \doteq 0$$

$$\Rightarrow (s - 26/9675)(s - 389/3180)(s - 745/529) \doteq 0 \Rightarrow$$

$$\lambda_1 = \frac{26}{9675}, \ \lambda_2 = \frac{389}{3180}, \ \lambda_3 = \frac{745}{529} \Rightarrow c[A(3)] = \frac{\lambda_3}{\lambda_1} = \frac{\frac{745}{529}}{\frac{26}{9675}} = 524.0566$$

Clearly, $c[A(3)^T A(3)] = (c[A(3)])^2$.

(c) Take $A(4)$.

$$A(4) = \begin{bmatrix} 1 & \frac{1}{2} & \frac{1}{3} & \frac{1}{4} \\ \frac{1}{2} & \frac{1}{3} & \frac{1}{4} & \frac{1}{5} \\ \frac{1}{3} & \frac{1}{4} & \frac{1}{5} & \frac{1}{6} \\ \frac{1}{4} & \frac{1}{5} & \frac{1}{6} & \frac{1}{7} \end{bmatrix}$$

$$A(4) = Q_0 R_0 = \begin{bmatrix} -0.83812 & 0.52265 & -0.15397 & -0.02631 \\ -0.41906 & -0.44171 & 0.72775 & 0.31568 \\ -0.27937 & -0.52882 & -0.13951 & -0.78920 \\ -0.20953 & -0.50207 & -0.65361 & 0.52613 \end{bmatrix}$$

$$\begin{bmatrix} -1.19320 & -0.67049 & -0.47493 & -0.36984 \\ 0.00000 & -0.11853 & -0.12566 & -0.11754 \\ 0.00000 & 0.00000 & -0.00622 & -0.00957 \\ 0.00000 & 0.00000 & 0.00000 & 0.00019 \end{bmatrix}$$

$$A(4)_1 = R_0 Q_0$$

$$= \begin{bmatrix} 1.49110 & 0.10941 & 0.0037426 & -3.9372 \times 10^{-5} \\ 0.10941 & 0.17782 & 0.0080931 & -9.4342 \times 10^{-5} \\ 0.00374 & 0.00809 & 0.0071205 & -0.00012282 \\ -3.9372 \times 10^{-5} & -9.4342 \times 10^{-5} & -0.00012282 & 9.8863 \times 10^{-5} \end{bmatrix}$$

$$A(4)_1 = Q_1 R_1 = \begin{bmatrix} -0.997320 & 0.073211 & -0.000868 & 2.1324 \times 10^{-6} \\ -0.073173 & -0.996260 & 0.046000 & -0.0002696 \\ -0.002503 & -0.045938 & -0.998790 & 0.0175510 \\ 2.6333 \times 10^{-5} & 0.000538 & 0.017545 & 0.9998500 \end{bmatrix}$$

$$\begin{bmatrix} -1.49520 & -0.12214 & -0.0043425 & 4.6479 \times 10^{-5} \\ 0.00000 & -0.16952 & -0.0081160 & 9.6801 \times 10^{-5} \\ 0.00000 & 0.00000 & -0.0067449 & 0.00012010 \\ 0.00000 & 0.00000 & 0.0000000 & 9.6718 \times 10^{-5} \end{bmatrix}$$

$$A(4)_2 = R_1 Q_1 = \begin{bmatrix} 1.500100 & 0.012424 & 1.6887 \times 10^{-5} & 2.5468 \times 10^{-9} \\ 0.012424 & 0.169260 & 0.0003099 & 5.1991 \times 10^{-8} \\ 1.6887 \times 10^{-5} & 0.000310 & 0.0067389 & 1.6969 \times 10^{-6} \\ 2.5468 \times 10^{-9} & 5.1991 \times 10^{-8} & 1.6969 \times 10^{-6} & 9.6703 \times 10^{-5} \end{bmatrix}$$

$$A(4)_2 = Q_2 R_2$$

$$= \begin{bmatrix} -0.999970 & 0.008282 & -3.9108 \times 10^{-6} & -1.3792 \times 10^{-10} \\ -0.008282 & -0.999960 & 0.0018313 & 1.5392 \times 10^{-7} \\ -1.1257 \times 10^{-5} & -0.001831 & -1.0000000 & -0.00025182 \\ -1.6977 \times 10^{-9} & -3.0723 \times 10^{-7} & -0.0002518 & 1.00000000 \end{bmatrix}$$

$$\begin{bmatrix} -1.50010 & -0.01383 & -1.9529 \times 10^{-5} & -2.9966 \times 10^{-9} \\ 0.00000 & -0.16915 & -0.0003221 & -5.5105 \times 10^{-8} \\ 0.00000 & 0.00000 & -0.0067383 & -1.7212 \times 10^{-6} \\ 0.00000 & 0.00000 & 0.0000000 & 9.6702 \times 10^{-5} \end{bmatrix}$$

$$A(4)_3 = R_2 Q_2$$

$$= \begin{bmatrix} 1.500200 & 0.001401 & 7.5850 \times 10^{-8} & -1.6405 \times 10^{-13} \\ 0.001401 & 0.169140 & 1.2340 \times 10^{-5} & -2.9710 \times 10^{-11} \\ 7.5850 \times 10^{-8} & 1.2340 \times 10^{-5} & 0.0067383 & -2.4351 \times 10^{-8} \\ -1.6417 \times 10^{-13} & -2.9710 \times 10^{-11} & -2.4351 \times 10^{-8} & 9.6702 \times 10^{-5} \end{bmatrix}$$

$A(4)_3 = Q_3 R_3$

$$= \begin{bmatrix} -1.0000000 & 0.0009338 & -1.7566 \times 10^{-8} & 8.8905 \times 10^{-15} \\ -0.0009338 & -1.0000000 & 7.2955 \times 10^{-5} & -8.7996 \times 10^{-11} \\ -5.0559 \times 10^{-8} & -7.2955 \times 10^{-5} & -1.0000 & 3.6138 \times 10^{-6} \\ 1.0943 \times 10^{-13} & 1.7565 \times 10^{-10} & 3.6138 \times 10^{-6} & 1.0000 \end{bmatrix}$$

$$\begin{bmatrix} -1.50020 & -0.00156 & -8.7713 \times 10^{-8} & 1.9304 \times 10^{-13} \\ 0.00000 & -0.16914 & -1.2831 \times 10^{-5} & 3.1503 \times 10^{-11} \\ 0.00000 & 0.00000 & -0.006738 & 2.4701 \times 10^{-8} \\ 0.00000 & 0.00000 & 0.000000 & 9.6702 \times 10^{-5} \end{bmatrix}$$

$A(4)_4 = R_3 Q_3$

$$= \begin{bmatrix} 1.500200 & 0.000158 & 3.4068 \times 10^{-10} & -1.0796 \times 10^{-16} \\ 0.000158 & 0.169140 & 4.9159 \times 10^{-7} & 1.7074 \times 10^{-14} \\ 3.4068 \times 10^{-10} & 4.9159 \times 10^{-7} & 0.0067383 & 3.4947 \times 10^{-10} \\ 1.0582 \times 10^{-17} & 1.6986 \times 10^{-14} & 3.4947 \times 10^{-10} & 9.6702 \times 10^{-5} \end{bmatrix}$$

$A(4)_4 = Q_4 R_4$

$$= \begin{bmatrix} -1.0000 & 0.00010528 & -7.8899 \times 10^{-11} & -5.7307 \times 10^{-19} \\ -0.00010528 & -1.0000 & 2.9064 \times 10^{-6} & 5.0310 \times 10^{-14} \\ -2.2709 \times 10^{-10} & -2.9064 \times 10^{-6} & -1.0000 & -5.1863 \times 10^{-8} \\ -7.0539 \times 10^{-18} & -1.0042 \times 10^{-13} & -5.1863 \times 10^{-8} & 1.0000 \end{bmatrix}$$

$$\begin{bmatrix} -1.50020 & -0.00018 & -3.9397 \times 10^{-10} & 1.0608 \times 10^{-16} \\ 0.00000 & -0.16914 & -5.1117 \times 10^{-7} & -1.8100 \times 10^{-14} \\ 0.00000 & 0.00000 & -0.0067383 & -3.5448 \times 10^{-10} \\ 0.00000 & 0.00000 & 0.0000000 & 9.6702 \times 10^{-5} \end{bmatrix}$$

$A(4)_5 = R_4 Q_4$

$$= \begin{bmatrix} 1.5002 & 1.7808 \times 10^{-5} & 1.5304 \times 10^{-12} & 1.1853 \times 10^{-16} \\ 1.7808 \times 10^{-5} & 0.16914 & 1.9584 \times 10^{-8} & -9.8322 \times 10^{-17} \\ 1.5302 \times 10^{-12} & 1.9584 \times 10^{-8} & 0.0067383 & -5.0152 \times 10^{-12} \\ -6.8213 \times 10^{-22} & -9.7112 \times 10^{-18} & -5.0153 \times 10^{-12} & 9.6702 \times 10^{-5} \end{bmatrix}$$

Thus, $\Lambda = \begin{bmatrix} 1.5002 & & & \\ & 0.16914 & & \\ & & 0.0067383 & \\ & & & 0.0000967 \end{bmatrix}$ and

$$c[A(4)] = \frac{1.5002}{0.0000967} = 15514.$$

Problems of Chapter 7

7.1
a) A zero dimensional polytope is a point.
b) One dimensional polytopes are line segments.
c) Two dimensional polytopes are n-gons:
triangle (3), rectangle (4), trapezoid (4), pentagon (5), ...

7.2 $\Delta_2 = \mathrm{conv}(e_1, e_2, e_3)$. See Figure S.6.

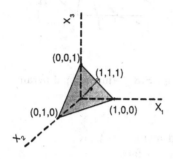

Fig. S.6. Δ_2 in $\mathbb{R}^3$

7.3 $C_3 = \mathrm{conv}((0,0,0)^T, (\alpha,0,0)^T, (0,\alpha,0)^T, (0,0,\alpha)^T, (\alpha,\alpha,0)^T,$
$$(\alpha,0,\alpha)^T, (0,\alpha,\alpha)^T, (\alpha,\alpha,\alpha)^T)$$

$$C_n = \{x \in \mathbb{R}^n : 0 \le x_i \le \alpha, \ i = 1,\ldots,n; \ \alpha \in \mathbb{R}_+\}.$$

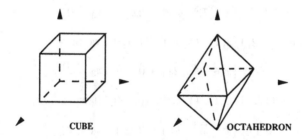

Fig. S.7. Cube and octahedron

$C_3^\Delta = \mathrm{conv}((\alpha,0,0)^T, (0,\alpha,0)^T, (0,0,\alpha)^T, (-\alpha,0,0)^T, (0,-\alpha,0)^T, (0,0,-\alpha)^T)$

$$C_n^\Delta = \left\{ x \in \mathbb{R}^n : \sum_{i=1}^{n} |x_i| \le \alpha, \ \alpha \in \mathbb{R}_+ \right\}.$$

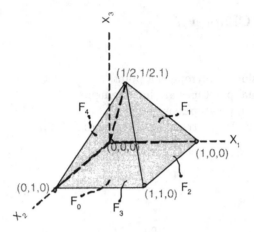

Fig. S.8. 3-dimensional pyramid

7.4

See Figure S.8 for a drawing of P_{n+1}.

Let a^i be the normal to face F_i, $i = 0, 1, 2, 3, 4$. Let $a^i x \leq b_i$ be the respective defining inequalities.

We know F_0 is the x_1-x_2 plane. Then, $F_0 = \left\{ x \in \mathbb{R}^3 : x_3 = 0 \right\}$.

We know that a^2 and a^4 are perpendicular to x_2-axis. Similarly, a^1 and a^3 are perpendicular to x_1 axis. Thus,

$$a^1 = (0, *, *)^T, \ a^2 = (*, 0, *)^T, \ a^3 = (0, *, *)^T, \ a^4 = (*, 0, *)^T.$$

Since F_1 contains $(1/2, 1/2, 1)$, $(1, 0, 0)$, $(0, 0, 0)$, what we have is

$$F_1 = \left\{ x \in \mathbb{R}^3 : 0x_1 - 2x_2 + 1x_3 = 0 \right\}.$$

Since F_2 contains $(1/2, 1/2, 1)$, $(1, 0, 0)$, $(1, 1, 0)$, we have

$$F_2 = \left\{ x \in \mathbb{R}^3 : 2x_1 + 0x_2 + 1x_3 = 2 \right\}.$$

Since F_3 contains $(1/2, 1/2, 1)$, $(1, 1, 0)$, $(0, 1, 0)$, it is

$$F_3 = \left\{ x \in \mathbb{R}^3 : 0x_1 + 2x_2 + 1x_3 = 2 \right\}.$$

And finally, $(1/2, 1/2, 1)$, $(0, 1, 0)$, $(0, 0, 0)$ are in F_4,

$$F_4 = \left\{ x \in \mathbb{R}^3 : -2x_1 + 0x_2 + 1x_3 = 0 \right\}.$$

Therefore,

$$P_3 = \{ x \in \mathbb{R}^3 : x_3 \geq 0, \ -2x_2 + x_3 \leq 0, \ 2x_1 + x_3 \leq 2,$$
$$2x_2 + x_3 \leq 2, \ -2x_1 + x_3 \leq 0 \}.$$

P_{n+1} is not a union of a cone at x_0 and a polytope.
P_{n+1} is a direct sum of a cone at x_0 and C_n.
P_{n+1} is an intersection of a cone at x_0 and C_{n+1} provided that $x_0 \in C_{n+1} \backslash C_n$.

7.5 See Figure S.9.

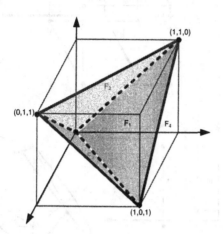

Fig. S.9. A tetrahedron

The diagonal ray $(1, 1, 1)^T$ of the cube is orthogonal to facet F_4. Thus, $F_4 = \{x \in \mathbb{R}^3 : x_1 + x_2 + x_3 = \alpha\}$. Since this facet contains $(0, 1, 1)^T$, $(1, 0, 1)^T$, $(1, 1, 0)^T$, the value of α is 2. Therefore,

$$F_4 = \{x \in \mathbb{R}^3 : x_1 + x_2 + x_3 = 2\}.$$

Since $(0, 0, 0)^T$ is on the tetrahedron, the following halfspace is valid and facet defining

$$H_4 = \{x \in \mathbb{R}^3 : x_1 + x_2 + x_3 \leq 2\}$$

Similarly,

$$F_1 = \{x \in \mathbb{R}^3 : x_1 - x_2 - x_3 = 0\},$$
$$F_2 = \{x \in \mathbb{R}^3 : -x_1 + x_2 - x_3 = 0\},$$
$$F_3 = \{x \in \mathbb{R}^3 : -x_1 - x_2 + x_3 = 0\}.$$

The following set describes the tetrahedron:

$$x_1 + x_2 + x_3 \leq 2,$$
$$x_1 - x_2 - x_3 \leq 0,$$
$$-x_1 + x_2 - x_3 \leq 0,$$
$$-x_1 - x_2 + x_3 \leq 0.$$

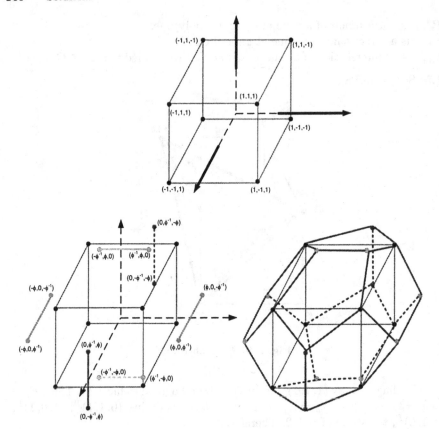

Fig. S.10. The dodecahedron, ϕ: golden ratio

7.6 See Figure S.10.

The polyhedron vertices of a dodecahedron can be given in a simple form for a dodecahedron of side length $a = \sqrt{5} - 1$ by

$$(0, \pm\phi^{-1}, \pm\phi)^T, \ (\pm\phi, 0, \pm\phi^{-1})^T, \ (\pm\phi^{-1}, \pm\phi, 0)^T \text{ and } (\pm1, \pm1, \pm1)^T;$$

where $\phi = \frac{1+\sqrt{5}}{2}$ is the golden ratio. We know $\phi - 1 = \frac{1}{\phi}$ and $\phi = 2\cos\frac{\pi}{5}$. See Figure S.11.

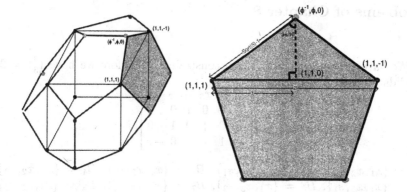

Fig. S.11. The extreme points of the dodecahedron, ϕ: golden ratio

Problems of Chapter 8

8.1
a) We have six variables and three constraints, therefore we have $\binom{6}{3} = 20$ candidate bases.

$$A = \begin{bmatrix} x_1 & x_2 & x_3 & s_1 & s_2 & s_3 \\ 2 & 1 & 0 & 1 & 0 & 0 \\ 0 & 0 & 1 & 0 & 1 & 0 \\ 0 & 1 & 0 & 0 & 0 & -1 \end{bmatrix}$$

$B_1 = \{x_1, x_2, x_3\}$, $B_2 = \{x_1, x_2, s_1\}$, $B_3 = \{x_1, x_2, s_2\}$, $B_4 = \{x_1, x_2, s_3\}$, $B_5 = \{x_1, x_3, s_1\}$, $B_6 = \{x_1, x_3, s_2\}$, $B_7 = \{x_1, x_3, s_3\}$, $B_8 = \{x_1, s_1, s_2\}$, $B_9 = \{x_1, s_1, s_3\}$, $B_{10} = \{x_1, s_2, s_3\}$, $B_{11} = \{x_2, x_3, s_1\}$, $B_{12} = \{x_2, x_3, s_2\}$, $B_{13} = \{x_2, x_3, s_3\}$, $B_{14} = \{x_2, s_1, s_2\}$, $B_{15} = \{x_2, s_1, s_3\}$, $B_{16} = \{x_2, s_2, s_3\}$, $B_{17} = \{x_3, s_1, s_2\}$, $B_{18} = \{x_3, s_1, s_3\}$, $B_{19} = \{x_3, s_2, s_3\}$, $B_{20} = \{s_1, s_2, s_3\}$.

$B_2, B_4, B_5, B_6, B_8, B_9, B_{12}, B_{15}, B_{17}, B_{19}$ are not bases since they form singular matrices. $B_7, B_{10}, B_{18}, B_{20}$ are infeasible since they do not satisfy non-negativity constraints. Thus, what we have is

$(x_1, x_2, x_3, s_1, s_2, s_3)^T = (3, 2, 10, 0, 0, 0)^T$ from $B_1 \hookrightarrow$ point F,
$(x_1, x_2, x_3, s_1, s_2, s_3)^T = (3, 2, 0, 0, 10, 0)^T$ from $B_3 \hookrightarrow$ point C,
$(x_1, x_2, x_3, s_1, s_2, s_3)^T = (0, 2, 10, 6, 0, 0)^T$ from $B_{11} \hookrightarrow$ point E,
$(x_1, x_2, x_3, s_1, s_2, s_3)^T = (0, 8, 10, 0, 0, 6)^T$ from $B_{13} \hookrightarrow$ point D,
$(x_1, x_2, x_3, s_1, s_2, s_3)^T = (0, 2, 0, 6, 10, 0)^T$ from $B_{14} \hookrightarrow$ point B,
$(x_1, x_2, x_3, s_1, s_2, s_3)^T = (0, 8, 0, 0, 10, 6)^T$ from $B_{16} \hookrightarrow$ point A.
See Figure S.12.

b)

1. *matrix form:*
   Let $x_B = (s_1, x_3, x_2)^T$, $x_N = (x_1, s_2, s_3)^T$. Then,

$$B = \begin{bmatrix} 1 & 0 & 1 \\ 0 & 1 & 0 \\ 0 & 0 & 1 \end{bmatrix} \Rightarrow B^{-1} = \begin{bmatrix} 1 & 0 & -1 \\ 0 & 1 & 0 \\ 0 & 0 & 1 \end{bmatrix}.$$

$$x_B = B^{-1}b = \begin{bmatrix} 1 & 0 & -1 \\ 0 & 1 & 0 \\ 0 & 0 & 1 \end{bmatrix} \begin{bmatrix} 8 \\ 10 \\ 2 \end{bmatrix} = \begin{bmatrix} 6 \\ 10 \\ 2 \end{bmatrix}.$$

We are on point E.

$$z = c_B^T x_B = [0, 2, 2] \begin{bmatrix} 6 \\ 10 \\ 2 \end{bmatrix} = 24.$$

$$c_N^T - c_B^T B^{-1} N = [1, 0, 0] - [0, 2, 2] \begin{bmatrix} 1 & 0 & -1 \\ 0 & 1 & 0 \\ 0 & 0 & 1 \end{bmatrix} \begin{bmatrix} 2 & 0 & 0 \\ 0 & 1 & 0 \\ 0 & 0 & -1 \end{bmatrix} = [1, -2, 2].$$

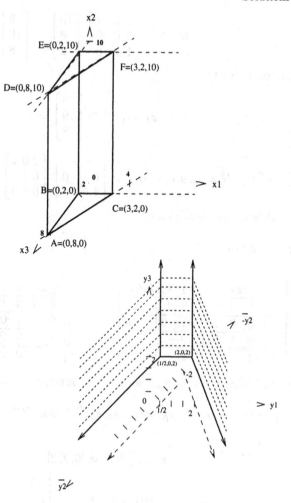

Fig. S.12. Exercise 8.1: Primal and dual polyhedra

Thus, s_3 enters.

$$B^{-1}N^{s_3} = \begin{bmatrix} 1 & 0 & -1 \\ 0 & 1 & 0 \\ 0 & 0 & 1 \end{bmatrix} \begin{bmatrix} 0 \\ 0 \\ -1 \end{bmatrix} = \begin{bmatrix} 1 \\ 0 \\ -1 \end{bmatrix} \begin{matrix} s_1 \\ x_3 \\ x_2 \end{matrix}$$

Thus, s_1 leaves.

New partition is $x_B = (s_3, x_3, x_2)^T$, $x_N = (x_1, s_2, s_1)^T$. Then,

$$B = \begin{bmatrix} 0 & 0 & 1 \\ 0 & 1 & 0 \\ -1 & 0 & 1 \end{bmatrix} \Rightarrow B^{-1} = \begin{bmatrix} 1 & 0 & -1 \\ 0 & 1 & 0 \\ 1 & 0 & 0 \end{bmatrix}.$$

$$x_B = B^{-1}b = \begin{bmatrix} 1 & 0 & -1 \\ 0 & 1 & 0 \\ 1 & 0 & 0 \end{bmatrix} \begin{bmatrix} 10 \\ 8 \\ 2 \end{bmatrix} = \begin{bmatrix} 6 \\ 10 \\ 8 \end{bmatrix}.$$

We are on point D.

$$z = c_B^T x_B = [0, 2, 2] \begin{bmatrix} 6 \\ 10 \\ 8 \end{bmatrix} = 36.$$

$$c_N^T - c_B^T B^{-1} N = [1, 0, 0] - [0, 2, 2] \begin{bmatrix} 1 & 0 & -1 \\ 0 & 1 & 0 \\ 1 & 0 & 0 \end{bmatrix} \begin{bmatrix} 2 & 0 & 1 \\ 0 & 1 & 0 \\ 0 & 0 & 0 \end{bmatrix} = [-3, -2, -2].$$

Thus, D is the optimal point.

2. *simplex tableau:*

$$
\begin{bmatrix}
 & x_1 & x_2 & x_3 & s_1 & s_2 & s_3 & RHS \\
s_1 & 2 & 0 & 0 & 1 & 0 & 1 & 6 \\
x_3 & 0 & 0 & 1 & 0 & 1 & 0 & 10 \\
x_2 & 0 & 1 & 0 & 0 & 0 & -1 & 2 \\
z & -1 & 0 & 0 & 0 & 2 & -2 & 24
\end{bmatrix}
\qquad
\begin{bmatrix}
 & x_1 & x_2 & x_3 & s_1 & s_2 & s_3 & RHS \\
s_3 & 2 & 0 & 0 & 1 & 0 & 1 & 6 \\
x_3 & 0 & 0 & 1 & 0 & 1 & 0 & 10 \\
x_2 & 2 & 1 & 0 & 0 & 0 & 0 & 8 \\
z & 3 & 0 & 0 & 2 & 2 & 0 & 36
\end{bmatrix}
$$

3. *revised simplex with product form of the inverse:*

Let $x_B = (s_1, x_3, x_2)^T$, $x_N = (x_1, s_2, s_3)^T$. Then, $B^{-1} = \begin{bmatrix} 1 & 0 & -1 \\ 0 & 1 & 0 \\ 0 & 0 & 1 \end{bmatrix}$.

$$w = c_B^T B^{-1} = [0, 2, 2].$$

$$r_{x_1} = c_{x_1} - wN^{x_1} = 1 - [0, 2, 2] \begin{bmatrix} 2 \\ 0 \\ 0 \end{bmatrix} = 1 > 0.$$

$$r_{s_2} = c_{s_2} - wN^{s_2} = 0 - [0, 2, 2] \begin{bmatrix} 0 \\ 1 \\ 0 \end{bmatrix} = -2 < 0.$$

$$r_{s_3} = c_{s_3} - wN^{s_3} = 0 - [0, 2, 2] \begin{bmatrix} 0 \\ 0 \\ -1 \end{bmatrix} = 2 > 0.$$

s_3 is the entering variable and s_1 leaves. $E_1^{-1} = \begin{bmatrix} \frac{1}{1} & 0 & 0 \\ -\frac{0}{1} & 1 & 0 \\ -\frac{-1}{1} & 0 & 1 \end{bmatrix}$.

$$x_B = E_1^{-1}\bar{b} = (6, 8, 10)^T.$$

$$w = [0, 2, 2]E_1^{-1}B^{-1} = [2, 2, 0].$$

$$w = c_B^T B^{-1} = [0, 2, 2]$$

$$r_{x_1} = c_{x_1} - wN^{x_1} = 1 - [2, 2, 0]\begin{bmatrix} 2 \\ 0 \\ 0 \end{bmatrix} = -3 < 0.$$

$$r_{s_2} = c_{s_2} - wN^{s_2} = 0 - [2, 2, 0]\begin{bmatrix} 0 \\ 1 \\ 0 \end{bmatrix} = -2 < 0.$$

$$r_{s_1} = c_{s_1} - wN^{s_1} = 0 - [0, 2, 2]\begin{bmatrix} 0 \\ 0 \\ -1 \end{bmatrix} = -2 < 0.$$

Optimal.

4. *revised simplex with $B = LU$ decomposition:*

Let $x_B = (s_1, x_3, x_2)^T$, $x_N = (x_1, s_2, s_3)^T$. Then, $B = \begin{bmatrix} 1 & 0 & 1 \\ 0 & 1 & 0 \\ 0 & 0 & 1 \end{bmatrix}$ is upper

triangular, $L = I_3$. Solve $Bx_B = b$ by back substitution.

$$x_2 = 2, \ x_3 = 10, \ s_1 = 8 - x_2 = 6.$$

Solve $wB = c_B$ by back substitution.

$$w_1 = 0, \ w_2 = 2, \ w_3 = 2 - w_1 = 2.$$

The rest is the same, s_3 enters and s_1 leaves.

New basis is $B = \begin{bmatrix} 0 & 0 & 1 \\ 0 & 1 & 0 \\ -1 & 0 & 1 \end{bmatrix}$.

$$PB = LU \Leftrightarrow \begin{bmatrix} 0 & 0 & 1 \\ 0 & 1 & 0 \\ 1 & 0 & 0 \end{bmatrix}\begin{bmatrix} 0 & 0 & 1 \\ 0 & 1 & 0 \\ -1 & 0 & 1 \end{bmatrix} = \begin{bmatrix} -1 & 0 & 1 \\ 0 & 1 & 0 \\ 0 & 0 & 1 \end{bmatrix} = I_3 U.$$

Solve $Bx_B = Pb = (2, 10, 8)^T$ by substitution.

$$x_2 = 2, \ x_3 = 10, \ s_3 = x_2 - 2 = 6.$$

Solve $wB = Pc_B = (2, 2, 0)^T$ by substitution.

$$w_1 = 0, \ w_2 = 2, \ w_3 = 2 - w_1 = 2.$$

The rest is the same.

5. *revised simplex with $B = QR$ decomposition:*

Let $x_B = (s_1, x_3, x_2)^T$, $x_N = (x_1, s_2, s_3)^T$. Then, $B = \begin{bmatrix} 1 & 0 & 1 \\ 0 & 1 & 0 \\ 0 & 0 & 1 \end{bmatrix}$ is upper

triangular, $Q = I_3$. The rest is the same as above, s_3 enters and s_1 leaves.

$$B = \begin{bmatrix} 0 & 0 & 1 \\ 0 & 1 & 0 \\ -1 & 0 & 1 \end{bmatrix} = \begin{bmatrix} 0 & 0 & 1 \\ 0 & 1 & 0 \\ -1 & 0 & 0 \end{bmatrix} \begin{bmatrix} 1 & 0 & -1 \\ 0 & 1 & 0 \\ 0 & 0 & 1 \end{bmatrix} = QR.$$

In order to solve $Bx_B = QRx_B = b = (8, 10, 2)^T = Q(Rx_B) = Qb'$,

$$b_3' = 8, \ b_2' = 10, \ b_1' = -2 \Rightarrow x_2 = 8, \ x_3 = 0, \ s_3 = x_2 - 2 = 6.$$

In order to solve $wB = c_B$, first solve $wQR = c_B = w'R$.

$$w_1' = 0, \ w_2' = 2, \ w_3' = 2 + w_1' = 2.$$

Then, solve $wQ = w'$

$$w_3 = 0, \ w_2 = 2, \ w_1 = 2.$$

The rest is the same.

c)

(D):

$$
\begin{aligned}
Min \ \ w = & 8y_1 + 10y_2 - 2y_3 \\
s.t. & \\
& 2y_1 \geq 1 \\
& y_1 - y_3 \geq 2 \\
& y_2 \geq 2 \\
& y_1, y_2, y_3 \geq 0.
\end{aligned}
$$

See Figure S.12.

8.2 The second constraint is redundant whose twice is exactly the last constraint plus the nonnegativity of x_1. Then,

$$A = \begin{bmatrix} x_1 & x_2 & x_3 & s_1 & s_3 \\ 2 & -1 & -1 & -1 & 0 \\ 1 & -2 & 2 & 0 & -1 \end{bmatrix}.$$

a) The bases are
$B_1 = \{x_1, x_2\}$, $B_2 = \{x_1, x_3\}$, $B_3 = \{x_1, s_1\}$, $B_4 = \{x_1, s_3\}$, $B_5 = \{x_2, x_3\}$,
$B_6 = \{x_2, s_1\}$, $B_7 = \{x_2, s_3\}$, $B_8 = \{x_3, s_1\}$, $B_9 = \{x_3, s_3\}$, $B_{10} = \{s_1, s_3\}$.

All bases except B_2, B_3 yield infeasible solutions since they do not satisfy the nonnegativity constraints. Thus, $(x_1, x_2, x_3, s_1, s_3)^T = (2, 0, 1, 0, 0)^T$ from B_2, and $(x_1, x_2, x_3, s_1, s_3)^T = (4, 0, 0, 5, 0)^T$ from B_3.

b)

Method 1:

At $(2,0,1,0,0)^T$, we have

$$B^{-1}N = \begin{bmatrix} x_2 & s_1 & s_3 \\ -\frac{4}{5} & -\frac{2}{5} & -\frac{1}{5} \\ -\frac{3}{5} & \frac{1}{5} & -\frac{2}{5} \end{bmatrix}, \quad B^{-1}b = \begin{bmatrix} 2 \\ 1 \end{bmatrix}.$$

If x_2 enters $\left. \begin{array}{r} x_1 - \frac{4}{5}x_2 = 2 \\ x_3 - \frac{3}{5}x_2 = 1 \\ x_2 \geq 0 \end{array} \right\} \hookrightarrow x_2 = \theta \Rightarrow r = (2 + \frac{4}{5}\theta, \theta, 1 + \frac{3}{5}\theta, 0, 0)^T$ is

feasible for $\theta > 0$. Thus, $r^1 = (\frac{4}{5}, 1, \frac{3}{5}, 0, 0)^T$ is an unboundedness direction and hence an extreme ray.

If s_3 enters $\left. \begin{array}{r} x_1 - \frac{1}{5}s_3 = 2 \\ x_3 - \frac{2}{5}s_3 = 1 \\ s_3 \geq 0 \end{array} \right\} \hookrightarrow s_3 = \theta \Rightarrow r = (2 + \frac{1}{5}\theta, 0, 1 + \frac{2}{5}\theta, 0, \theta)^T$

is feasible for $\theta > 0$. Thus, $r^2 = (\frac{1}{5}, 0, \frac{2}{5}, 0, 1)^T$ is another unboundedness direction and hence an extreme ray.

At $(4,0,0,5,0)^T$, we have

$$B^{-1}N = \begin{bmatrix} x_2 & x_3 & s_3 \\ -3 & 2 & -1 \\ -3 & 5 & -2 \end{bmatrix}, \quad B^{-1}b = \begin{bmatrix} 4 \\ 5 \end{bmatrix}.$$

If x_2 enters $\left. \begin{array}{r} x_1 - 2x_2 = 4 \\ s_1 - 3x_2 = 5 \\ x_2 \geq 0 \end{array} \right\} \hookrightarrow x_2 = \theta \Rightarrow r = (4 + 2\theta, \theta, 0, 5 + 3\theta, 0)^T$ is

feasible for $\theta > 0$. Thus, $r^3 = (2, 1, 0, 3, 0)^T$ is an unboundedness direction and hence an extreme ray.

If s_3 enters $\left. \begin{array}{r} x_1 - s_3 = 4 \\ s_1 - 2s_3 = 5 \\ s_3 \geq 0 \end{array} \right\} \hookrightarrow s_3 = \theta \Rightarrow r = (4 + \theta, 0, 0, 2\theta, \theta)^T$ is feasible

for $\theta > 0$. Thus, $r^4 = (1, 0, 0, 2, 1)^T$ is another unboundedness direction and hence an extreme ray.

Method 2:

Try to find some nonnegative vectors in $\mathcal{N}(A)$.

$$\theta \leq r^1 = \left(\frac{4}{5}, 1, \frac{3}{5}, 0, 0\right)^T \in \mathcal{N}(A).$$

$$\theta \leq r^2 = \left(\frac{1}{5}, 0, \frac{2}{5}, 0, 1\right)^T \in \mathcal{N}(A).$$

$$\theta \leq r^3 = (2, 1, 0, 3, 0)^T \in \mathcal{N}(A).$$

$$\theta \leq r^4 = (1, 0, 0, 2, 1)^T \in \mathcal{N}(A).$$

So, they are rays. Since every pair of the above vectors have zeros in different places, we cannot express one ray as a linear combination of the others, they are extreme rays.

c)

1. $x_1 + x_2 + x_3$:
$$c^1 = (1,1,1,0,0)^T \Rightarrow$$

$$\begin{cases} (c^1)^T r^1 = \frac{4}{5} + 1 + \frac{3}{5} + 0 + 0 = \frac{12}{5} > 0 \hookrightarrow \text{unbounded} \\ (c^1)^T r^2 = \frac{1}{5} + 0 + \frac{2}{5} + 0 + 0 = \frac{3}{5} > 0 \hookrightarrow \text{unbounded} \\ (c^1)^T r^3 = 2 + 1 + 0 + 0 + 0 = 3 > 0 \hookrightarrow \text{unbounded} \\ (c^1)^T r^4 = 1 + 0 + 0 + 0 + 0 = 1 > 0 \hookrightarrow \text{unbounded} \end{cases}$$

Thus, there is no finite solution.

2. $-2x_1 - x_2 - 3x_3$:
$$c^2 = (-2,-1,-3,0,0)^T \Rightarrow$$

$$\begin{cases} (c^2)^T r^1 = -\frac{8}{5} - 1 - \frac{9}{5} + 0 + 0 = -\frac{22}{5} \not> 0 \hookrightarrow \text{bounded} \\ (c^2)^T r^2 = -\frac{2}{5} - 0 - \frac{6}{5} + 0 + 0 = -\frac{8}{5} \not> 0 \hookrightarrow \text{bounded} \\ (c^2)^T r^3 = -4 - 1 + 0 + 0 + 0 = -5 \not> 0 \hookrightarrow \text{bounded} \\ (c^2)^T r^4 = -2 + 0 + 0 + 0 + 0 = -2 \not> 0 \hookrightarrow \text{bounded} \end{cases}$$

Thus, there is finite solution.

3. $-x_1 - 2x_2 + 2x_3$:
$$c^3 = (-1,-2,2,0,0)^T \Rightarrow$$

$$\begin{cases} (c^3)^T r^1 = -\frac{4}{5} - 2 + \frac{6}{5} + 0 + 0 = -\frac{8}{5} \not> 0 \hookrightarrow \text{bounded} \\ (c^3)^T r^2 = -\frac{1}{5} + 0 + \frac{4}{5} + 0 + 0 = \frac{3}{5} > 0 \hookrightarrow \text{unbounded} \\ (c^3)^T r^3 = -2 - 2 + 0 + 0 + 0 = -4 \not> 0 \hookrightarrow \text{bounded} \\ (c^3)^T r^4 = -1 + 0 + 0 + 0 + 0 = -1 \not> 0 \hookrightarrow \text{unbounded} \end{cases}$$

Thus, there is no finite solution.

d) $x_1 = 6$, $x_2 = 1$, $x_3 = \frac{1}{2} \Rightarrow s_1 = \frac{15}{2}$, $s_3 = 1$

$$\begin{bmatrix} 6 \\ 1 \\ \frac{1}{2} \\ \frac{15}{2} \\ 1 \end{bmatrix} = \alpha \begin{bmatrix} 2 \\ 0 \\ 1 \\ 0 \\ 0 \end{bmatrix} + (1-\alpha) \begin{bmatrix} 4 \\ 0 \\ 0 \\ 5 \\ 0 \end{bmatrix} + \mu_1 \begin{bmatrix} \frac{4}{5} \\ 1 \\ \frac{3}{5} \\ 0 \\ 0 \end{bmatrix} + \mu_2 \begin{bmatrix} \frac{1}{5} \\ 0 \\ \frac{2}{5} \\ 0 \\ 1 \end{bmatrix} + \mu_3 \begin{bmatrix} 2 \\ 1 \\ 0 \\ 3 \\ 0 \end{bmatrix} + \mu_4 \begin{bmatrix} 1 \\ 0 \\ 0 \\ 2 \\ 1 \end{bmatrix}.$$

$$\alpha, \mu_1, \mu_2, \mu_3, \mu_4 \geq 0$$

We have 5 unknowns and 5 equations. The solution is

$$\begin{bmatrix} 6 \\ 1 \\ \frac{1}{2} \\ \frac{15}{2} \\ 1 \end{bmatrix} = \frac{1}{2}\begin{bmatrix} 2 \\ 0 \\ 1 \\ 0 \\ 0 \end{bmatrix} + \frac{1}{2}\begin{bmatrix} 4 \\ 0 \\ 0 \\ 5 \\ 0 \end{bmatrix} + 0\begin{bmatrix} \frac{4}{5} \\ 1 \\ \frac{3}{5} \\ 0 \\ 0 \end{bmatrix} + 0\begin{bmatrix} \frac{1}{5} \\ 0 \\ \frac{2}{5} \\ 0 \\ 1 \end{bmatrix} + 1\begin{bmatrix} 2 \\ 1 \\ 0 \\ 3 \\ 0 \end{bmatrix} + 1\begin{bmatrix} 1 \\ 0 \\ 0 \\ 2 \\ 1 \end{bmatrix}.$$

$$\underbrace{\hspace{5cm}}_{\substack{\text{convex combination of} \\ \text{extreme points}}} \qquad \underbrace{\hspace{5cm}}_{\substack{\text{canonical combination of} \\ \text{extreme rays}}}$$

e)

1.

	x_1	x_2	x_3	s_1	s_3	RHS
x_1	1	$-\frac{4}{5}$	0	$-\frac{2}{5}$	$-\frac{1}{5}$	2
x_3	0	$-\frac{3}{5}$	1	$\frac{1}{5}$	$-\frac{2}{5}$	1
$-z$	0	4	0	0	1	-4

2.

$$B^{-1}(b - \Delta b) = \begin{bmatrix} \frac{2}{5} & \frac{1}{5} \\ -\frac{1}{5} & \frac{2}{5} \end{bmatrix}\left(\begin{bmatrix} 3 \\ 4 \end{bmatrix} - \begin{bmatrix} 3 \\ 1 \end{bmatrix}\right) = \begin{bmatrix} 2 \\ 1 \end{bmatrix} - \begin{bmatrix} \frac{7}{5} \\ -\frac{1}{5} \end{bmatrix} = \begin{bmatrix} \frac{3}{5} \\ \frac{6}{5} \end{bmatrix}$$

The values of basic variables will change but not the optimal basis.

3. The solution above is $(\frac{3}{5}, 0, \frac{6}{5})^T$ which satisfies the new constraint, no problem!

8.3 a)

1. $\mathcal{B} = \{s_1, s_2, s_3\} \Rightarrow B = I$, $c_B = \theta$, $\mathcal{N} = \{x_1, x_2, x_3, x_4\}$ at their lower bounds and $c_N^T = (2, 3, 1, 4)$.

$$x_B = B^{-1}b - B^{-1}Nx_N = \begin{bmatrix} 30 \\ 13 \\ 20 \end{bmatrix} - \begin{bmatrix} 1 & 2 & 3 & 5 \\ 1 & 1 & 0 & 0 \\ 0 & 0 & 3 & 4 \end{bmatrix}\begin{bmatrix} 1 \\ 0 \\ 3 \\ 0 \end{bmatrix}$$

$$= \begin{bmatrix} 30 \\ 13 \\ 20 \end{bmatrix} - \begin{bmatrix} 10 \\ 1 \\ 9 \end{bmatrix} = \begin{bmatrix} 20 \\ 12 \\ 11 \end{bmatrix} = \begin{bmatrix} s_1 \\ s_2 \\ s_3 \end{bmatrix}$$

$$\Rightarrow z = c_B^T x_B + c_N^T x_N = 2 + 0 + 3 + 0 = 5.$$

$c_N^T - c_B^T B^{-1}N = (2, 3, 1, 4)$. Then, Bland's rule (lexicographical order) marks the first variable. Since the reduced cost of x_1 is positive and x_1 is at its lower bound; as x_1 is increased, so is z. Hence, x_1 enters.

$$\begin{bmatrix} 0 \\ 0 \\ 0 \end{bmatrix} \le \begin{bmatrix} s_1 \\ s_2 \\ s_3 \end{bmatrix} = \begin{bmatrix} 20 \\ 12 \\ 11 \end{bmatrix} - \begin{bmatrix} 1 \\ 1 \\ 0 \end{bmatrix} \alpha,$$

$$\alpha \le 6 - 1 = 5(\text{bounds of } x_1)$$

$$\Rightarrow \alpha = \min\{20, 12, 5\} = 5.$$

x_1 leaves immediately at its upper bound, $x_1 = 6$.

2. $B = I$, $c_B = \theta$, $c_N^T = (2, 3, 1, 4)$, $x_B^T = (15, 7, 11)$ $z = 12 + 0 + 3 + 0 = 15$, $c_N^T - c_B^T B^{-1} N = (2, 3, 1, 4)$. Then, Bland's rule marks the second variable. Since the reduced cost of x_2 is positive and x_2 is at its lower bound; as x_2 is increased, so is z. Hence, x_2 enters.

$$\begin{bmatrix} 0 \\ 0 \\ 0 \end{bmatrix} \le \begin{bmatrix} s_1 \\ s_2 \\ s_3 \end{bmatrix} = \begin{bmatrix} 15 \\ 7 \\ 11 \end{bmatrix} - \begin{bmatrix} 2 \\ 1 \\ 0 \end{bmatrix} \alpha,$$

$$\alpha \le 10 - 0 = 10(\text{bounds of } x_2)$$

$$\Rightarrow \alpha = \min\left\{\frac{15}{2}, 7, 10\right\} = 7.$$

Thus, s_2 leaves.

3.

$$\mathcal{B} = \{s_1, x_2, s_3\} \Rightarrow B = \begin{bmatrix} 1 & 2 & 0 \\ 0 & 1 & 0 \\ 0 & 0 & 1 \end{bmatrix} \Rightarrow B^{-1} = \begin{bmatrix} 1 & -2 & 0 \\ 0 & 1 & 0 \\ 0 & 0 & 1 \end{bmatrix}$$

$$x_B = \begin{bmatrix} s_1 \\ x_2 \\ s_3 \end{bmatrix} = \begin{bmatrix} 1 & -2 & 0 \\ 0 & 1 & 0 \\ 0 & 0 & 1 \end{bmatrix} \begin{bmatrix} 30 \\ 13 \\ 20 \end{bmatrix} - \begin{bmatrix} 1 & -2 & 0 \\ 0 & 1 & 0 \\ 0 & 0 & 1 \end{bmatrix} \begin{bmatrix} 1 & 0 & 3 & 5 \\ 1 & 1 & 0 & 0 \\ 0 & 0 & 3 & 4 \end{bmatrix} \begin{bmatrix} 6 \\ 0 \\ 3 \\ 0 \end{bmatrix}$$

$$= \begin{bmatrix} 4 \\ 13 \\ 20 \end{bmatrix} - \begin{bmatrix} -1 & -2 & 3 & 5 \\ 1 & 1 & 0 & 0 \\ 0 & 0 & 3 & 4 \end{bmatrix} \begin{bmatrix} 6 \\ 0 \\ 3 \\ 0 \end{bmatrix} = \begin{bmatrix} 4 \\ 13 \\ 20 \end{bmatrix} - \begin{bmatrix} 3 \\ 6 \\ 9 \end{bmatrix} = \begin{bmatrix} 1 \\ 7 \\ 11 \end{bmatrix}$$

$$\Rightarrow z = (0, 3, 0) \begin{bmatrix} 1 \\ 7 \\ 11 \end{bmatrix} + (2, 0, 1, 4) \begin{bmatrix} 6 \\ 0 \\ 3 \\ 0 \end{bmatrix} = 21 + (12 + 3) = 36.$$

$$c_N^T - c_B^T B^{-1} N = (2, 0, 1, 4) - (0, 3, 0) \begin{bmatrix} -1 & -2 & 3 & 5 \\ 1 & 1 & 0 & 0 \\ 0 & 0 & 3 & 4 \end{bmatrix}$$

$$= (2, 0, 1, 4) - (3, 3, 0, 0) = (-1, -3, 1, 4),$$

where $\mathcal{N} = \{x_1, s_1, x_3, x_4\}$. Then, Bland's rule (lexicographical order) marks the first variable. Since the reduced cost of x_1 is negative and x_1 is at its upper bound; as x_1 is decreased, z is increased. Hence, x_1 enters.

$$\begin{bmatrix} 0 \\ 0 \\ 0 \end{bmatrix} \le \begin{bmatrix} s_1 \\ x_2 \\ s_3 \end{bmatrix} = \begin{bmatrix} 1 \\ 7 \\ 11 \end{bmatrix} - \begin{bmatrix} -1 \\ 1 \\ 0 \end{bmatrix} \alpha \le \begin{bmatrix} \infty \\ 10 \\ \infty \end{bmatrix},$$

$$\alpha \le 6 - 1 = 5 (\text{bounds of } x_1)$$

$$\Rightarrow \alpha = \min \{1, 10 - 7, 5\} = 1.$$

Thus, s_1 leaves.

4.

$$\mathcal{B} = \{x_1, x_2, s_3\} \Rightarrow B = \begin{bmatrix} 1 & 2 & 0 \\ 1 & 1 & 0 \\ 0 & 0 & 1 \end{bmatrix} \Rightarrow B^{-1} = \begin{bmatrix} -1 & 2 & 0 \\ 1 & -1 & 0 \\ 0 & 0 & 1 \end{bmatrix}$$

$$x_B = \begin{bmatrix} x_1 \\ x_2 \\ s_3 \end{bmatrix} = \begin{bmatrix} -1 & 2 & 0 \\ 1 & -1 & 0 \\ 0 & 0 & 1 \end{bmatrix} \begin{bmatrix} 30 \\ 13 \\ 20 \end{bmatrix} - \begin{bmatrix} -1 & 2 & 0 \\ 1 & -1 & 0 \\ 0 & 0 & 1 \end{bmatrix} \begin{bmatrix} 1 & 0 & 3 & 5 \\ 0 & 1 & 0 & 0 \\ 0 & 0 & 3 & 4 \end{bmatrix} \begin{bmatrix} 0 \\ 0 \\ 3 \\ 0 \end{bmatrix}$$

$$= \begin{bmatrix} -4 \\ 17 \\ 20 \end{bmatrix} - \begin{bmatrix} -1 & 2 & -3 & -5 \\ 1 & -1 & 3 & 5 \\ 0 & 0 & 3 & 4 \end{bmatrix} \begin{bmatrix} 0 \\ 0 \\ 3 \\ 0 \end{bmatrix} = \begin{bmatrix} -4 \\ 17 \\ 20 \end{bmatrix} - \begin{bmatrix} -9 \\ 9 \\ 9 \end{bmatrix} = \begin{bmatrix} 5 \\ 8 \\ 11 \end{bmatrix}$$

$$\Rightarrow z = (2, 3, 0) \begin{bmatrix} 5 \\ 8 \\ 11 \end{bmatrix} + (0, 0, 1, 4) \begin{bmatrix} 0 \\ 0 \\ 3 \\ 0 \end{bmatrix} = (10 + 24) + 3 = 37.$$

$$c_N^T - c_B^T B^{-1} N = (0, 0, 1, 4) - (2, 3, 0) \begin{bmatrix} -1 & 2 & -3 & -5 \\ 1 & -1 & 3 & 5 \\ 0 & 0 & 3 & 4 \end{bmatrix}$$

$$= (0, 0, 1, 4) - (1, 1, 3, 5) = (-1, -1, -2, -1),$$

where $\mathcal{N} = \{s_1, s_2, x_3, x_4\}$. All of the reduced costs are negative for all the nonbasic variables that all are at their lower bounds. Hence, $x^* = (x_1, x_2, x_3, x_4)^T = (5, 8, 3, 0)^T$ is the optimum solution, where $z^* = 37$.

b)(P):

$$\max 2x_1 + 3x_2 + x_3 + 4x_4$$

s.t.

$$x_1 + 2x_2 + 3x_3 + 5x_4 \le 30 \quad (y_1)$$

$$x_1 + x_2 \le 13 \quad (y_2)$$

$$3x_3 + x_4 \leq 20 \quad (y_3)$$

$$-x_1 \leq -1 \quad (y_4)$$

$$x_1 \leq 6 \quad (y_5)$$

$$x_2 \leq 10 \quad (y_6)$$

$$-x_3 \leq -3 \quad (y_7)$$

$$x_3 \leq 9 \quad (y_8)$$

$$x_4 \leq 5 \quad (y_9)$$

$$x_1, x_2, x_3, x_4 \geq 0$$

(D):

$$\min 30y_1 + 13y_2 + 20y_3 - y_4 + 6y_5 + 10y_6 - 3y_7 + 9y_8 + 5y_9$$

s.t.

$$y_1 + y_2 - y_4 + y_5 \geq 2 \quad (x_1)$$

$$2y_1 + y_2 + y_6 \geq 3 \quad (x_2)$$

$$3y_1 + 3y_2 - y_7 + y_8 \geq 1 \quad (x_3)$$

$$5y_1 + y_3 + y_9 \geq 4 \quad (x_4)$$

$$y_1, y_2, y_3, y_4, y_5, y_6, y_7, y_8, y_9 \geq 0$$

The optimal primal solution, $x^* = (x_1, x_2, x_3, x_4)^T = (5, 8, 3, 0)^T$, satisfies constraints (y_1, y_2, y_7) as binding, i.e. the corresponding slacks are zero. By complementary slackness, the dual variables y_1, y_2, y_7 might be in the optimal dual basis. The other primal constraints have positive surplus values at the optimality, therefore $y_3^* = y_4^* = y_5^* = y_6^* = y_8^* = y_9^* = 0$. Moreover, the reduced costs of the surplus variables at the optimal primal solution are both 1 for s_1 and s_2, which are the optimal values of $y_1^* = y_2^* = 1$. Since the optimal primal basis contains the nonzero valued x_1 and x_2, the corresponding dual constraints are binding: $1 + 1 - 0 + 0 = 2\sqrt{}$ and $2 + 1 + 0 = 3\sqrt{}$. Furthermore, the optimal primal solution has nonbasic variables x_3 and x_4, then the corresponding dual surplus variables may be in the dual basis: $3+3-y_7^* + 0 \geq 1$, and $5 + 0 + 0 > 4 \Rightarrow$ the corresponding surplus, say $t_4^* = 1$ in the dual optimal basis. The optimal primal objective function value is $z^* = 37$, which is equal to the optimal dual objective function value by the strong duality theorem. Then, $37 = 30(1) + 13(1) + 20(0) - (0) + 6(0) + 10(0) - 3y_7^* + 9(0) + 5(0) + 0t_1^* + 0t_2^* 0t_3^* + 0t_4^*$, yielding $y_7^* = 2$.

8.4

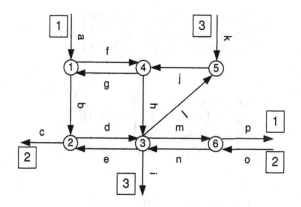

Fig. S.13. A multi-commodity flow instance

Let us take the instance given in Figure S.13, where $K = 3$ and

$$V = \{1, 2, 3, 4, 5, 6, A, C, I, K, OP\},$$

$$A = \{a, b, c, d, e, f, g, h, i, j, k, l, m, n, o, p\}.$$

Let us fix all capacities at 10 and all positive supplies/demands at 10 with unit carrying costs.

a)

$$Min \ \sum_k \sum_a c_{ka} x_{ka}$$

s.t.

$$\sum_{a \in T(i)} x_{ka} - \sum_{a \in H(i)} x_{ka} = d_{ki}$$

$$\sum_k x_{ka} \le U_a$$

$$x_{ka} \le u_{ka}$$

$$x_{ka} \ge 0 \ (\text{integer})$$

In general, we have mK variables, $m + nK$ constraints and mK simple bounds other than the nonnegativity constraints. In our example instance, we have

$$Min \ (x_{1a} + x_{2a} + x_{3a}) + \cdots + (x_{1p} + x_{2p} + x_{3p})$$

s.t.

$$\left.\begin{array}{l} (x_{1a} + x_{1g}) - (x_{1b} + x_{1f}) = 0 \\ (x_{2a} + x_{2g}) - (x_{2b} + x_{2f}) = 0 \\ (x_{3a} + x_{3g}) - (x_{3b} + x_{3f}) = 0 \end{array}\right\} \ \text{node 1}$$

$$\vdots$$

$$\left.\begin{array}{l} (x_{1p}) - (x_{1o}) = 10 \\ (x_{2p}) - (x_{2o}) = -10 \\ (x_{3p}) - (x_{3o}) = 0 \end{array}\right\} \ \text{node OP}$$

$$x_{1a} + x_{2a} + x_{3a} \leq 10$$

$$\vdots$$

$$x_{1p} + x_{2p} + x_{3p} \leq 10$$

$$x_{1a}, \cdots, x_{3p} \geq 0 \ \text{(integer)}$$

b)

$$Min \ \sum_k \sum_{P \in \mathcal{P}^k} C_{kP} f_P$$

s.t.

$$\sum_{P \in \mathcal{P}^k} f_P = D_k$$

$$\sum_k \sum_{P \in \mathcal{P}^k} I_{aP} f_P \leq U_a$$

$$f_P \leq \mu_P$$

$$f_P \geq 0 \ \text{(integer)}$$

We have (huge number of) $K2^m$ variables, $m + K$ constraints and $K2^m$ simple bounds other than the nonnegativity constraints. The following sets the relation between the decision variables of the two formulations whose constraints are isomorphic:

$$x_{ak} = \sum_{P \in \mathcal{P}^k} I_{aP} f_P, \quad f_P = \min_{a \in P} \sum_k x_{ak} \ \text{(applied recursively)}.$$

In our example instance, s_1 is node A and t_3 is node I. If we enumerate paths (some of them is given in Figure S.14), we have

Comm.	Path#	Path
	1	$a \mapsto f \mapsto h \mapsto m \mapsto p$
	2	$a \mapsto b \mapsto d \mapsto m \mapsto p$
1	3	$a \mapsto f \mapsto h \mapsto e \mapsto d \mapsto m \mapsto p$
	4	$a \mapsto f \mapsto g \mapsto b \mapsto d \mapsto m \mapsto p$
	5	$a \mapsto b \mapsto d \mapsto l \mapsto j \mapsto h \mapsto m \mapsto p$
	6	$a \mapsto f \mapsto g \mapsto b \mapsto d \mapsto l \mapsto j \mapsto h \mapsto m \mapsto p$
	7	$o \mapsto n \mapsto e \mapsto c$
	8	$o \mapsto n \mapsto l \mapsto j \mapsto h \mapsto e \mapsto c$
2	9	$o \mapsto n \mapsto l \mapsto j \mapsto g \mapsto b \mapsto c$
	10	$o \mapsto n \mapsto l \mapsto j \mapsto g \mapsto f \mapsto h \mapsto e \mapsto c$
	11	$o \mapsto n \mapsto e \mapsto d \mapsto l \mapsto j \mapsto g \mapsto b \mapsto c$
	12	$k \mapsto j \mapsto h \mapsto i$
	13	$k \mapsto j \mapsto g \mapsto b \mapsto d \mapsto i$
	14	$k \mapsto j \mapsto h \mapsto e \mapsto d \mapsto i$
	15	$k \mapsto j \mapsto h \mapsto m \mapsto n \mapsto i$
3	16	$k \mapsto j \mapsto h \mapsto m \mapsto p \mapsto o \mapsto n \mapsto i$
	17	$k \mapsto j \mapsto h \mapsto e \mapsto d \mapsto m \mapsto n \mapsto i$
	18	$k \mapsto j \mapsto g \mapsto b \mapsto d \mapsto m \mapsto n \mapsto i$
	19	$k \mapsto j \mapsto g \mapsto b \mapsto d \mapsto m \mapsto p \mapsto o \mapsto n \mapsto i$
	20	$k \mapsto j \mapsto h \mapsto e \mapsto d \mapsto m \mapsto p \mapsto o \mapsto n \mapsto i$

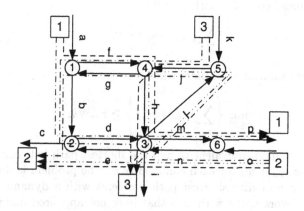

Fig. S.14. Some paths in our multi-commodity flow instance

and the formulation will be

$$Min \ 5f_1 + \cdots + 10f_{20}$$
$$s.t.$$
$$f_1 + \cdots + f_6 = 10$$
$$f_7 + \cdots + f_{11} = 10$$
$$f_{12} + \cdots + f_{20} = 10$$
$$f_1 + f_2 + f_3 + f_4 + f_5 + f_6 \leq 10$$
$$\vdots$$
$$f_1 + f_2 + f_3 + f_4 + f_5 + f_6 + f_{16} + f_{19} + f_{20} \leq 10$$
$$f_1, \cdots, f_{20} \geq 0 \ (\text{integer})$$

The first three constraints make the capacity constraints for arcs a, c, i and k redundant.

c)

$$w_a \leftrightarrow \sum_k \sum_{P \in \mathcal{P}^k} I_{aP} f_P \leq U_a$$

$$\pi_k \leftrightarrow \sum_{P \in \mathcal{P}^k} f_P = D_k$$

Then, the reduced cost of a path P will be

$$\sum_{a \in P} (c_{ka} + w_a) - \pi_k,$$

and the current solution is optimal when

$$\min_{P \in \mathcal{P}^k} \left\{ \sum_{a \in P} (c_{ka} + w_a) \right\} \geq \pi_k, \ \forall k.$$

The above problem is equivalent to find the shortest path between s_k and t_k using arc costs $c_{ka} + w_a$ for each commodity k. The problem is decomposed into K single commodity shortest path problems with a dynamic objective function that favors paths with arcs that have not appeared many times in current paths.

d)
$$x^T =$$
$$[f_1, f_2, f_3, f_4, f_5, f_6, f_7, f_8, f_9, f_{10}, f_{11}, f_{12}, f_{13}, f_{14}, f_{15}, f_{16}, f_{17}, f_{18}, f_{19}, f_{20} |$$
$$| s_b, s_d, s_e, s_f, s_g, s_h, s_j, s_l, s_m, s_n, s_o, s_p].$$

$$c = \begin{bmatrix} 5 & 5 & 7 & 7 & 8 & 9 & 4 & 7 & 7 & 9 & 9 & 4 & 6 & 6 & 6 & 8 & 8 & 8 & 10 & 10 & | & 0 & 0 & 0 & 0 & 0 & 0 & 0 & 0 & 0 & 0 & 0 & 0 \end{bmatrix}$$

$$A = \left[\begin{array}{cccccccccccccccccccc|cccccccccccc}
1&1&1&1&1&1&0&0&0&0&0&0&0&0&0&0&0&0&0&0&0&0&0&0&0&0&0&0&0&0&0&0 \\
0&0&0&0&0&0&1&1&1&1&1&0&0&0&0&0&0&0&0&0&0&0&0&0&0&0&0&0&0&0&0&0 \\
0&0&0&0&0&0&0&0&0&0&0&1&1&1&1&1&1&1&1&1&0&0&0&0&0&0&0&0&0&0&0&0 \\
\hline
0&1&0&0&1&1&0&0&1&0&1&0&1&0&0&0&0&1&1&0&1&0&0&0&0&0&0&0&0&0&0&0 \\
0&1&1&1&1&1&0&0&0&0&1&0&1&1&0&0&1&1&1&1&0&1&0&0&0&0&0&0&0&0&0&0 \\
0&0&1&0&0&0&1&1&0&1&1&0&0&1&0&0&1&0&0&1&0&0&1&0&0&0&0&0&0&0&0&0 \\
1&0&1&1&0&1&0&0&0&1&0&0&0&0&0&0&0&0&0&0&0&0&0&1&0&0&0&0&0&0&0&0 \\
0&0&0&1&0&1&0&0&1&1&1&0&1&0&0&0&0&1&1&0&0&0&0&0&1&0&0&0&0&0&0&0 \\
1&0&1&0&1&1&0&1&0&1&0&1&0&1&1&1&1&0&0&1&0&0&0&0&0&1&0&0&0&0&0&0 \\
0&0&0&0&1&1&0&1&1&1&1&1&1&1&1&1&1&1&1&1&0&0&0&0&0&0&1&0&0&0&0&0 \\
0&0&0&0&1&1&0&1&1&1&0&0&0&0&0&0&0&0&0&0&0&0&0&0&0&0&0&1&0&0&0&0 \\
1&1&1&1&1&1&0&0&0&0&0&0&0&0&0&1&1&1&1&1&0&0&0&0&0&0&0&0&1&0&0&0 \\
0&0&0&0&0&0&1&1&1&1&0&0&0&1&1&1&1&1&1&0&0&0&0&0&0&0&0&0&0&1&0&0 \\
0&0&0&0&0&0&1&1&1&1&0&0&0&0&1&0&0&1&1&1&0&0&0&0&0&0&0&0&0&0&1&0 \\
1&0&0&0&0&1&0&0&0&0&0&0&0&0&0&1&0&0&1&1&0&0&0&0&0&0&0&0&0&0&0&1 \\
\end{array}\right]$$

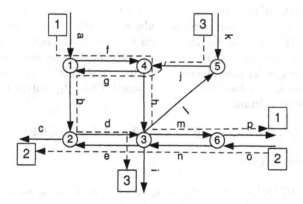

Fig. S.15. Starting bfs solution for our multi-commodity flow instance: repeated

$$B_1 = \begin{bmatrix} 1&0&0&0&0&0&0&0&0&0&0&0&0&0&0 \\ 0&1&0&0&0&0&0&0&0&0&0&0&0&0&0 \\ 0&0&1&0&0&0&0&0&0&0&0&0&0&0&0 \\ 0&0&1&1&0&0&0&0&0&0&0&0&0&0&0 \\ 0&0&1&0&1&0&0&0&0&0&0&0&0&0&0 \\ 0&1&0&0&0&1&0&0&0&0&0&0&0&0&0 \\ 1&0&0&0&0&0&1&0&0&0&0&0&0&0&0 \\ 0&0&1&0&0&0&0&1&0&0&0&0&0&0&0 \\ 1&0&0&0&0&0&0&0&1&0&0&0&0&0&0 \\ 0&0&1&0&0&0&0&0&0&1&0&0&0&0&0 \\ 0&0&0&0&0&0&0&0&0&0&1&0&0&0&0 \\ 1&0&0&0&0&0&0&0&0&0&0&1&0&0&0 \\ 0&1&0&0&0&0&0&0&0&0&0&0&1&0&0 \\ 0&1&0&0&0&0&0&0&0&0&0&0&0&1&0 \\ 1&0&0&0&0&0&0&0&0&0&0&0&0&0&1 \end{bmatrix}.$$

$$c_{B_1}^T = \begin{bmatrix} 5\,4\,6\,0\,0\,0\,0\,0\,0\,0\,0\,0\,0\,0\,0 \end{bmatrix}.$$

$$y_1 = c_{B_1}^T B_1^{-1} = [\pi | w] = [5,4,6 | \theta].$$

Then, the lengths of arcs are $c_{ka} + w_a = 1 + 0 = 1$, $\forall$ arcs.

For commodity one, the minimum shortest path ($P_2 : \ a \mapsto b \mapsto d \mapsto m \mapsto p$) other than P_1 has length 5 which is equal to the corresponding dual variable $\pi_1 = 5$. For commodity two, the minimum shortest path has length 6 which is strictly greater than the corresponding dual variable $\pi_2 = 4$. However, $P_{12} : \ k \mapsto j \mapsto h \mapsto i$ has length $4 < 6 = \pi_3$! Thus, f_{12} enters to the basis with the updated column

$$(B_1^{-1} A^{12})^T = \begin{bmatrix} 0\,0\,1\,{-1}\,{-1}\,0\,0\,{-1}\,1\,0\,0\,0\,0\,0\,0 \end{bmatrix}$$

and the updated RHS is

$$x_{B_1}^T = (B_1^{-1} b)^T = \begin{bmatrix} f_1 & f_7 & f_{13} & s_b & s_d & s_e & s_f & s_g & s_h & s_j & s_l & s_m & s_n & s_o & s_p \end{bmatrix}$$

$$x_{B_1}^T = (B_1^{-1} b)^T = \begin{bmatrix} 10 & 10 & 10 & 0 & 0 & 0 & 0 & 0 & 0 & 10 & 0 & 0 & 0 & 0 \end{bmatrix},$$

therefore the slack variable corresponding to arc h, s_h, leaves.

$$B_2 = \begin{bmatrix} 1 0 0 0 0 0 0 0 0 0 0 0 0 0 0 \\ 0 1 0 0 0 0 0 0 0 0 0 0 0 0 0 \\ 0 0 1 0 0 0 0 1 0 0 0 0 0 0 0 \\ 0 0 1 1 0 0 0 0 0 0 0 0 0 0 0 \\ 0 0 1 0 1 0 0 0 0 0 0 0 0 0 0 \\ 0 1 0 0 0 1 0 0 0 0 0 0 0 0 0 \\ 1 0 0 0 0 0 1 0 0 0 0 0 0 0 0 \\ 0 0 1 0 0 0 0 1 0 0 0 0 0 0 0 \\ 1 0 0 0 0 0 0 0 1 0 0 0 0 0 0 \\ 0 0 1 0 0 0 0 0 1 1 0 0 0 0 0 \\ 0 0 0 0 0 0 0 0 0 0 1 0 0 0 0 \\ 1 0 0 0 0 0 0 0 0 0 0 1 0 0 0 \\ 0 1 0 0 0 0 0 0 0 0 0 0 1 0 0 \\ 0 1 0 0 0 0 0 0 0 0 0 0 0 1 0 \\ 1 0 0 0 0 0 0 0 0 0 0 0 0 0 1 \end{bmatrix}$$

$$c_{B_2}^T = \begin{bmatrix} 5\ 4\ 6\ 0\ 0\ 0\ 0\ 0\ 4\ 0\ 0\ 0\ 0\ 0\ 0 \end{bmatrix}$$

$$y_2 = \begin{bmatrix} 7\ 4\ 6 | 0\ 0\ 0\ 0\ 0\ -2\ 0\ 0\ 0\ 0\ 0\ 0 \end{bmatrix}$$

Then, the lengths of arcs are $c_{ka} + w_a = 1 + 0 = 1$, $\forall$ arcs except arc h, whose length is $1 - 2 = -1$.

For commodity one, the minimum shortest path $P_2 : a \mapsto b \mapsto d \mapsto m \mapsto p$ has length 5, which is strictly less than the corresponding dual variable $\pi_1 = 7$. Thus, f_2 enters to the basis with the updated column

$$(B_2^{-1} A^2)^T = \begin{bmatrix} 1\ 0\ 1\ 0\ 0\ 0\ -1\ -1\ -1\ 0\ 0\ 0\ 0\ 0\ -1 \end{bmatrix}$$

and the updated RHS is

$$x_{B_2}^T = (B_2^{-1} b)^T = \begin{bmatrix} f_1\ f_7\ f_{13}\ s_b\ s_d\ s_e\ s_f\ s_g\ f_{12}\ s_j\ s_l\ s_m\ s_n\ s_o\ s_p \end{bmatrix}$$

$$x_{B_2}^T = (B_2^{-1} b)^T = \begin{bmatrix} 10\ 10\ 10\ 0\ 0\ 0\ 0\ 0\ 0\ 0\ 10\ 0\ 0\ 0\ 0 \end{bmatrix};$$

therefore, either f_1 or f_7 leaves. We choose f_1!

$$B_3 = \begin{bmatrix} 1 & 0 & 0 & 0 & 0 & 0 & 0 & 0 & 0 & 0 & 0 & 0 & 0 & 0 & 0 & 0 \\ 0 & 1 & 0 & 0 & 0 & 0 & 0 & 0 & 0 & 0 & 0 & 0 & 0 & 0 & 0 & 0 \\ 0 & 0 & 1 & 0 & 0 & 0 & 0 & 1 & 0 & 0 & 0 & 0 & 0 & 0 & 0 & 0 \\ 1 & 0 & 1 & 1 & 0 & 0 & 0 & 0 & 0 & 0 & 0 & 0 & 0 & 0 & 0 & 0 \\ 1 & 0 & 1 & 0 & 1 & 0 & 0 & 0 & 0 & 0 & 0 & 0 & 0 & 0 & 0 & 0 \\ 0 & 1 & 0 & 0 & 0 & 1 & 0 & 0 & 0 & 0 & 0 & 0 & 0 & 0 & 0 & 0 \\ 0 & 0 & 0 & 0 & 0 & 0 & 1 & 0 & 0 & 0 & 0 & 0 & 0 & 0 & 0 & 0 \\ 0 & 0 & 1 & 0 & 0 & 0 & 0 & 1 & 0 & 0 & 0 & 0 & 0 & 0 & 0 & 0 \\ 0 & 0 & 0 & 0 & 0 & 0 & 0 & 0 & 1 & 0 & 0 & 0 & 0 & 0 & 0 & 0 \\ 0 & 0 & 1 & 0 & 0 & 0 & 0 & 0 & 1 & 1 & 0 & 0 & 0 & 0 & 0 & 0 \\ 0 & 0 & 0 & 0 & 0 & 0 & 0 & 0 & 0 & 1 & 0 & 0 & 0 & 0 & 0 & 0 \\ 1 & 0 & 0 & 0 & 0 & 0 & 0 & 0 & 0 & 0 & 1 & 0 & 0 & 0 & 0 & 0 \\ 0 & 1 & 0 & 0 & 0 & 0 & 0 & 0 & 0 & 0 & 0 & 1 & 0 & 0 & 0 & 0 \\ 0 & 1 & 0 & 0 & 0 & 0 & 0 & 0 & 0 & 0 & 0 & 0 & 1 & 0 & 0 & 0 \\ 0 & 0 & 0 & 0 & 0 & 0 & 0 & 0 & 0 & 0 & 0 & 0 & 0 & 0 & 0 & 1 \end{bmatrix}$$

$$c_{B_3}^T = \begin{bmatrix} 5 & 4 & 6 & 0 & 0 & 0 & 0 & 4 & 0 & 0 & 0 & 0 & 0 & 0 \end{bmatrix}$$

$$y_3 = \begin{bmatrix} 5 & 4 & 6 | 0 & 0 & 0 & 0 & 0 & -2 & 0 & 0 & 0 & 0 & 0 & 0 \end{bmatrix}.$$

Then, the lengths of arcs are $c_{ka} + w_a = 1 + 0 = 1$, $\forall$ arcs except arc h, whose length is $1 - 2 = -1$. For all the three commodities, the minimum shortest distances between the source and the sink nodes are greater and equal to the corresponding dual variables. Therefore, the current solution given below is optimal.

$$x_{B_3}^T = (B_3^{-1}b)^T = \begin{bmatrix} f_2 & f_7 & f_{13} & s_b & s_d & s_e & s_f & s_g & f_{12} & s_j & s_l & s_m & s_n & s_o & s_p \end{bmatrix}$$

$$x_{B_3}^T = (B_3^{-1}b)^T = \begin{bmatrix} 10 & 10 & 0 & 0 & 0 & 0 & 10 & 10 & 10 & 0 & 10 & 0 & 0 & 0 & 10 \end{bmatrix}$$

The optimum solution is depicted in Figure S.16.

e) When the number of variables (columns of A) is huge, the following question is asked: Can one generate column A^j by some oracle that can answer the question, Does there exist a column with with reduced cost < 0? If so, the oracle returns one. So, the sketch of so called "A Column Generation Algorithm" is given below:

S1. Solve $LP(J)$:

$$\min \left\{ \sum_{j \in J} c_j x_j : \sum_{j \in J} A^j x_j = b, \ x \geq 0 \right\}$$

for some $J \subseteq I = \{1, \ldots, n\}$.

S2. Using dual variables π that are optimal for $LP(J)$, ask the oracle if there exists $j \notin J$ such that $c_j \pi A^j < 0$. If so, add it to J and perform pivot(s) to solve new $LP(J)$; Go back S1. If not, we have the optimal solution to LP over all columns.

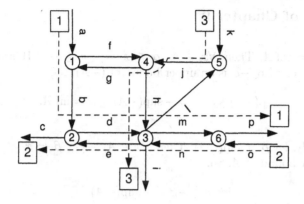

Fig. S.16. The optimum solution for our multi-commodity flow instance

In a sense, we partition the optimization problem into two levels: Main / Subproblem, or Master / Slave, or Superior / Inferior; where the subproblem has a structure that can be exploited easily. The main problem generates dual variables and the subproblem generates new primal variables; and the loop stops when primal-dual conditions are satisfied.

The dual to the above column generation approach gives rise to the separation problem, where we are about to solve LP with large number of rows (equations). We first solve over restricted subset of rows (analogous to solving over subset of columns) and ask oracle if other rows are satisfied. If so, we are done; if not, we ask the oracle to return a separating hyperplane that has current rows satisfied in one half space and a violation in the other. This approach leads to Bender's decomposition.

Problems of Chapter 9

9.1 Let $\alpha = \inf A$. Then, $\forall x \in A$, $\alpha \leq x \Leftrightarrow -x \leq -\alpha$. Hence, $(-A)$ in bounded above. Also, $-\alpha$ is an upper bound of $(-A)$. So,

$$\sup(-A) \leq -\alpha \Leftrightarrow -\sup(-A) \geq \alpha = \inf A.$$

Conversely, let $\beta = \sup(-A)$. Then, $\forall x \in A$, $-x \leq \beta \Leftrightarrow x \geq -\beta$. Hence, $-\beta$ is a lower bound of A. So,

$$\inf A \geq -\beta = -\sup(-A).$$

Thus, $\inf A = -\sup(-A)$.

9.2
a) If $m = 0$, $(b^m)^{1/n} = (b^0)^{1/n} = 1^{1/n} = 1$ (see (c)).

If $m > 0$, $(b^m)^{1/n} = \dfrac{\overbrace{(b \cdots b)^{1/n}}^{m \text{ times}}}{} = \overbrace{b^{1/n} \cdots b^{1/n}}^{m \text{ times}} = (b^{1/n})^m$.

If $m < 0$, let $m' = -m > 0$. Then,
$(b^m)^{1/n} = (b^{-m'})^{1/n} = (\frac{1}{b^{m'}})^{1/n} = \frac{1}{(b^{m'})^{1/n}} = \frac{1}{(b^{1/n})^{m'}} = \frac{1}{(b^{1/n})^{-m}} = (b^{1/n})^m$.

b) If $m = 0$, all terms are 1.

If $m > 0$, $(b^m)^n = \underbrace{\overbrace{b^m \cdots b^m}}_{n} = \overbrace{b \cdots b}^{m} \cdots \overbrace{b \cdots b}^{m} \Big\} n = b^{mn}$.

Similarly, $(b^m)^n = b^{mn}$.
If $m < 0$, let $m' = -m > 0$. Then,
$(b^m)^n = (b^{-m'})^n = (\frac{1}{b^{m'}})^n = \frac{1}{(b^{m'})^n} = \frac{1}{b^{m'n}} = \frac{1}{b^{-(mn)}} = b^{mn}$.

c) Let $1^{1/n} = x$ where $x \succ 0$. Then, $x^n = 1$. Also, $1^{1/n} = 1$. Since the positive n^{th} root of 1 is unique, we get $x = 1$.

d) Let $b^{1/nq} = \alpha$, and $(b^{1/n})^{1/q} = \beta$ where $\alpha, \beta \succ 0$.
Then, $b = \alpha^{nq}$ and $b^{1/n} = \beta^q \Rightarrow b = (\beta^q)^n = \beta^{qn} = \beta^{nq} = \alpha^{nq}$. Since the positive nq^{th} root of b is unique, we get $\alpha = \beta$, i.e. $b^{1/nq} = (b^{1/n})^{1/q}$. Similarly, $b^{1/nq} = (b^{1/q})^{1/n}$.

e) If $p = 0$, then $b^{p+q} = b^{0+q} = b^q = b^0 b^q = b^p b^q$. Similarly, if $q = 0$, $b^{p+q} = b^p b^q$. So assume $p \neq 0$, $q \neq 0$.

$$\begin{cases} Case\ 1: \ p > 0, \ q > 0, \ b^{p+q} = \underbrace{b\cdots b}_{p+q} = \underbrace{b\cdots b}_{p}\underbrace{b\cdots b}_{q} = b^p b^q; \\ Case\ 2: \ p < 0, \ q > 0, \ \text{Let } p' = -p > 0. \ \text{So, } b^{p+q} = b^{-p'+q}; \\ \quad \begin{cases} Case\ 2a: \ p' = q \Rightarrow b^{-p'+q} = b^0 = 1 = \frac{b^q}{b^q} = \frac{b^q}{b^{p'}} = b^p b^q. \\ Case\ 2b: \ p' < q \Rightarrow b^{-p'+q} = \underbrace{b\cdots b}_{q-p'} = \underbrace{b\cdots b}_{q-p'}\frac{\overbrace{b\cdots b}^{p'}}{\underbrace{b\cdots b}_{p'}} = \frac{b^q}{b^{p'}} = b^p b^q. \\ Case\ 2c: \ p' > q \Rightarrow b^{-p'+q} = b^{-(p'-q)} = \frac{1}{b^{p'-q}} = \frac{1}{b^{p'}b^{-q}} = \frac{b^q}{b^{p'}} = b^p b^q. \end{cases} \\ Case\ 3: \ p > 0, \ q < 0, \ \text{similar to Case 2;} \\ Case\ 4: \ p < 0, \ q < 0, \ \text{then, } p+q < 0 \Rightarrow b^{p+q} = \frac{1}{b^{-(p+q)}} = \frac{1}{b^{-p}b^{-q}} = b^p b^q. \end{cases}$$

9.3

a) Let $\alpha = (b^m)^{1/n}$, $\beta = (b^p)^{1/q}$ where $\alpha, \beta > 0$.

$$\begin{cases} Case\ 1: \ m = 0, \Rightarrow p = 0. \ \text{So, } \alpha = \beta = 1. \\ Case\ 2: \ m > 0, \Rightarrow p > 0. \ \alpha = (b^m)^{1/n} \Rightarrow \alpha^n = b^m \Rightarrow b = (\alpha^n)^{\frac{1}{m}}. \\ \qquad\qquad \text{Similarly, } b = (\beta^q)^{\frac{1}{p}} \Rightarrow b^{mp} = \alpha^{np} = \beta^{qm}. \\ \qquad\qquad \text{Thus, } np = qm \Rightarrow \alpha^{np} = \beta^{np}. \\ \qquad\qquad \text{Since the positive } (np)^{th} \text{ root is unique, } \alpha = \beta. \\ Case\ 3: \ m < 0, \Rightarrow p < 0. \ \text{Let } m' = -m, \ p' = -p \Rightarrow m', p' > 0.\text{Case 2!} \\ \qquad\qquad (b^m)^{1/n} = (b^{-m'})^{1/n} = (\frac{1}{b^{-m'}})^{1/n} = \frac{1}{(b^{-m'})^{1/n}} = \\ \qquad\qquad = \frac{1}{(b^{-p'})^{1/q}} = (b^p)^{1/q}. \end{cases}$$

So, b^r, $r \in \mathbb{Q}$ are well defined.

b) Let $r = \frac{m}{n}$, $s = \frac{p}{q}$ where $n, q > 0$.

$$b^{r+s} = (b^{mq+np})^{\frac{1}{nq}} = (b^{mq}b^{np})^{\frac{1}{nq}} = (b^{mq})^{\frac{1}{nq}}(b^{np})^{\frac{1}{nq}} =$$

$$= ((b^m)^q)^{\frac{1}{nq}}((b^p)^n)^{\frac{1}{nq}} = (((b^m)^q)^{1/q})^{1/n}(((b^p)^n)^{1/n})^{1/q} =$$

$$(b^m)^{1/n}(b^p)^{1/q} = b^r b^s.$$

c) Let $b^t \in B(r)$. Then, $t \in \mathbb{Q}$, $t \le r \Rightarrow r - t \ge 0$, $r - t \in \mathbb{Q}$. Since $b > 1$ and $r - t$ is a nonnegative rational number, we get $b^{r-t} \ge 1$.

    Claim: Let $b > 1$, $s \in \mathbb{Q}_+$. $b^s \ge 1$.

    Proof: If $s = 0 \Rightarrow b^s = b^0 = 1$. Assume $s > 0$. Then, $s = \frac{p}{q}$ where $p, q > 0$.
$b^s = (b^p)^{1/q}$. $b > 1 \Rightarrow a = b^p > 1 \Rightarrow b^s = a^{1/q} > 1$.

    Hence, $1 \ge b^{r-t} = b^r b^{-t} = \frac{b^r}{b^t} \Rightarrow b^t \le b^r$. That is $\forall b^t \in B(r)$, $b^t \le b^r$; i.e. b^r is an upper bound for $B(r)$. Then, $\sup(B(r)) \le b^r$. If $r \in \mathbb{Q}$, $b^r \in B(r)$. So, $b^r \le \sup(B(r))$. Thus, $b^r = \sup(B(r))$.

    Now, we can safely define $b^x = \sup(B(x))$, $\forall x \in \mathbb{R}$.

d) Fix b^r arbitrary in $B(x)$ and fix b^s arbitrary in $B(y)$: $r, s \in \mathbb{Q}$, $r \le x$, $s \le y$.

Then, $r + s \in \mathbb{Q}$, $r + s \le x + y \Rightarrow b^{r+s} = b^s b^r \in B(x + y) \Rightarrow b^r b^s \le b^{x+y}$. Keep s fixed. $b^r \le \frac{b^{x+y}}{b^s}$, $\forall b^r \in B(x)$. Thus, $\frac{b^{x+y}}{b^s}$ is an upper bound for $B(x)$. Hence, $b^x = \sup(B(x)) \le \frac{b^{x+y}}{b^s} \Leftrightarrow b^s \le \frac{b^{x+y}}{b^x}$. Similarly, $b^r \le \frac{b^{x+y}}{b^y}$.

Now vary s. $\forall b^s \in B(y)$, $b^s \le \frac{b^{x+y}}{b^x}$. Thus, $\frac{b^{x+y}}{b^x}$ is an upper bound for $B(y)$.

$$b^y = \sup(B(y)) \le \frac{b^{x+y}}{b^x} \Rightarrow b^x b^y \le b^{x+y}.$$

Claim: $b^x b^y \ge b^{x+y}$.

Proof: Suppose not. $b^x b^y < b^{x+y}$ for some $x, y \in \mathbb{R}$. $\exists a \in \mathbb{Q} \subset \mathbb{R} \ni b^x b^y < a < b^{x+y}$, by Archimedean property. $b^x, b^y > 0 \Rightarrow a > 0$. Since $a < b^{x+y}$, a is NOT an upper bound of $B(x + y)$. So, $\exists b^r \in B(x + y) \ni a > b^r$. Let $t = \frac{b^r}{a} > 1$. If $n > \frac{b-1}{t-1}$ (see problem 9.4–c)) $b^{1/n} < t = \frac{b^r}{a} \Rightarrow a < b^r b^{-1/n} = b^{r-1/n}$ (true for rationals). Also $r - 1/n < r \le x + y$. So, $r - \frac{1}{n} - x < y$. $\exists v \in \mathbb{Q} \ni r - \frac{1}{n} - x < v < y$. Then, $v < y$ and $r - \frac{1}{n} - v < x$. Thus, $b^v \in B(y)$ and $b^{r-\frac{1}{n}-v} \in B(x)$. That is, $b^v \le b^y$ and

$$b^{r-\frac{1}{n}-v} \le b^x \Leftrightarrow b^{r-\frac{1}{n}} = b^{r-\frac{1}{n}-v} b^v \le b^x b^y < a < b^{r-\frac{1}{n}}.$$

We have a contradiction from the first and the last terms of the above relation.

9.4

a) $b^n - 1 = (b - 1)(b^{n-1} + b^{n-2} + \cdots + b + 1) > (b - 1)(1 + \cdots + 1) \ge (b - 1)n$.

b) Let $t = b^{1/n}$. Apply part a) for t: $t^n - 1 \ge n(t - 1) \Rightarrow b - 1 \ge n(b^{1/n} - 1)$.

c) $\frac{b-1}{t-1} < n \Rightarrow \frac{b-1}{n} < t - 1 \Rightarrow \frac{b-1}{n} + 1 < t$. We have $\frac{b-1}{n} \ge b^{1/n} - 1$. Thus, $b^{1/n} < t$.

d) Let $t = \frac{y}{b^w} > 1$. Use part c), $b^{1/n} < t = \frac{y}{b^w} \Rightarrow b^{w+1/n} = b^w b^{1/n} < y$ if $n > \frac{b-1}{t-1}$.

e) $y > 0 \Rightarrow t = \frac{b^w}{y} > 1$. If $n > \frac{b-1}{t-1}$, use c), $b^{1/n} < t = \frac{b^w}{y} \Rightarrow y < \frac{b^w}{b^{1/n}} = b^{w-1/n}$.

f)

Claim: A is bounded above.

Proof: If not, $\forall \beta > 0$, $\exists w \in A \ni w > \beta$. In particular, $\forall n \in \mathbb{N}$, $\exists w \in A \ni w > n$. Hence, $\forall n \in \mathbb{N}$, $\exists w \in A \ni b^n < b^w < y$, i.e. $\forall n \in \mathbb{N}$, $b^n < y$. If $0 < y \le 1$, we have a Contradiction since $b^n > 1$. Assume $y > 1$, use (c) $\forall n \ni n > \frac{y-1}{b-1}$, $y^{1/n} < b \Rightarrow y < b^n$. Hence, $\forall n \ni n > \frac{y-1}{b-1}$ we have $b^n < y < b^n$, Contradiction.

Let $x = \sup(A) = \sup\{w \in \mathbb{R} : b^w \prec y\}$.

<u>Claim:</u> $b^x = y$.

<u>Proof:</u> If not, $b^x < y$ or $b^x > y$. If $b^x < y$, by (d) $\forall n \in \mathbb{N}$, $b^{x+1/n} <$ y, $x + 1/n \in A$. Contradiction to the upper bound $x > x + 1/n$. If $b^x > y$, then <u>Claim:</u> if $u < x$, $u \in A$. <u>Proof:</u> $u < x \Rightarrow u$ is nor an upper bound of A. $\exists w \in A \ni u < w \Rightarrow w - u > 0 \Rightarrow b^{w-u} > 1 \Rightarrow \frac{b^w}{b^u} > 1 \Rightarrow b^w > b^u$. So, $u \in A$. $y < b^x \Rightarrow$ use (e) $\forall n \in \mathbb{N} \ni y < b^{x-1/n}$; so $x + 1/n \notin A$. Thus, $x \le x - 1/n$ ($u < x \Rightarrow u \in A$), Contradiction.

Hence, $b^x = y$.

g) Let $b > 1$, $y > 0$ be fixed. Suppose $x \ne x' \ni b^x = y = b^{x'}$.
Without loss of generality, we may assume that , $x < x' \Rightarrow x' - x > 0 \Rightarrow$ $b^{x-x'} \Rightarrow b^x > b^{x'}$, Contradiction.

9.5
a) $\forall z \in F$, $z^2 \succeq 0$ (if $z = 0 \Rightarrow z^2 = 0$. If $z \succ 0 \Rightarrow z^2 \succ 0$). Assume that $x \ne 0 \Rightarrow x^2 \succ 0$. If $y^2 \succeq 0 \Rightarrow x^2 + y^2 \succ 0$, Contradiction. So $x = 0$, then $x^2 + y^2 = 0 + y^2 = 0 \Rightarrow y = 0$.

b) Trivial by induction.

9.6 Note that "$a \sim b$ if $a - b$ is divisible by m" is different from saying "$\frac{a-b}{m}$ is an integer", since the above one is defined for all fixed $m \in \mathbb{Z}$ including $m = 0$, but the latter one is defined for all $m \ne 0$.

a) $a \sim a$, $\forall a \in \mathbb{Z}$ (take $k = 0$). Then, $\sim$ is reflexive.
$a \sim b \Rightarrow \exists k \in \mathbb{Z} \ni a - b = mk$. Then, $b - a = m(-k)$ where $-k \in \mathbb{Z}$. Thus, $b \sim a$, yielding that $\sim$ is symmetric.
$a \sim b$ and $b \sim c \Rightarrow \exists k_1, k_2 \in \mathbb{Z} \ni a - b = k_1 m$, $b - c = k_2 m$. Then, $a - c = (k_1 + k_2)m$ where $k_1 + k_2 \in \mathbb{Z}$. Hence, $a \sim c$, meaning that $\sim$ is transitive.

Thus, $\sim$ is an equivalence relation.

b) <u>Case 1:</u> $m = 0$. Then, $a \sim b \Leftrightarrow a = b$. So, $[a] = \{a\}$, and the number of equivalence classes is ∞.
<u>Case 2:</u> $m \ne 0$. Then, $a \sim b \Leftrightarrow \exists k \in \mathbb{Z} \ni a = b + mk$. Hence,

$$[a] = \{a, a + m, a - m, a + 2m, a - 2m, \cdots\},$$

and the number of distinct equivalence classes is $|m|$.

9.7
a) $x \sim y \Rightarrow x \in [0, 1]$ and $y \in [0, 1] \Rightarrow y \sim x$ (i.e. symmetric).
$x \sim y$ and $y \sim z \Rightarrow x \in [0, 1]$ and $y \in [0, 1]$ and $z \in [0, 1] \Rightarrow x \sim z$ (i.e. transitive).
If $x \notin [0, 1]$, then $x \sim x$ does not hold. For reflexibility we want $x \sim x$ to hold

$\forall x \in \mathbb{R}$. Hence, $\sim$ is not reflexive.

b) The statement

$$x \sim y \Rightarrow y \sim x, \ x \sim y \text{ and } y \sim x \Rightarrow x \sim x; \text{ therefore, } x \sim x, \ \forall x \in X$$

starts with the following assumption: $\forall x \in X, \ \exists y \in X \ni x \sim y$. If $\sim$ is symmetric and transitive and also has this additional property, then it is necessarily reflexive. But if it does not have this property, then it is not reflexive.

9.8

a) We will make the proof by induction on n. If $n = 1$, $X = X_1$ is countable by hypothesis. Assume that the proposition is true for $n = k$, i.e. $X_1 \times \cdots \times X_k$ is countable. We will prove the proposition for $n = k + 1$, i.e. prove that $X = X_1 \times \cdots \times X_k \times X_{k+1}$ is countable. Let $Y = X_1 \times \cdots \times X_k$. Then, $X = Y \times X_{k+1}$ and Y is countable by the induction hypothesis. Then, the elements of Y and X_{k+1} can be listed as sequences $Y = \{y_1, y_2, \ldots\}$, $X_{k+1} = \{x_1, x_2, \ldots\}$. Now, for $X = Y \times X_{k+1}$, we use Cantor's counting scheme and see that X is countable.

b) Let X be countable. Then, $X = \{x_1, x_2, \ldots\}$. Let $A = \{x_2, x_3, \ldots\}$. Then, A is a proper subset of X and $f : X \mapsto A$ defined by $f(x_n) = x_{n+1}$, $n = 1, 2, \ldots$ is one-to-one and onto. Thus, every countable set is numerically equivalent to a proper subset of itself.

c) If $f : X \mapsto Y$ is onto, then $\exists g : Y \mapsto X \ni f \circ g = id_Y$. Moreover, g is one-to-one. Let $A = g(Y)$, then $A \subset X$ and $g : Y \mapsto A$ is one-to-one and onto. So, $A \sim Y$. Since $A \subset X$ and X is countable, A is either finite or countable. To see that A cannot be uncountable, we express $X = \{x_1, x_2, \ldots\}$. If A is not finite, then $A = \{x_{i_1}, x_{i_2}, \ldots\}$, where i_n's are positive integers and $i_n \neq i_m$ for $n \neq m$. Now, we define $f : \mathbb{N} \mapsto A$ by $f(n) = x_{i_n}$. Then, f is one-to-one and onto. If A is finite, $A \sim Y \Rightarrow Y$ is finite; if A is countable, $A \sim Y \Rightarrow Y$ is countable. Thus, Y is at most countable.

Problems of Chapter 10

10.1 Fix $x, y \in \mathbb{R}^k$ arbitrary.

$$d_2(x, y) = [\sum_{i=1}^{k} (x_i - y_i)^2]^{1/2}, \quad d_1(x, y) = \sum_{i=1}^{k} |x_i - y_i|,$$

$$d_\infty(x, y) = \max_i \{|x_i - y_i|\} = |x_j - y_j|.$$

$d_1 \sim d_\infty$:

$$d_\infty(x, y) = |x_j - y_j| \leq \sum_{i=1}^{k} |x_i - y_i| = d_1(x, y) \Rightarrow A = 1.$$

$$d_\infty(x, y) = |x_j - y_j| \geq |x_i - y_i|, \ \forall i = 1, 2, \ldots, k$$

$$\Rightarrow k d_\infty(x, y) = k|x_j - y_j| \geq \sum_{i=1}^{k} |x_i - y_i| \Rightarrow B = k.$$

$d_2 \sim d_\infty$:

$$[d_\infty(x, y)]^2 = (x_j - y_j)^2 \leq \sum_{i=1}^{k} (x_i - y_i)^2 \Rightarrow d_\infty(x, y) \leq d_2(x, y) \Rightarrow A = 1.$$

$$[d_\infty(x, y)]2 = (x_j - y_j)^2 \geq |x_i - y_i|, \ \forall i = 1, 2, \ldots, k$$

$$\Rightarrow k[d_\infty(x, y)]^2 \geq [d_2(x, y)]^2 \Rightarrow B = \sqrt{k}.$$

$d_1 \sim d_2$: $d_1 \sim d_\infty$ and $d_2 \sim d_\infty \Rightarrow d_1 \sim d_2$.

10.2

Consider the discrete metric $d(p, q) = \begin{cases} 0, & \text{if } p = q, \\ r, & \text{if } p \neq q \end{cases}$ on X.

$$B_r(p) = \{p\}, \ B_r[p] = X, \ \overline{B_r(p)} = \{p\} \neq X.$$

10.3

($\Leftarrow$:)
Let $\emptyset \neq A \subsetneq X$. A is both open and closed. Let $B = A^c$, B is also both open and closed. $A \cup B = X$. If A is closed then B is open, we have $A \cap \bar{B} = A \cap B = \emptyset$. If B is closed then A is open, we have $B \cap \bar{A} = A \cap B = \emptyset$. Thus X is disconnected.

($\Rightarrow$:)
X is disconnected. $\exists A \neq \emptyset, \exists B \neq \emptyset \ni X = A \cup B$ and $(A \cap \bar{B}) \cap (\bar{A} \cap B) = \emptyset \Rightarrow A \cap B = \emptyset$. Thus $A^c = B \neq \emptyset \Rightarrow A \subsetneq X$.
$A \cup B = X \Rightarrow A \cup \bar{B} = X$, $A \cap \bar{B} = \emptyset \Rightarrow A = (\bar{B})^c$, i.e. A is open.

$A \cup B = X \Rightarrow \bar{A} \cup B = X$, $\bar{A} \cap B = \emptyset \Rightarrow B = (\bar{A})^c$, i.e. B is open.
A and B are separated and $A \cup B = X \Rightarrow A = B^c$, so A is closed. Similarly,
B is closed.

10.4

Let us place the origin at the lower left corner of the PCB. Then,

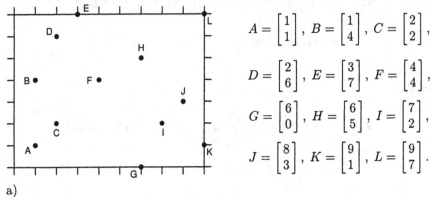

$$A = \begin{bmatrix} 1 \\ 1 \end{bmatrix}, \quad B = \begin{bmatrix} 1 \\ 4 \end{bmatrix}, \quad C = \begin{bmatrix} 2 \\ 2 \end{bmatrix},$$

$$D = \begin{bmatrix} 2 \\ 6 \end{bmatrix}, \quad E = \begin{bmatrix} 3 \\ 7 \end{bmatrix}, \quad F = \begin{bmatrix} 4 \\ 4 \end{bmatrix},$$

$$G = \begin{bmatrix} 6 \\ 0 \end{bmatrix}, \quad H = \begin{bmatrix} 6 \\ 5 \end{bmatrix}, \quad I = \begin{bmatrix} 7 \\ 2 \end{bmatrix},$$

$$J = \begin{bmatrix} 8 \\ 3 \end{bmatrix}, \quad K = \begin{bmatrix} 9 \\ 1 \end{bmatrix}, \quad L = \begin{bmatrix} 9 \\ 7 \end{bmatrix}.$$

a)

Use l_1 norm:

l_1	A	B	C	D	E	F	G	H	I	J	K	L
A	0	3	2	6	8	6	6	9	7	9	8	14
B	3	0	3	3	5	3	9	6	8	8	11	11
C	2	3	0	4	6	4	6	7	5	7	8	12
D	6	3	4	0	2	4	10	5	9	9	12	8
E	8	5	6	2	0	4	10	5	9	9	12	6
F	6	3	4	4	4	0	6	3	5	5	8	8
G	6	9	6	10	10	6	0	5	3	5	4	10
H	9	6	7	5	5	3	5	0	4	4	7	5
I	7	8	5	9	9	5	3	4	0	2	3	7
J	9	8	7	9	9	5	5	4	2	0	3	5
K	8	11	8	12	12	8	4	7	3	3	0	6
L	14	11	12	8	6	8	10	5	7	5	6	0

Nearest neighbor (in l_1 metric):

$$A \mapsto C \mapsto B \mapsto D(D \text{ or } F) \mapsto E \mapsto F \mapsto H$$

$$\mapsto I(I \text{ or } J) \mapsto J \mapsto K \mapsto G \mapsto L \mapsto A$$

Initial tour length is 54. See Figure S.17.

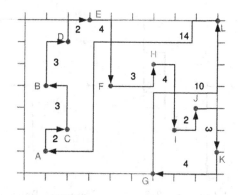

Fig. S.17. Nearest neighbor (in l_1 metric): initial solution

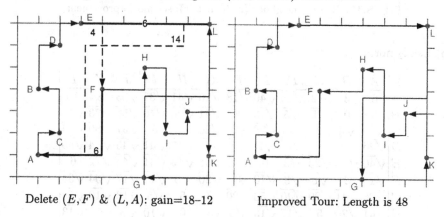

Delete (E, F) & (L, A): gain=18–12   Improved Tour: Length is 48

Fig. S.18. Nearest neighbor (in l_1 metric): first improvement

The gain values are tabulated below. See Figure S.18.

GAIN	(B,D)	(D,E)	(E,F)	(F,H)	(H,I)	(I,J)	(J,K)	(K,G)	(G,L)	(L,A)
(A,C)	-2	-8	-6	-8	-8	-10	-12	-8	-6	
(C,B)		-4	-2	-4	-8	-8	-12	-14	-4	2
(B,D)			-2	-2	-8	-12	-14	-14	-4	0
(D,E)				-4	-8	-14	-16	-16	-4	0
(E,F)					-2	-8	-10	-10	-4	6
(F,H)						-4	-6	-6	2	0
(H,I)							0	-2	2	6
(I,J)								-2	4	0
(J,K)									2	4
(K,G)										6

The maximum gain is 6, due to the deletion of (E, F) and (L, A). The situation after this step is illustrated in Figure S.19.

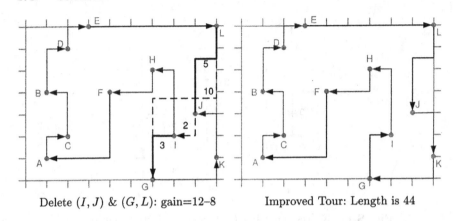

Delete (I, J) & (G, L): gain=12−8   Improved Tour: Length is 44

Fig. S.19. Nearest neighbor (in l_1 metric): second improvement

b) Use l_2 norm:

l_2	A	B	C	D	E	F	G	H	I	J	K	L
A	0	3	$\sqrt{2}$	$\sqrt{26}$	$\sqrt{40}$	$\sqrt{18}$	$\sqrt{26}$	$\sqrt{41}$	$\sqrt{37}$	$\sqrt{53}$	8	10
B	3	0	$\sqrt{5}$	$\sqrt{5}$	$\sqrt{13}$	3	$\sqrt{41}$	$\sqrt{26}$	$\sqrt{40}$	$\sqrt{50}$	$\sqrt{73}$	$\sqrt{73}$
C	$\sqrt{2}$	$\sqrt{5}$	0	4	$\sqrt{26}$	$\sqrt{8}$	$\sqrt{20}$	5	5	$\sqrt{37}$	$\sqrt{50}$	$\sqrt{74}$
D	$\sqrt{26}$	$\sqrt{5}$	4	0	$\sqrt{2}$	$\sqrt{8}$	$\sqrt{40}$	$\sqrt{17}$	$\sqrt{41}$	$\sqrt{45}$	$\sqrt{74}$	$\sqrt{50}$
E	$\sqrt{40}$	$\sqrt{13}$	$\sqrt{26}$	$\sqrt{2}$	0	$\sqrt{10}$	$\sqrt{58}$	$\sqrt{13}$	$\sqrt{41}$	$\sqrt{41}$	$\sqrt{72}$	6
F	$\sqrt{18}$	3	$\sqrt{8}$	$\sqrt{8}$	$\sqrt{10}$	0	$\sqrt{20}$	$\sqrt{5}$	$\sqrt{13}$	$\sqrt{17}$	$\sqrt{34}$	$\sqrt{34}$
G	$\sqrt{26}$	$\sqrt{41}$	$\sqrt{20}$	$\sqrt{40}$	$\sqrt{58}$	$\sqrt{20}$	0	5	$\sqrt{5}$	$\sqrt{13}$	$\sqrt{10}$	$\sqrt{58}$
H	$\sqrt{41}$	$\sqrt{26}$	5	$\sqrt{17}$	$\sqrt{13}$	$\sqrt{5}$	5	0	$\sqrt{10}$	$\sqrt{8}$	5	$\sqrt{13}$
I	$\sqrt{37}$	$\sqrt{40}$	5	$\sqrt{41}$	$\sqrt{41}$	$\sqrt{13}$	$\sqrt{5}$	$\sqrt{10}$	0	$\sqrt{2}$	$\sqrt{5}$	$\sqrt{29}$
J	$\sqrt{53}$	$\sqrt{50}$	$\sqrt{37}$	$\sqrt{45}$	$\sqrt{41}$	$\sqrt{17}$	$\sqrt{13}$	$\sqrt{8}$	$\sqrt{2}$	0	$\sqrt{5}$	$\sqrt{17}$
K	8	$\sqrt{73}$	$\sqrt{50}$	$\sqrt{74}$	$\sqrt{72}$	$\sqrt{34}$	$\sqrt{10}$	5	$\sqrt{5}$	$\sqrt{5}$	0	6
L	10	$\sqrt{73}$	$\sqrt{74}$	$\sqrt{50}$	6	$\sqrt{34}$	$\sqrt{58}$	$\sqrt{13}$	$\sqrt{29}$	$\sqrt{17}$	6	$\sqrt{0}$

Nearest neighbor (in l_2 metric):

$$A \mapsto C \mapsto B \mapsto D \mapsto E \mapsto F \mapsto H \mapsto J$$

$$\mapsto I \mapsto G(G \text{ or } K) \mapsto K \mapsto L \mapsto A$$

Initial tour length is 38.3399. See Figure S.20.

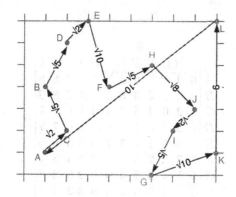

Fig. S.20. Nearest neighbor (in l_2 metric): initial solution

The gain values are tabulated below. See Figure S.21 for the improvement.

$GAIN$	(B,D)	(D,E)	(E,F)	(F,H)	(H,I)	(I,J)	(J,K)	(K,G)	(G,L)	(L,A)
(A,C)	−3.35	−7.37	−4.58	−5.59	−8.24	−9.45	−6.90	−7.59	−9.19	
(C,B)		−3.96	−2.70	−3.46	−7.01	−8.76	−6.93	−7.62	−7.38	0.63
(B,D)			−1.04	−2.65	−6.74	−9.82	−9.06	−9.61	−7.38	−1.41
(D,E)				−2.78	−6.28	−10.28	−10.37	−11.12	−7.19	−1.98
(E,F)					−1.74	−5.43	−5.48	−7.12	−5.15	2.92
(F,H)						−3.64	−4.13	−4.07	−1.20	0.00
(H,I)							−1.70	−1.25	−0.29	1.94
(I,J)								−1.27	−0.21	1.21
(J,K)									−1.62	1.75
(K,G)										−2.45

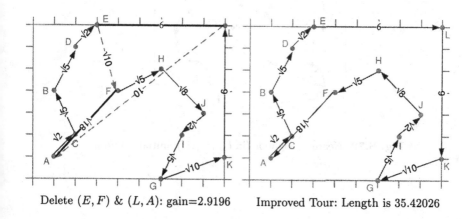

Delete (E,F) & (L,A): gain=2.9196   Improved Tour: Length is 35.42026

Fig. S.21. Nearest neighbor (in l_2 metric): improvement

c) Use l_∞ norm:

l_∞	A	B	C	D	E	F	G	H	I	J	K	L
A	0	3	1	5	6	3	5	5	6	7	8	8
B	3	0	2	2	3	3	5	5	6	7	8	8
C	1	2	0	4	5	2	4	4	5	6	7	7
D	5	2	4	0	1	2	6	4	5	6	7	7
E	6	3	5	1	0	3	7	3	5	5	6	6
F	3	3	2	2	3	0	4	2	3	4	5	5
G	5	5	4	6	7	4	0	5	2	3	3	7
H	5	5	4	4	3	2	5	0	3	2	4	3
I	6	6	5	5	5	3	2	3	0	1	2	5
J	7	7	6	6	5	4	3	2	1	0	2	4
K	8	8	7	7	6	5	3	4	2	2	0	6
L	8	8	7	7	6	5	7	3	5	4	6	0

Nearest neighbor (in l_∞ metric):

$$A \mapsto C \mapsto B \mapsto D \mapsto E \mapsto F \mapsto H \mapsto J$$

$$\mapsto I \mapsto G(G \text{ or } K) \mapsto K \mapsto L \mapsto A$$

Initial tour length is 33. See Figure S.22.

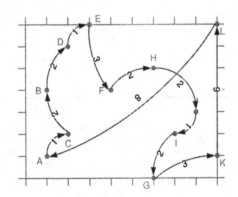

Fig. S.22. Nearest neighbor (in l_∞ metric): initial solution

The gain values are tabulated below. See Figure S.23 for the improvement.

GAIN	(B,D)	(D,E)	(E,F)	(F,H)	(H,I)	(I,J)	(J,K)	(K,G)	(G,L)	(L,A)
(A,C)	−4	−8	−4	−4	−8	−10	−7	−8	−8	
(C,B)		−4	−3	−3	−7	−9	−6	−7	−7	0
(B,D)			0	−3	−7	−9	−8	−7	−7	−3
(D,E)				−2	−6	−9	−9	−8	−6	−4
(E,F)					−2	−4	−4	−6	−2	2
(F,H)						−4	−6	−6	2	0
(H,I)							−2	−2	0	0
(I,J)								−1	0	−1
(J,K)									−1	0
(K,G)										−4

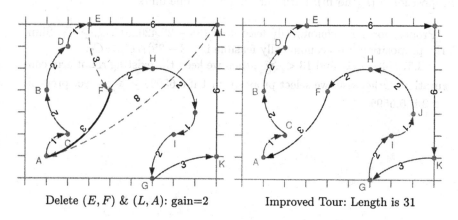

Delete (E,F) & (L,A): gain=2   Improved Tour: Length is 31

Fig. S.23. Nearest neighbor (in l_∞ metric): improvement

d)

Case 1: we need to complete the tour for the consecutive PCB's:

Current situation (l_1 norm): Tour duration is 44 time units.
Proposition 1 (l_2 norm): Tour duration is 35.42026 time units.
Proposition 2 (l_∞ norm): Tour duration is 31 time units.

Proposition 1 is economically feasible if $(44 - 35.42026)NC_o \geq C_1$. Similarly, proposition 2 is economically feasible if $(44 - 31)NC_o \geq C_2$.

Case 2: we may delete the most costly connection:

For the odd numbered PCBs among $1, \ldots, N$;

l_1 norm: $L \mapsto J \mapsto K \mapsto G \mapsto I \mapsto H \mapsto F \mapsto A \mapsto C \mapsto B \mapsto D \mapsto E$ with length 38;

l_2 norm: $K \mapsto G \mapsto I \mapsto J \mapsto H \mapsto F \mapsto A \mapsto C \mapsto B \mapsto D \mapsto E \mapsto L$ with length 29.42026;

l_∞ norm: $K \mapsto G \mapsto I \mapsto J \mapsto H \mapsto F \mapsto A \mapsto C \mapsto B \mapsto D \mapsto E \mapsto L$ with length 25.

For the even numbered PCBs, we reverse the order as

l_1 norm: $E \mapsto D \mapsto B \mapsto C \mapsto A \mapsto F \mapsto H \mapsto I \mapsto G \mapsto K \mapsto J \mapsto L$;

l_2 norm: $L \mapsto E \mapsto D \mapsto B \mapsto C \mapsto A \mapsto F \mapsto H \mapsto J \mapsto I \mapsto G \mapsto K$;

l_∞ norm: $L \mapsto E \mapsto D \mapsto B \mapsto C \mapsto A \mapsto F \mapsto H \mapsto J \mapsto I \mapsto G \mapsto K$.

Current situation (l_1 norm): Path duration is 38 time units.
Proposition 1 (l_2 norm): Path duration is 29.42026 time units.
Proposition 2 (l_∞ norm): Path duration is 25 time units.

Proposition 1 is economically feasible if $(38 - 29.42026)NC_o \geq C_1$. Similarly, proposition 2 is economically feasible if $(38 - 25)NC_o \geq C_2$.

If $8.57974 < \frac{C_1}{NC_o}$ and $13 < \frac{C_2}{NC_o}$, then we keep the existing robot arm configuration. Otherwise, we select proposition 1 if $0.65998 \geq \frac{C_1}{C_2}$; select proposition 2 if $0.65998 \leq \frac{C_1}{C_2}$.

Problems of Chapter 11

11.1
a) ($\Rightarrow$): Let $\varepsilon > 0$ and x_0 be given. Let $b = f(x_0) - \varepsilon$. Then, by assumption, the set $B = \{x \in X : f(x) > f(x_0) - \varepsilon\}$ is open. Moreover, $x_0 \in B$ since $f(x_0) > f(x_0) - \varepsilon$. So, $\exists \delta > 0 \ni B_\delta(x_0) \subset B$; that is, $x \in B_\delta(x_0) \Rightarrow x \in B \Rightarrow f(x) > f(x_0) - \varepsilon$.

($\Leftarrow$): Let $b \in \mathbb{R}$ be given. We will show that the set $A = \{x \in X : f(x) > b\}$ is open. If $A = \emptyset$, then A is open. Assume $A \neq \emptyset$, show that every point of A is an interior point. Let $x_0 \in A$. Then, $f(x_0) > b$. Let $\varepsilon = f(x_0) - b$. Then, by our assumption, $\exists \delta > 0 \ni x \in B_\delta(x_0) \Rightarrow f(x) > f(x_0) - \varepsilon = b \Rightarrow x \in A$. Hence, $B_\delta(x_0) \subset A$, that is $x_0 \in int A$.

b) Similar as above.

11.2 f is continuous and X is compact $\Rightarrow f(X) = B$ is compact in Y. $q \in \overline{f(X)} = \overline{B} = B$ since B is compact, therefore closed. So, by $q \in B = f(X)$, we have $\exists p \in X \ni q = f(p)$. Next, we will show that $p_n \to p$. $f : X \mapsto B$ is continuous, one-to-one and onto. Since X is compact, $f^{-1} : B \mapsto X$ is continuous. Moreover, $f(p_n), q \in B$ and $f(p_n) \to q$. Then,

$$\underbrace{f^{-1}(f(p_n))}_{p_n} \to \underbrace{f^{-1}(q)}_{p}.$$

11.3 Let the wire be the circle $C_r = \{(x, y) : x^2 + y^2 = r^2\}$. For $\alpha = (x, y) \in C_r$, let $T(\alpha)$ be the temperature at α and let $f : C_r \mapsto \mathbb{R}$ be such that $f(\alpha) = T(\alpha) - T(-\alpha)$. Note that α and $-\alpha$ are diametrically opposite points. Then, T, and hence, f are continuous.
<u>Claim:</u> $\exists \alpha \in C_r \ni f(\alpha) = 0$.
<u>Proof:</u> Assume not, $\forall \alpha \in C_r$, $T(\alpha) \neq T(-\alpha)$. Define $A = \{\alpha \in C_r : f(\alpha) > 0\}$, $B = \{\alpha \in C_r : f(\alpha) < 0\}$. Then, A and B are both open in C_r. Why? (since they are the inverse images of the open sets $(0, +\infty)$ and $(-\infty, 0)$ under the continuous function f.) $A \cap B \neq \emptyset$, because of the heated wire; $A \cup B = C_r$, since we assumed $\forall \alpha \in C_r$, $T(\alpha) \neq T(-\alpha)$; moreover, $A \neq \emptyset$, there is at least one point (the point where heat is applied). Suppose not, then $C_r = B$, $\forall \alpha \in C_r$, $f(\alpha) < 0 \Leftrightarrow T(\alpha) < T(-\alpha)$. But, then $T(-\alpha) < T(-(-\alpha)) = T(\alpha)$, Contradiction. Hence $A \neq \emptyset$. Similarly, with the same argument, $B \neq \emptyset$, think of the opposite point to where heat is applied. So, A is nonempty, proper ($A^c = B \neq \emptyset$) subset of C_r which is both open and closed (A^c is open). Thus, C_r is disconnected. Contradiction.

Another way of proving the statement is the following: Let $x \in A$ and $y \in B$, and we know that f is continuous as well as $f(x) > 0 > f(y)$. Apply the intermediate value theorem (Corollary 11.4.2) to conclude that $\exists \alpha \in C_r \ni f(\alpha) = 0$.

Problems of Chapter 12

12.1 Use the Mean Value Theorem: $h : \mathbb{R} \mapsto \mathbb{R}$ is nondecreasing if $h'(x) \geq 0$.

$$y \leq x \Rightarrow h(x) - h(y) = h'(c)(x - y) \geq 0 \Rightarrow h(x) \geq h(y).$$

$g'(x) = \frac{x f'(x) - f(x)}{x^2}$, $f(x) = f(x) - f(0) = f'(c)x \leq x f'(x)$, $0 < c < x$.
So $g'(x) \geq 0$, $\forall x \Rightarrow g$ is nondecreasing.

12.2 Use the Mean Value Theorem: $f_i(y) - f_i(x) = f_i'(c_i)(y - x)$. $f' = 0 \Rightarrow$
$f_i' = 0$, $\forall i$; thus $f_i(y) = f_i(x)$ which means f is constant.

12.3 $\frac{\partial f}{\partial x}(0,0) = \cos(0 + 2 \cdot 0) = 1$, $\frac{\partial f}{\partial y}(0,0) = 2\cos(0 + 2 \cdot 0) = 2$;
$\frac{\partial^2 f}{\partial x^2}(0,0) = 0$, $\frac{\partial^2 f}{\partial y^2}(0,0) = 0$, $\frac{\partial^2 f}{\partial x^2}(0,0) = 0$ and $\frac{\partial^2 f}{\partial x \partial y}(0,0) = 0$.

$$f(x,y) = x + 2y + R_2(x,y)(0,0),$$

where $\frac{R_2(x,y)}{|(x,y)|^2}(0,0) \to 0$ as $(x,y) \to (0,0)$.

12.4
a) Let us take the first order Taylor's approximation for any nonzero direction
h,

$$f(x^* + h) = f(x^*) + \nabla f(x^*)^T h + R_1(x^*, h), \quad \frac{R_1(x^*, h)}{\|h\|^1} \to 0 \text{ as } h \to \theta.$$

Since $\frac{R_1(x^*,h)}{\|h\|^1} \approx \frac{1}{2} h^T \nabla^2 f(\xi) h$, where $\xi = x^* + \alpha h$, $0 < \alpha < 1$, we say that

$$f(x^* + h) \approx f(x^*) + \nabla f(x^*)^T h.$$

Since x^* is a local minimizer, $f(x^*) \leq f(x^* + h)$, $\forall h$ small. Therefore, for
all feasible directions $\nabla f(x^*)^T h \geq 0$, where the left hand side is known as
the directional derivative of the function. Since we have an unconstrained
minimization problem, all directions h (and so are inverse directions $-h$) are
feasible,

$$\nabla f(x^*)^T h \geq 0 \geq -\nabla f(x^*)^T h = \nabla f(x^*)^T(-h) \geq 0, \forall h \neq \theta.$$

Thus, we must have $\nabla f(x^*) = \theta$.

b) Let us take the second order Taylor's approximation for any nonzero (but
small in magnitude) direction h,

$$f(x^* + h) \approx f(x^*) + \nabla f(x^*)^T h + \frac{1}{2} h^T \nabla^2 f(x^*) h.$$

Since $\nabla f(x^*) = \theta$, we have

$$f(x^* + h) \approx f(x^*) + \frac{1}{2}h^T\nabla^2 f(x^*)h.$$

Suppose that $\nabla^2 f(x^*)$ is not positive semi-definite. Then,

$$\exists v \in \mathbb{R}^n \ni v^T\nabla^2 f(x^*)v < 0;$$

even for the remainder term, $v^T\nabla^2 f(\xi)v < 0$ if $\|x^* - \xi\|$ is small enough. If we take h as being along v, we should have $f(x^*) > f(x^* + h)$, Contradiction to the local minimality of $f(x^*)$. Thus, $\nabla^2 f(x^*)$ is positive semi-definite.

If we combine the first order necessary condition and the second order necessary condition after deleting the term -semi-, we will arrive at the sufficiency condition for x^* being the strict local minimizer.

c) At every iteration, we will approximate $f(x)$ by a quadratic function $Q(p)$ using the first three terms of its Taylor series about the point x_{k-1}:

$$f(x_{k-1} + p) \approx f(x_{k-1}) + \nabla f(x_{k-1})^T p + \frac{1}{2}p^T\nabla^2 f(x_{k-1})p \doteq Q(p);$$

and we will minimize Q as a function of p, then we will finally set $x_k = x_{k-1} + p_k$.

Let us take the derivative of Q:

$$\frac{dQ}{dp} = \nabla f(x_{k-1}) + \nabla^2 f(x_{k-1})^T p \approx \nabla f(x_{k-1} + p).$$

Since we expect $\theta = \nabla f(x_{k-1} + p_k) \approx \nabla f(x_{k-1}) + \nabla^2 f(x_{k-1})^T p_k$,

$$\nabla^2 f(k-1)^T p_k = -\nabla f(x_{k-1}) \Leftrightarrow p_k = -[\nabla^2 f(x_{k-1})]^{-1}\nabla f(x_{k-1}).$$

This method of finding a root of a function is known as Newton's method, which has a quadratic rate of convergence except in some degenerate cases. Newton's method for finding $\nabla f(x) = \theta$ is simply to iterate as $x_k = x_{k-1} - [\nabla^2 f(x_{k-1})]^{-1}\nabla f(x_{k-1})$.

d) See Figure S.24 for the plot of the bivariate function, $f(x_1, x_2) = x_1^4 + 2x_1^3 + 24x_1^2 + x_2^4 + 12x_2^2$, in the question.

$$\nabla f\left(\begin{bmatrix} x_1 \\ x_2 \end{bmatrix}\right) = \begin{bmatrix} 4x_1^3 + 6x_1^2 + 48x_1 \\ 4x_2^3 + 24x_2 \end{bmatrix},$$

$$\nabla^2 f\left(\begin{bmatrix} x_1 \\ x_2 \end{bmatrix}\right) = \begin{bmatrix} 12x_1^2 + 12x_1 + 48 & 0 \\ 0 & 12x_2^2 + 24 \end{bmatrix}.$$

Let $x_{(0)} = \begin{bmatrix} 1 \\ 1 \end{bmatrix} \Rightarrow \nabla f\left(\begin{bmatrix} 1 \\ 1 \end{bmatrix}\right) = \begin{bmatrix} 58 \\ 28 \end{bmatrix}$, $\nabla^2 f\left(\begin{bmatrix} 1 \\ 1 \end{bmatrix}\right) = \begin{bmatrix} 72 & 0 \\ 0 & 36 \end{bmatrix}$. Then,

$$x_{(1)} = \begin{bmatrix} 1 \\ 1 \end{bmatrix} - \begin{bmatrix} \frac{1}{72} & \\ & \frac{1}{36} \end{bmatrix}\begin{bmatrix} 58 \\ 28 \end{bmatrix} = \begin{bmatrix} 1 - \frac{58}{72} \\ 1 - \frac{28}{36} \end{bmatrix} = \begin{bmatrix} \frac{14}{72} \\ \frac{4}{36} \end{bmatrix}.$$

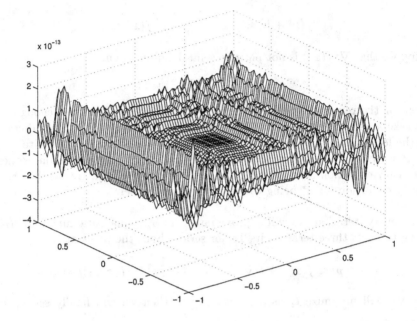

Fig. S.24. Plot of $f(x_1, x_2) = x_1^4 + 2x_1^3 + 24x_1^2 + x_2^4 + 12x_2^2$

$$\nabla f \left(\begin{bmatrix} 0.194444 \\ 0.222222 \end{bmatrix} \right) = \begin{bmatrix} 9.589592 \\ 5.377229 \end{bmatrix},$$

$$\nabla^2 f \left(\begin{bmatrix} 0.194444 \\ 0.222222 \end{bmatrix} \right) = \begin{bmatrix} 50.78704 & 0 \\ 0 & 24.59259 \end{bmatrix}$$

Then,

$$x_{(2)} = \begin{bmatrix} 0.194444 \\ 0.222222 \end{bmatrix} - \begin{bmatrix} \frac{1}{50.78704} & \\ & \frac{1}{24.59259} \end{bmatrix} \begin{bmatrix} 9.589592 \\ 5.377229 \end{bmatrix} = \begin{bmatrix} 0.005625 \\ 0.003570 \end{bmatrix}.$$

$$\nabla f \left(\begin{bmatrix} 0.005625 \\ 0.003570 \end{bmatrix} \right) = \begin{bmatrix} 0.270179 \\ 0.085676 \end{bmatrix},$$

$$\nabla^2 f \left(\begin{bmatrix} 0.005625 \\ 0.003570 \end{bmatrix} \right) = \begin{bmatrix} 48.06788 & 0 \\ 0 & 24.00015 \end{bmatrix}$$

Then,

$$x_{(3)} = \begin{bmatrix} 0.005625 \\ 0.003570 \end{bmatrix} - \begin{bmatrix} \frac{1}{48.06788} & \\ & \frac{1}{24.00015} \end{bmatrix} \begin{bmatrix} 0.270179 \\ 0.085676 \end{bmatrix} = \begin{bmatrix} 0.00000398 \\ 0.00000002 \end{bmatrix}.$$

$$\nabla f \left(\begin{bmatrix} 0.00000398 \\ 0.00000002 \end{bmatrix} \right) = \begin{bmatrix} 0.000191000 \\ 0.000000364 \end{bmatrix},$$

$$\nabla^2 f\left(\begin{bmatrix} 0.00000398 \\ 0.00000002 \end{bmatrix}\right) = \begin{bmatrix} 48.00005 & 0 \\ 0 & 24 \end{bmatrix}.$$

Then,

$$x_{(4)} = \begin{bmatrix} 0.00000398 \\ 0.00000002 \end{bmatrix} - \begin{bmatrix} \frac{1}{48.00005} & \\ & \frac{1}{24} \end{bmatrix} \begin{bmatrix} 0.0001910000 \\ 0.000000364 \end{bmatrix} = \begin{bmatrix} 1.98 \times 10^{-12} \\ 0 \end{bmatrix}.$$

Finally, $\nabla f\left(\begin{bmatrix} 1.98 \times 10^{-12} \\ 0 \end{bmatrix}\right) = \begin{bmatrix} 9.5 \times 10^{-11} \\ 0 \end{bmatrix}$, which is close to θ.

Thus, $x^* = \begin{bmatrix} 0 \\ 0 \end{bmatrix} \Rightarrow \nabla f(x^*) = \theta,\ \nabla^2 f(x^*) = \begin{bmatrix} 48 & 0 \\ 0 & 24 \end{bmatrix}$. Since $\nabla^2 f(x^*)$ is diagonal with positive entries, it is positive definite. Therefore, $x^* = \theta$ is a local minimizer with $f(x^*) = 0$.

Problems of Chapter 13

13.1

Let there be given two series

$$A = \sum_0^\infty u_k \text{ and } B = \sum_0^\infty v_k$$

with nonnegative terms.

(a) If $u_k \le v_k$, $\forall k$, the convergence of series B implies the convergence of series A and the divergence of series A implies the divergence of series B. Suppose that B is convergent. Let $S = \sum_0^\infty v_k$ be finite.

$$\sum_0^n u_k \le \sum_0^n v_k \le S, \ n = 0, 1, \ldots$$

thus partial sum of A is bounded, hence it is convergent.
Suppose that A is divergent. Thus its nth partial sum increases indefinitely together with n.

$$\sum_0^n u_k \le \sum_0^n v_k, \ n = 0, 1, \ldots$$

Thus, n^{th} partial sum of B increases indefinitely together with n, too. That is, B is divergent.

(b) If $\lim_{k\to\infty} \frac{u_k}{v_k} = \alpha > 0$, then series A and B are simultaneously convergent and divergent.
$\lim_{k\to\infty} \frac{u_k}{v_k} = \alpha > 0$, $v_k \ge 0$, $\forall k$. Then,

$$\forall \epsilon > 0 \ \exists N \ni \alpha - \epsilon < \frac{u_k}{v_k} < \alpha + \epsilon, \ \forall k > N.$$

$\Rightarrow v_k(\alpha - \epsilon) < u_k < v_k(\alpha + \epsilon)$. If B is convergent, so is $\sum_0^\infty v_k(\alpha - \epsilon)$. Thus, A is convergent by (a). If B is divergent, so is $\sum_0^\infty v_k(\alpha - \epsilon)$. Thus, A is divergent by (a).

13.2

a) $\sum_0^\infty \frac{x^k}{k!}$:

It is convergent for $x = 0$. Let us assume that $x > 0$.

$$\frac{u_{k+1}}{u_k} = \frac{\frac{x^{k+1}}{(k+1)!}}{\frac{x^k}{k!}} = \frac{x}{k+1} \to 0 \text{ as } k \to \infty.$$

Thus, it is convergent.

b) $\sum_1^\infty \frac{x^k}{k^\alpha}$, where $\alpha > 0$:

It is convergent for $x = 0$. Let us assume that $0 < x < 1$.

$$\frac{u_{k+1}}{u_k} = \frac{\frac{x^{k+1}}{(k+1)^\alpha}}{\frac{x^k}{k^\alpha}} = x\left(\frac{k}{k+1}\right)^\alpha \to x \text{ as } k \to \infty.$$

Thus, it is convergent. If $x > 1$, it is divergent since $\frac{u_{k+1}}{u_k} \to x$ as $k \to \infty$. If $x = 1$, we have $\sum_1^\infty k^{-\alpha}$, $\alpha > 0$. Then,

$$\lim_{n \to \infty} \frac{u_{n+1}}{u_n} = \left(\frac{k}{k+1}\right)^{-\alpha}.$$

The series is convergent when $\alpha > 1$ and divergent when $\alpha < 1$. In the special case where $\alpha = 1$, it is (f), the harmonic series which is divergent.

c) $\sum_1^\infty (e^{\frac{1}{k}} - 1)$:

$e^{\frac{1}{k}} - 1 \approx \frac{1}{k}$ as $k \to \infty$. Thus, it is divergent (see part f) below).

d) $\sum_1^\infty \ln\left(1 + \frac{1}{k}\right)$:

$\ln\left(1 + \frac{1}{k}\right) \approx \frac{1}{k}$ as $k \to \infty$. Thus, it is divergent (see part f) below).

e) $\sum_1^\infty q^{k+\sqrt{k}}$, where $q > 0$:

$\sqrt[k]{u_k} = q^{1+k^{-0.5}} \to q$ as $k \to \infty$. Thus, it is convergent for $0 \leq q < 1$ and divergent for $q > 1$. If $q = 1$, then $u_k = 1$ and $\sum 1$ is divergent.

f) $\sum_1^\infty \frac{1}{n}$:

$$\frac{u_{k+1}}{u_k} = \frac{\frac{1}{k+1}}{\frac{1}{k}} = \frac{k}{k+1} \to 1 \text{ as } k \to \infty.$$

The Harmonic series is divergent!

13.3

a) For each object $i = 1, \ldots, n$, either it is selected or not; that is $x_i \in S_i = \{0, 1\}$. Then,

$$g(x) = \prod_{i=1}^n (x^0 + x^1) = (1+x)^n,$$

Without loss of generality, we may assume that $r = \sum x_i$ objects are selected. We know from Problem 1.3.a) that the number of distinct ways of selecting

$r \leq n$ objects out of n objects is $\binom{n}{r}$. Thus, $a_r = \binom{n}{r}$. We cannot choose more than n objects; that is $a_r = 0$, $r > n$. Therefore,

$$g(x) = (1+x)^n = \sum_{r=0}^{n} \binom{n}{r} x^r.$$

Let us prove the power expansion as a corollary to the Binomial theorem.

$$(x+y)^n = \sum_{i=0}^{n} \binom{n}{i} x^i y^{n-i}$$

The Binomial theorem states that $(1+z)^n = \sum_{i=0}^{n} \binom{n}{i} z^i$. Let $z = \frac{x}{y}$. Then,

$$\left(1 + \frac{x}{y}\right)^n = \sum_{i=0}^{n} \binom{n}{i} \left(\frac{x}{y}\right)^i \Leftrightarrow \left(\frac{x+y}{y}\right)^n = \frac{(x+y)^n}{y^n} = \sum_{i=0}^{n} \binom{n}{i} \left(\frac{x^i}{y^i}\right)$$

$$\Leftrightarrow (x+y)^n = \sum_{i=0}^{n} \binom{n}{i} x^i y^{n-i}.$$

Let us prove the multinomial theorem as a corollary to the Binomial theorem by induction on k.

$$(x_1 + \cdots x_k)^n = \sum_{\substack{i_1, \ldots, i_k \in \mathbb{Z}_+ \\ i_1 + \cdots + i_k = n}} \binom{n}{i_1, \ldots, i_k} x_1^{i_1} \cdots x_k^{i_k}$$

Let $l = 2$ and $x_1 = x$, $x_2 = y$. We use the power expansion to state that the induction base ($k = l = 2$) is true. Let use assume as induction hypothesis that

$$(x_1 + \cdots x_l)^n = \sum_{\substack{i_1, \ldots, i_l \in \mathbb{Z}_+ \\ i_1 + \cdots + i_l = n}} \binom{n}{i_1, \ldots, i_l} x_1^{i_1} \cdots x_l^{i_l}$$

holds.

$$(x_1 + \cdots + x_l + x_{l+1})^n =$$

$$\sum_{\substack{i_1, \ldots, i_l, i_{l+1} \in \mathbb{Z}_+ \\ i_1 + \cdots + i_l + i_{l+1} = n}} \binom{n}{i_1, \ldots, i_l, i_{l+1}} x_1^{i_1} \cdots x_l^{i_l} x_{l+1}^{i_{l+1}}$$

needs to be shown.

Let $x = x_1 + \cdots + x_l$ and $y = x_{l+1}$ in the power expansion.

$$(x+y)^n = \sum_{i=0}^{n} \binom{n}{i}(x_1 + \cdots + x_l)^i x_{l+1}^{n-i}$$

$$= \sum_{i=0}^{n} \binom{n}{i} \sum_{\substack{i_1,\ldots,i_l \in \mathbb{Z}_+ \\ i_1 + \cdots + i_l = i}} \binom{i}{i_1,\ldots,i_l} x_1^{i_1} \cdots x_l^{i_l} x_{l+1}^{n-i}$$

$$= \sum_{\substack{i_1,\ldots,i_l,i_{l+1} \in \mathbb{Z}_+ \\ i_1 + \cdots + i_l + i_{l+1} = n}} \binom{n}{i_1,\ldots,i_l,i_{l+1}} x_1^{i_1} \cdots x_l^{i_l} x_{l+1}^{i_{l+1}}$$

b) For each object $i = 1,\ldots,n$, either it is not selected or selected once, twice, thrice, and so on; that is $x_i \in S_i = \mathbb{Z}_+$. Then,

$$g(x) = \prod_{i=1}^{n}(x^0 + x^1 + x^2 + \cdots) = (1 + x + x^2 + \cdots)^n,$$

Without loss of generality, we may assume that $r = \sum x_i$ objects are selected. We know from 14.4 that the number of distinct ways of selecting r objects out of n objects with replacement is $\binom{n+r-1}{n-1} = \binom{n-1+r}{r}$. Thus, $a_r = \binom{n-1+r}{r}$. Therefore,

$$g(x) = (1 + x + x^2 + \cdots)^n = \sum_{r=0}^{n} \binom{n-1+r}{r} x^r.$$

c)

$$x_1 + x_2 + x_3 + x_4 = 13, \quad x_i = 1,2,3,4,5,6 \ \forall i \Rightarrow$$

$$g(x) = (x + x^2 + x^3 + x^4 + x^5 + x^6)^4 = x^4(1 + x + x^2 + x^3 + x^4 + x^5)^4$$

We are interested in the coefficient of x^{13} of $g(x)$, which is the coefficient of x^9 of $h(x) = (1 + x + x^2 + x^3 + x^4 + x^5)^4$.

$$\begin{aligned}
p(x) &= 1 + x + x^2 + x^3 + x^4 + \cdots \\
xp(x) &= \quad\ \ x + x^2 + x^3 + x^4 + \cdots \\
\hline
(1-x)p(x) &= 1 \ \Rightarrow p(x) = \frac{1}{1-x}
\end{aligned}$$

$$p(x) = 1 + x + x^2 + x^3 + x^4 + x^5 + x^6 + x^7 + \cdots$$

Similarly, $\dfrac{x^6 p(x) = \quad x^6 + x^5 + \cdots}{}$

$$(1 - x^6)p(x) = 1 + x + x^2 + x^3 + x^4 + x^5 = \frac{1 - x^6}{1 - x}$$

Then,

$$h(x) = (1 - x^6)^4 [p(x)]^4 = (1 - x^6)^4 (1 + x + x^2 + x^3 + x^4 + \cdots)^4 = k(x)l(x);$$

by the Binomial theorem

$$k(x) = \binom{4}{0} - \binom{4}{1}x^6 + \binom{4}{2}x^{12} - \binom{4}{3}x^{18} \binom{4}{4}x^{24},$$

and by the multiset problem

$$l(x) = \binom{3}{0} + \binom{4}{1}x + \binom{5}{2}x^2 + \binom{6}{3}x^3 + \binom{7}{4}x^4 + \cdots + \binom{12}{9}x^9 + \cdots$$

The ninth convolution of $k(x)l(x)$ is the answer:

$$\binom{4}{0}\binom{12}{9} - \binom{4}{1}\binom{6}{3} = 220 - 4(20) = 140$$

Therefore, the probability is

$$P(\text{having a sum of 13}) = \frac{140}{6^4} = 0.1080247$$

d)

$$a_n - 5a_{n-1} + 6a_{n-2} = 0, \ \forall n = 2, 3, 4, \ldots \Leftrightarrow$$

$$a_n x^n - 5a_{n-1}x^n + 6a_{n-2}x^n = 0, \ \forall n = 2, 3, 4, \ldots$$

Summing the above equation for all n, we get

$$\sum_{n=2}^{\infty} a_n x^n - 5\sum_{n=2}^{\infty} a_{n-1}x^n + 6\sum_{n=2}^{\infty} a_{n-2}x^n = 0$$

$$[g(x) - a_1 x - a_0] - 5x[g(x) - a_0] + 6x^2[g(x)] = 0$$

Using the boundary conditions ($a_0 = 2$ and $a_1 = 5$) we have

$$g(x) = \frac{a_0 + a_1 x - 5a_0 x}{6x^2 - 5x + 1} = \frac{2 - 5x}{(3x - 1)(2x - 1)} = \frac{1}{1 - 2x} + \frac{1}{1 - 3x}$$

$$g(x) = (1 + 2x + 4x^2 + \cdots + 2^i x^i + \cdots) + (1 + 3x + 9x^2 + \cdots + 3^i x^i + \cdots)$$

$$\Rightarrow a_n = 2^n + 3^n.$$

13.4

a) The left hand side of the following constraint represents the complementary survival probability of a threat,

$$1 - \prod_j (1 - p_{ji})^{x_{ji}} \ge d_i, \ \forall i.$$

Then,

$$1 - d_i \ge \prod_j (1 - p_{ji})^{x_{ji}} \Leftrightarrow \zeta \log(1 - d_j) \ge \sum_j [\zeta \log(1 - p_{ji})] x_{ji}, \forall \zeta \ge 0.$$

With a suitable choice of ζ, and let $-b_i = \zeta log(1 - d_i)$, $-a_{ji} = \zeta \log(1 - p_{ji})$, we will have

$$\sum_j a_{ji} x_{ji} \ge b_i, \ \forall i.$$

Let $\alpha_{ji} = \lfloor a_{ji} \rfloor$ and $\beta_i = \lfloor b_i \rfloor$ (with a suitable choice of $\zeta \ge 0$), yielding

$$\sum_j \alpha_{ji} x_{ji} \ge \beta_i, \ \forall i.$$

b) The first three objective functions are equivalent to each other, so are the last two. The flaw lies in the equivalence of the third and the fourth objective functions: $\max \beta_{iz} \not\equiv \min(1 - \beta_{iz})$. In particular,

$$\max y_1 + y_2 + y_3 \equiv \min(1 - y_1) + (1 - y_2) + (1 - y_3)$$

is true. However,

$$\max y_1 y_2 y_3 \equiv \min(1 - y_1)(1 - y_2)(1 - y_3) = 1 - \cdots - y_1 y_2 y_3$$

is false because of the cross terms.

Problems of Chapter 14

14.1

$$y''(t) - y(t) = e^{2t} \Leftrightarrow s^2\eta(s) - 2s - \eta(s) = \frac{1}{s-2} \Leftrightarrow \eta(s)(s^2-1) = \frac{1}{s-2} + 2s.$$

$$\Rightarrow \eta(s) = \frac{1}{(s-2)(s^2-1)} + \frac{2s}{s^2-1}.$$

1. If $\eta(s) = \frac{A}{s-2} + \frac{B}{s+1} + \frac{C}{s-1} = \frac{1}{(s-2)(s^2-1)}$. Solve for A, B, C:

$$\left. \begin{array}{c} A + B + C = 0 \\ 3B + C = 0 \\ -A + 2B - 2C = 1 \end{array} \right\} \Rightarrow A = \frac{2}{6}, \ B = \frac{1}{6}, \ C = \frac{-3}{6}.$$

Thus, $\eta(s) = \frac{2}{6(s-2)} + \frac{1}{6(s+1)} - \frac{3}{6(s-1)}$.

2. If $\eta(s) = \frac{E}{s-1} + \frac{F}{s+1} = \frac{2s}{(s+1)(s-1)} = \frac{2s}{s^2-1}$. Solve for E, F:

$$\left. \begin{array}{c} E + F = 2 \\ E - F = 0 \end{array} \right\} \Rightarrow E = 1, \ F = 1.$$

Thus, $\eta(s) = \frac{1}{s-1} + \frac{1}{s+1}$.

Then, we have

$$\eta(s) = \frac{1}{3(s-2)} + \frac{7}{6(s+1)} + \frac{1}{2(s-1)} \Leftrightarrow y(t) = \frac{1}{3}e^{2t} + \frac{7}{6}e^{-t} + \frac{1}{2}e^{t}.$$

14.2

$$y(k+1) = y(k) + 2e^k \Leftrightarrow z\eta(z) - z = \eta(z) + 2\frac{z}{z-e^1} \Leftrightarrow \eta(z)(z-1) = z + 2\frac{z}{z-e}.$$

$$\Rightarrow \eta(z) = \frac{z}{z-1} + 2\frac{z}{(z-1)(z-e)}.$$

If $\eta(z) = \frac{2}{(z-1)(z-e)} = \frac{A}{z-1} + \frac{B}{z-e}$ then $A = \frac{2}{1-e}$ and $B = \frac{2}{e-1}$.
Therefore,

$$\eta(z) = \frac{z}{z-1} + z\left[\frac{2}{1-e}\left(\frac{1}{z-1}\right) + \frac{2}{e-1}\left(\frac{1}{z-e}\right) \right]$$

$$\Leftrightarrow y(k) = 1 + \frac{2}{1-e}(1 - e^k).$$

14.3

$$\frac{dx}{dt} = -0.2y, \ \frac{dy}{dt} = -0.3x - 0.1y, \ x(0) = 50, \ y(0) = 100.$$

$$\frac{dy}{dt} = -0.3x - 0.1y \Rightarrow \frac{dy^2}{dt^2} = -0.3\frac{dx}{dt} - 0.1\frac{dy}{dt} \Rightarrow \frac{dy^2}{dt^2} - 0.06y + 0.1\frac{dy}{dt} = 0 \ (\maltese)$$

$$\eta\left[\frac{dy^2}{dt^2}\right] = s^2\eta(s) - sy(0) - y(0) = s^2\eta(s) - 100s + 25,$$

since $\frac{dy}{dt}\big|_{t=0} = -0.3(50) - 0.1(100) = -25$. Moreover,

$$\eta\left[\frac{dy}{dt}\right] = s\eta(s) - y(0) = s\eta(s) - 100$$

$(\maltese)$: $[s^2\eta(s) - 100s + 25] - 0.06\eta(s) + 0.1[s\eta(s) - 100] = 0$

$$\Leftrightarrow \eta(s)(s^2 - 0.06 + 0.1s) = 100s - 25 + 10 \Leftrightarrow$$

$$\eta(s) = \frac{100s - 15}{(s + 0.3)(s - 0.2)} = \frac{A}{s + 0.3} + \frac{B}{s - 0.2}$$

$A + B = 100, -0.2A + 0.3B = -15 \Rightarrow A = 90, B = 10 \Rightarrow$

$(\maltese)$: $\eta(s) = \dfrac{90}{s + 0.3} + \dfrac{10}{s - 0.2} \Rightarrow y(t) = 90e^{-0.3t} + 10e^{0.2t}$

$$\begin{bmatrix} y(0) \\ y(1) \\ y(2) \\ y(3) \\ y(4) \end{bmatrix} = \begin{bmatrix} 100.0000 \\ 78.88767 \\ 64.31129 \\ 54.81246 \\ 49.36289 \end{bmatrix}$$

14.4

Let $x(n) = F_{n+1}$, and the initial conditions are $x(0) = 1$, $x(1) = 1$.

$$x(n + 1) = x(n) + x(n - 1), \ n = 2, 3, \ldots \ (\yen)$$

$\eta[x(n + 1)] = z\eta(z) - zx(0) = z\eta(z) - z$ and $\eta[x(n - 1)] = \frac{1}{z}\eta(z)$.

$(\yen)$: $x(n + 1) = x(n) + x(n - 1) \Leftrightarrow z\eta(z) - z - \eta(z) - \dfrac{1}{z}\eta(z) = 0$

$$\Leftrightarrow \eta(z) = \frac{z}{z - 1 - \frac{1}{z}} = \frac{z}{z(1 - \frac{1}{z} - \frac{1}{z^2})} \Leftrightarrow$$

$$\eta(z) = \frac{1}{1 - \frac{1}{z} - \frac{1}{z^2}} = \frac{1}{\left(1 - \frac{1-\sqrt{5}}{2z}\right)\left(1 - \frac{1+\sqrt{5}}{2z}\right)} = \frac{A}{1 - \frac{1-\sqrt{5}}{2z}} + \frac{B}{1 - \frac{1+\sqrt{5}}{2z}}$$

$$A + B = 1$$

$$A\left(\frac{1 - \sqrt{5}}{2z}\right) + B\left(\frac{1 + \sqrt{5}}{2z}\right) = 0 \Rightarrow A = \frac{5 + \sqrt{5}}{10}, \ B = \frac{5 - \sqrt{5}}{10} \Rightarrow$$

$(\yen)$: $\eta(z) = \dfrac{5 + \sqrt{5}}{10}\left(\dfrac{1}{1 - \frac{1-\sqrt{5}}{2z}}\right) + \dfrac{5 - \sqrt{5}}{10}\left(\dfrac{1}{1 - \frac{1+\sqrt{5}}{2z}}\right)$

Since $\mathcal{Z}^{-1}\left(\frac{1}{1 - \frac{a}{z}}\right) = a^n y(n)$, we have

$$x(n) = \frac{5+\sqrt{5}}{10}\left(\frac{1+\sqrt{5}}{2}\right)^{n} + \frac{5-\sqrt{5}}{10}\left(\frac{1+\sqrt{5}}{2}\right)^{n}, \quad n = 2, 3, \ldots$$

Thus,

$$F_n = x(n-1) = \frac{5+\sqrt{5}}{10}\left(\frac{1+\sqrt{5}}{2}\right)^{n-1} + \frac{5-\sqrt{5}}{10}\left(\frac{1+\sqrt{5}}{2}\right)^{n-1}, \quad n = 1, 2, \ldots$$

Finally,

$$F_{100} = x(99) = 354\,224\,848\,179\,261\,915\,075$$

Index

Early Titles in the
INTERNATIONAL SERIES IN
OPERATIONS RESEARCH & MANAGEMENT SCIENCE
(Continued)

* A list of the more recent publications in the series is at the front of the book *